AF443465

Topics in Applied Physics

Volume 131

Series editors

Mildred S. Dresselhaus, Cambridge, USA
Young Pak Lee, Seoul, Korea, Republic of (South Korea)
Paolo M. Ossi, Milan, Italy

Topics in Applied Physics is a well-established series of review books, each of which presents a comprehensive survey of a selected topic within the area of applied physics. Edited and written by leading research scientists in the field concerned, each volume contains review contributions covering the various aspects of the topic. Together these provide an overview of the state of the art in the respective field. Topics in Applied Physics is addressed to all scientists at universities and in industry who wish to obtain an overview and to keep abreast of advances in applied physics. The Managing Editors are open to any suggestions for topics coming from the community of applied physicists no matter what the field and encourage prospective book editors to approach them with ideas. 2015 Impact Factor: 0.889

More information about this series at http://www.springer.com/series/560

Byung-Eun Park · Hiroshi Ishiwara
Masanori Okuyama · Shigeki Sakai
Sung-Min Yoon
Editors

Ferroelectric-Gate Field Effect Transistor Memories

Device Physics and Applications

 Springer

Editors

Byung-Eun Park
School of Electrical and Computer
  Engineering
University of Seoul
Seoul
Korea, Republic of (South Korea)

Hiroshi Ishiwara
Frontier Collaborative Research Center
Tokyo Institute of Technology
Yokohama
Japan

Masanori Okuyama
Graduate School of Engineering Science
Osaka University
Osaka
Japan

Shigeki Sakai
National Institute of Advanced Industrial
  Science and Technology (AIST)
Tsukuba
Japan

Sung-Min Yoon
Department of Advanced Materials
  Engineering for Information
  and Electronics
Kyunghee University
Yongin
Korea, Republic of (South Korea)

ISSN 0303-4216   ISSN 1437-0859   (electronic)
Topics in Applied Physics
ISBN 978-94-024-0839-3   ISBN 978-94-024-0841-6   (eBook)
DOI 10.1007/978-94-024-0841-6

Library of Congress Control Number: 2016944338

Printed on acid-free paper

This Springer imprint is published by Springer Nature
The registered company is Springer Science+Business Media B.V. Dordrecht

Preface

The applications of ferroelectric materials have been manifold and overspreading, covering all areas of our workplaces and homes over the present century. By far the largest number of applications in ferroelectric materials remained to be associated with bulk ceramics, but a trend toward thin films for specified applications, which provides additional benefits including lower operating voltage, higher speed operations, and scaling ability, has been energetically developed. Among various applications of ferroelectric thin films, the development of nonvolatile ferroelectric random access memory (FeRAM) has been most actively progressed since the late 1980s and reached modest mass production for specific application since 1995. These activities have been mainly motivated by the physical limitation and technological drawbacks of the Flash memory. There are two types of memory cells in ferroelectric nonvolatile memories. One is the capacitor-type (1T1C-type) FeRAM, in which a ferroelectric material is used as in the storage capacitor of a dynamic random access memory (DRAM) structure. The other is the field effect transistor (FET)-type (1T-type) FeRAM, in which the conventional gate insulator is replaced with a ferroelectric thin film. Although the FET-type FeRAM claims the ultimate scalability and nondestructive readout characteristics, the capacitor-type FeRAMs have been the main interest for the major semiconductor memory companies, because the ferroelectric FETs have fatal handicaps of cross-talk for random accessibility and short retention time. The present main applications of the commercialized FeRAMs are low-density nonvolatile embedded memory. Eventually, the huge expectations for the FeRAMs turned out to be somewhat different from what they had in prospect, to the contrary, the Flash memories keep the overwhelmingly dominant position in the nonvolatile memory-related industries thanks to their fascinating technology improvements and tremendously increasing needs for them.

Unlike these Si-based electronics demanding an ultra-high performance and an aggressive device scaling, the requirements for the nonvolatile memory components embedded into the large-area electronics implemented on glass, plastic or paper substrates are considerably different. If we can additionally obtain such features as mechanical flexibility, transparency to the visible light, lower power operation and higher device reliability with a simpler process at lower temperature, it would have a great impact on the related industries. The ferroelectric gate field effect transistor (FeFET) is a very promising candidate because it can be reproducibly operated with a definitely designable principle and be prepared by a very simple process.

There are already lots of books dealing with the capacitor-type FeRAMs. Thus, this book aims to provide the readers with development history, technical issues, fabrication methodologies, and promising applications of FET-type ferroelectric memory devices, which have not been comprehensively discussed in the published books. The book is composed of chapters written by remarkable leading researchers. Authors are organized with top experts, who have been engaged with in ferroelectric materials and related device technologies, including oxide and organic ferroelectric thin films for a long time.

In Chap. 1, this book presents a comprehensive review of past and present technologies for the FET-type ferroelectric memories through historical development background. The feature, principle, theoretical analysis and improvement of FET-type ferroelectric memories are also overviewed. The Chap. 2 contains practical characteristics of FET-type ferroelectric memory devices. With the important aspects of channel layers, the practical properties of silicon-based and thin film-based channel layer ferroelectric gate field effect transistors are discussed in Chap. 2. In Chap. 3, the properties of organic ferroelectric materials such as poly(vinylidene fluoride) (PVDF), polyvinylidene fluoride trifluoroethylene [P(VDF-TrFE)] and polyvinylidene fluoride tetrafluoroethylene [P(VDF-TeFE)] in ferroelectric gate field effect transistors are studied. This chapter is also contains discussions on both silicon-based and thin film-based channel layer of ferroelectric gate field effect transistors. Utilizing advantage of the organic materials, flexible ferroelectric gate field effect transistors using plastic and paper substrates are fabricated and discussed in Chap. 3. Finally, future prospects for novel applications such as NAND flash memory circuits, relaxor, adaptive-learning synaptic devices and high-k gate oxide thin film transistors are discussed in Chap. 4.

The technology field of FET-type ferroelectric memory has great impacts on both fundamental device physics and commercial industrial opportunities. Notwithstanding, there have hardly been appropriate technical books dealing with an insightful review to all the key technologies and applications of the FET-type ferroelectric memory devices. This book provides a comprehensive coverage of material characteristics, process technologies, and device operations for the memory field effect transistors employing inorganic or organic ferroelectric thin films, which have not been discussed in previously published books. These issues would be big helps for further advances in large-area electronics implemented on glass, plastic or paper substrate as well as in conventional silicon (Si) electronics. Thus, this book is a useful source of information for graduate students of material science and electronic engineering, device and process engineers in industries.

We would like to express our sincere thanks to many other colleagues (in alphabetical order), in particular to Profs. Norifumi Fujimura, Cheol Seong Hwang, Takeshi Kanashima, Thomas Mikolajick, Cheol-Min Park, Tatsuya

Shimoda, Eisuke Tokumitsu, and Takeshi Yoshimura for their invaluable contributions. We are also very thankful to Dr. Yoshihisa Fujisaki, Dr. Yukihiro Kaneko, Dr. Richard H. Kim, Dr. Min Hyuk Park, Dr. Uwe Schroeder, Dr. Stefan Slesazeck, and Dr. Mitsue Takahashi for the continuous support. Last but not least, we would like to thank to Mrs. Cindy Zitter and Ms. Annie Kang for their tremendous efforts in arranging the manuscripts for this book.

Seoul, Korea, Republic of (South Korea)
Tokyo, Japan
Osaka, Japan
Tsukuba, Japan
Suwon, Korea, Republic of (South Korea)

Byung-Eun Park
Hiroshi Ishiwara
Masanori Okuyama
Shigeki Sakai
Sung-Min Yoon

Contents

Part VI Practical Characteristics of Organic Ferroelectric-Gate FETs : Ferroelectric-Gate Field Effect Transistors with Flexible Substrates

11 Mechanically Flexible Non-volatile Field Effect Transistor Memories with Ferroelectric Polymers

Richard H. Kim and Cheolmin Park

12 Non-volatile Paper Transistors with Poly(vinylidene fluoride-trifluoroethylene) Thin Film Using a Solution Processing Method

Byung-Eun Park

Contributors

Norifumi Fujimura Department of Physics and Electronics, Graduate School of Engineering, Osaka Prefecture University, Sakai, Osaka, Japan

Yoshihisa Fujisaki YourFriend, Hachioji, Tokyo, Japan

Cheol Seong Hwang Department of Materials Science and Engineering and Inter-University Semiconductor Research Center, College of Engineering, Seoul National University, Seoul, Korea, Republic of (South Korea)

Hiroshi Ishiwara Tokyo Institute of Technology, Midori-ku, Yokohama, Japan

Takeshi Kanashima Graduate School of Engineering Science, Osaka University, Toyonaka, Osaka, Japan

Yukihiro Kaneko Advanced Research Division, Panasonic Corporation, Seika-cho, Soraku-gun, Kyoto, Japan

Richard H. Kim Department of Materials Science and Engineering, Yonsei University, Seodaemun-gu, Seoul, Korea, Republic of (South Korea)

Thomas Mikolajick Nanoelectronic Materials Laboratory—NaMLab gGmbH, Dresden, Germany; IHM, Technische Universitaet Dresden, Dresden, Germany

Masanori Okuyama Institute for Nanoscience Design, Osaka University, Toyonaka, Osaka, Japan

Byung-Eun Park School of Electrical and Computer Engineering, University of Seoul, Dongdaemun-ku, Seoul, Korea, Republic of (South Korea)

Cheolmin Park Department of Materials Science and Engineering, Yonsei University, Seodaemun-gu, Seoul, Korea, Republic of (South Korea)

Min Hyuk Park Department of Materials Science and Engineering and Inter-University Semiconductor Research Center, College of Engineering, Seoul National University, Seoul, Korea, Republic of (South Korea)

Shigeki Sakai National Institute of Advanced Industrial Science and Technology, Tsukuba, Ibaraki, Japan

Uwe Schroeder Nanoelectronic Materials Laboratory—NaMLab gGmbH, Dresden, Germany

Tatsuya Shimoda Green Devices Research Center, Japan Advanced Institute of Science and Technology, Nomi, Ishikawa, Japan; School of Materials Science, Japan Advanced Institute of Science and Technology, Nomi, Ishikawa, Japan

Stefan Slesazeck Nanoelectronic Materials Laboratory—NaMLab gGmbH, Dresden, Germany

Mitsue Takahashi National Institute of Advanced Industrial Science and Technology, Tsukuba, Ibaraki, Japan

Eisuke Tokumitsu Green Devices Research Center, Japan Advanced Institute of Science and Technology, Nomi, Ishikawa, Japan

Sung-Min Yoon Department of Advanced Materials Engineering for Information and Electronics, Kyung Hee University, Yongin, Gyeonggi-Do, Korea, Republic of (South Korea)

Takeshi Yoshimura Graduate School of Engineering, Department of Physics and Electronics, Osaka Prefecture University, Sakai, Osaka, Japan

Part I
Introduction

Chapter 1
Features, Principles and Development of Ferroelectric–Gate Field-Effect Transistors

Masanori Okuyama

Abstract Ferroelectric-gate field effect transistor (FeFET) memories are overviewed. The FeFET shows excellent features as an integrated memory such as nonvolatality, better scalability, higher read-write speeds, lower dissipation powers, higher tamper resistances and higher radioactivity tolerance. But, memory retention was the most critical problem for its practical realization. Mechanisms of degradation of the retention are discussed in metal-ferroelectric-insulator-semiconductor (MFIS) gate structure in which the insulator is inserted between the ferroelectric and the semiconductor to avoid interface damages suffered during the device preparation at high temperature. It is concluded from careful discussion that leakage currents through insulator-semiconductor and metal-ferroelectric junctions store charges in the interface between the ferroelectric and the insulator layers, which reduce apparent dielectric polarization and promote the degradation of the retention. Electronic property of the interfaces and the ferroelectric layer in the MFIS structure has been improved by nitrogen radical treatment and thermal annealing, and the retention of MFIS capacitance is shown to extend very much. Moreover, several kinds of improved MFIS FETs are introduced and the memory retention has been extended very much to be useful for the practical realization of excellent memory devices.

1.1 Background of Ferroelectric Memories

1.1.1 Historical Background

In 1963, Moll and Tarui [1] proposed a novel solid state memory resistor that controls the conductance of a semiconductor by using a ferroelectric material. The structure consisted of thin trigricine sulfate (TGS) sandwiched between an Au

M. Okuyama (✉)
Institute for Nanoscience Design, Osaka University, Machikaneyama 1-3,
Toyonaka, Osaka 560-8531, Japan
e-mail: okuyama@insd.osaka-u.ac.jp

© Springer Science+Business Media Dordrecht 2016
B.-E. Park et al. (eds.), *Ferroelectric-Gate Field Effect Transistor Memories*,
Topics in Applied Physics 131, DOI 10.1007/978-94-024-0841-6_1

bottom electrode and a thin film transistor. The transistor was composed of two counter–electrodes on the TGS, covered with CdS/SiO layers and a top Al electrode. In this structure, a positively (negatively) polarizing voltage on the Al electrode decreased (increased) the conductance between the two electrodes. The measured ON–OFF resistance ratio was 40. Moll and Tarui's structure triggered a flurry of interest, and many memory devices consisting of semiconducting-film field effect transistors (FETs) on ferroelectric crystals and ceramics were proposed in the late 1960s [2–6]. However, devices based on semiconductor thin films cannot be miniaturized because of unsatisfactory memory characteristics. This situation was improved by stacking a ferroelectric thin film on the crystalline semiconductor of an FET, a structure that is currently used integrated devices. Wu fabricated an FET memory by layering a ferroelectric $Bi_4Ti_3O_{12}$ thin film on the Si surface of the FET channel region [7]. However, in this device, the drain current is not directly controlled by the polarization charge of the ferroelectric; rather it is indirectly controlled by the bound charges in the ferroelectric.

To enable direct control of the switching and memory effects in the drain current by ferroelectric polarization, Sugibuchi et al. [8] constructed a Si crystalline FET with a ferroelectric $Bi_4Ti_3O_{12}/SiO_2$ gate structure. Due to its metal/ferroelectric/insulator/semiconductor gate structure, this FET is called an MFIS FET. MFIS FETs are fabricated by sputtering a $Bi_4Ti_3O_{12}$ film onto a SiO_2 layer and heat-treating the layered structure. The hysteresis in the drain current–gate voltage characteristics of an MFIS FET is consistent with polarization–electric field hysteresis. Matsui et al. deposited $Pb_{1-x}La_xTiO_3$(PLT) and $Pb_{1-x}La_x(Zr_yTi_{1-x})O_3$(PLZT) ferroelectric thin films onto the channel region of a GaAs FET and onto the SiO_2 gate of a Si MOSFET substrate [9–11]. Again, the hystereses observed in the drain current–gate voltage characteristics were confirmed to be due to polarization hysteresis. However, sputtering the ferroelectric thin films at high temperature damages the SiO_2/Si interface, degrading the stability and long-term retention of the memorized states. Since then, maximizing the MFIS FET characteristics by improving the electronic properties of ferroelectric/oxide/Si interfaces has become a research priority, and good interfaces have become available in recent years. Consequently, as described in later chapters and literature in general, FETs with ferroelectric thin-film gate currently exhibit strong nonvolatile memory characteristics.

In the late 1980s, researchers constructed a new type of ferroelectric memory by wiring a storage capacitor (composed of ferroelectric thin films) to an addressing FET [12, 13]. This memory device was inspired by the remarkable polarization hysteresis behavior of ferroelectric oxide films, which yield splendid switching and memory characteristics. Since its inception, various ferroelectric memories have been markedly improved and commercialized as a nonvolatile memory device [14–16].

1.1.2   Classification of Non-volatile Ferroelectric Memories

As mentioned in the previous section and illustrated in Fig. 1.1, devices holding ferroelectric memories are broadly classifiable into two categories: FETs with storage capacitors consisting of ferroelectric thin films, and FETs with ferroelectric gate. Compared with other memory devices such as DRAM, SRAM and flash memory, ferroelectric memory devices realize better nonvolatility (retention of the memorized states when the electrical supply is turned off), better scalability, higher read-write speeds, lower dissipation powers, higher tamper resistances (difficulty of reading the memorized information from outside the devices) and higher radioactivity tolerances.

We first discuss the early FETs that used ferroelectric capacitors. The acronym for these devices is 1T1C-type FeRAM (**Fe**rroelectric **R**andom **A**ccess **M**emory) or simply FeRAM. In these devices, the ferroelectric capacitor memorizes ON and OFF states when positive and negative pulses with amplitudes exceeding the coercive voltage are applied. The memorized state is read out by applying a single pulse of the recognized polarity. When the polarity of the pulse voltage opposes that of the polarization of the ferroelectric capacitor, polarization reversal occurs, releasing a large current flow into the output circuit. In contrast, when the pulse has the same polarity as the ferroelectric polarization, the current output is very small. The intensity of the measured current indicates the memorized state. In the early years, the memorized weak polarization states were robustly judged by combining two pairs of FETs and ferroelectric capacitors (2T2C), which ensured stable memorization and read-out. However, the 1T1C FeRAM destroys the memorized polarization states after the read-out pulse application, because the polarization is aligned in the direction of the read-out voltage. Losing memory means destructive read-out and so the memory should be recovered by a pulse that rewrites the memory after the read-out. Moreover, because the charge change during polarization reversal is proportional to the area of the ferroelectric capacitor, the read-out current decrease proportionally to the area. Therefore, when such memory devices are scaled down, large remanent polarization is required. Lead zirconate titanate, $PbZr_xTi_{1-x}O_3$ (PZT) thin films have long been adopted in ferroelectric, pyroelectric and piezoelectric applications as they exhibit large remanent polarizations, low coercive forces, large pyroelectric and piezoelectric coefficients. However, frequent

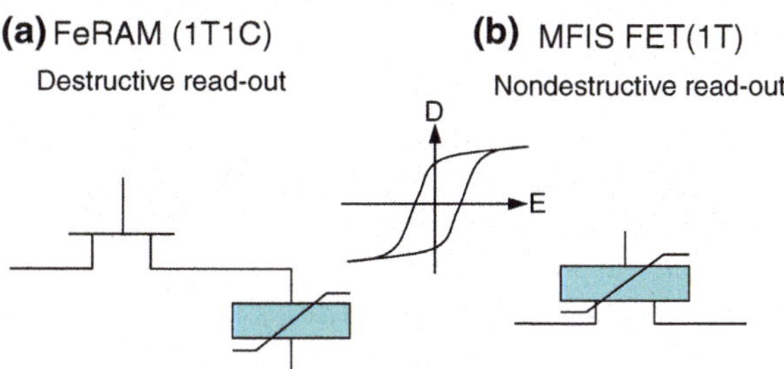

Fig. 1.1 Schematic of **a** 1T1C-type FeRAM and **b** 1T-type ferroelectric gate FET memory

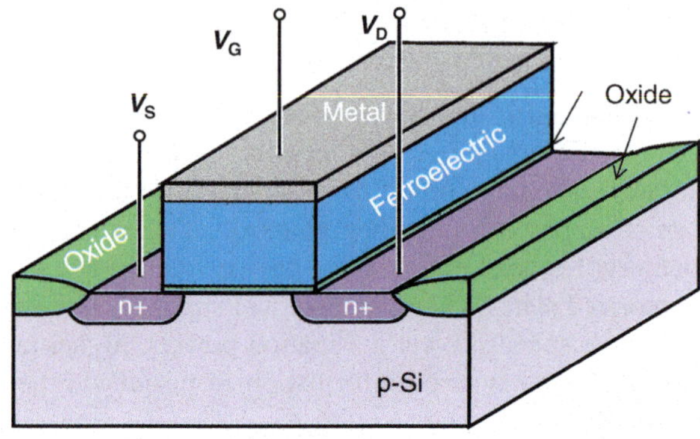

Fig. 1.2 Device structure of ferroelectric gate FET having MFIS structure. V_S, V_D and V_G show voltages of source, drain and gate, respectively

voltage-induced polarization reversal leads to a condition called *fatigue*, or degradation of the remanent polarization. Fatigue increases drastically after 10^6–10^7 cycles of reversals. Modern thin films based on bismuth-layered structures $SrBi_2Ta_2O_9$(SBT), $Bi_4Ti_3O_{12}$(BIT) and $Bi_{4-x}La_xTi_3O_{12}$(BLT) [17, 18] and the adoption of Ir/IrO_2 and $SrRuO_3$ electrodes on PZT films [19] have improved the number of the voltage application repetitions over 10^{12} cycles which is acceptable for conventional memory. Now, 1T1C FeRAMs are commercially manufactured by several companies, and are incorporated in such diverse applications as IC cards, train tickets, smart meters, radio frequency identifiers and communication devices.

The second category is ferroelectric-gate FET memory, known as 1T-type or FeFET (Fig. 1.2). FeFET devices receive pulses below the coercive voltage and read out drain currents without memorized polarization changes. Thus, unlike in the earlier 1T1C devices, the memorized states are preserved after the read-out and no rewriting process is required. The remanent polarization is required only to control the Si surface potential of the FET; moreover, it must be below a threshold polarization (determined by multiplying the dielectric constant times the breakdown field of the oxide thin film between the ferroelectric thin film and channel region of the FET). In contrast, 1T1C memory requires a large polarization to keep the readout charge at a readable level. Therefore, unlike their 1T1C-type counterparts, ferroelectric FETs can be easily miniaturized and densely integrated (i.e., they have high scalability), rendering them more suitable for material development.

To improve their tolerances to severe conditions, the ferroelectric film and Si channel regions are usually separated by an insulating thin film of SiO_2 (or some other metal oxide). This layer confers protection against high temperatures (~ 500–$700\ °C$), impingement of high energy ions and electrons on the substrate in plasma environments, and chemical contamination in highly reactive solutions during ferroelectric thin film deposition and device fabrication. Consequently, the Si surfaces are protected from surface degradation. Figure 1.3a shows the capacitance–gate voltage characteristics of the MFIS structures of SBT/HfO_2 and SBT/HfO_2 [20]. The observed counterclockwise hysteresis is a ferroelectric hysteretic effect. SBT/HfO_2 and BLT/HfO_2 structures have also been incorporated into ferroelectric gate FETs. The drain current–gate voltage characteristics of MFIS FETs are plotted

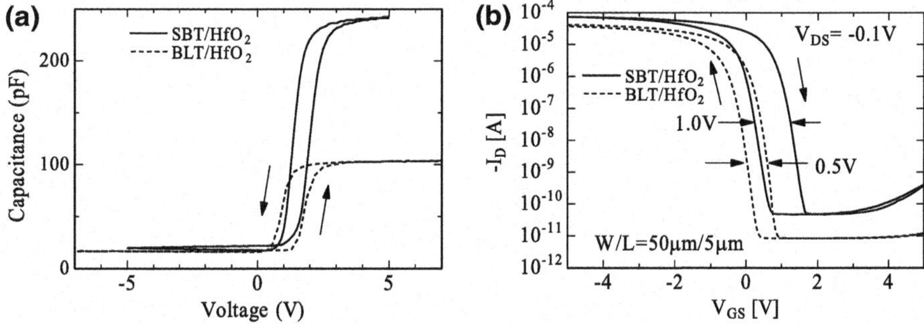

Fig. 1.3 Representative gate voltage dependences of **a** capacitance of MFIS structures and **b** drain current of MFIS FETs using SBT/HfO$_2$ and BLT HfO$_2$ gate structures. Copyright 2005 The Japan Society of Applied Physics

in Fig. 1.3b. The ON and OFF currents of the MFIS FET correspond to the low and high capacitance states of the MFIS structure, respectively. However, the performances of the MFIS FETs using SiO$_2$ layer were degraded by low endurance, disturbance, and poor retention of the memorized states. Memory retention is an especially critical problem, being affected by reduction and/or charge compensation of the polarization of the ferroelectric thin film. As discussed in the next section, the memorized states become deteriorated by the poor properties of the ferroelectric and insulator films, degradation of the interfaces between the insulator and ferroelectric and semiconductors, and depolarization field in the ferroelectric layer.

1.2 Degradation and Improvement of Memorized States in MFIS Structures

1.2.1 Degradation of Memorized States

As mentioned in the previous section, memory retention is the most critical problem in ferroelectric gate FETs. Figure 1.4 illustrates the capacitance aging of the MFIS structure. Among the best candidates for nonvolatile memories is a ferroelectric layer of SBT thin film deposited on an SiO$_2$/n-Si substrate by pulsed laser deposition (PLD). SBT exhibits less-fatigue properties and sufficient remanent polarization [17]. Figure 1.4 also shows the band profiles of the MFIS structure, neglecting some realistic properties such as space charges and the internal field at zero bias voltage. These profiles are classified into three typical states. The OFF state is obtained immediately after the metal electrode receives a short positive pulse voltage that exceeds the coercive voltage. The ON state is obtained immediately after application of a negative pulse bias below the coercive voltage. The aged state of the capacitance vs. hold time characteristics occurs when the structure

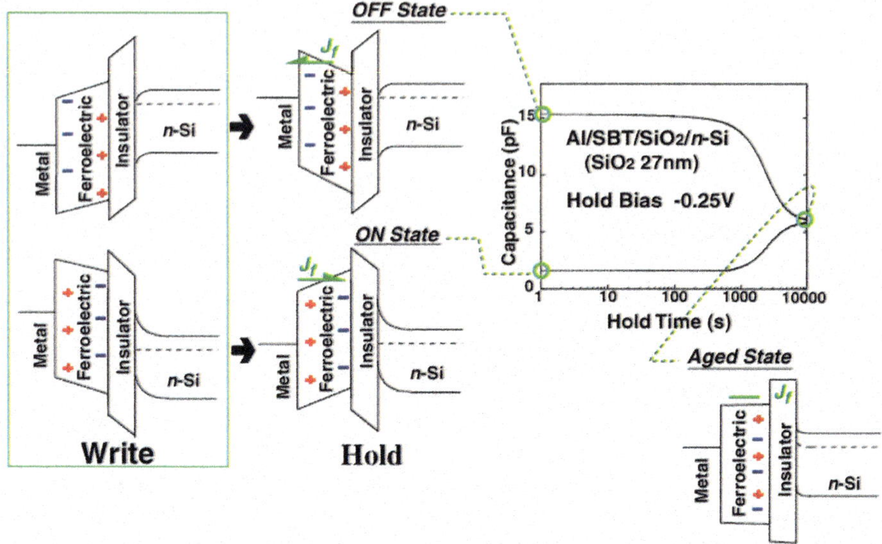

Fig. 1.4 Capacitance aging characteristics of MFIS structure and band profiles at 'ON', 'OFF' and 'aged' states

has been held at zero or a certain voltage for sufficient time to lose the retained data. In the aged state, the polarization in the ferroelectric layer can no longer control the Si surface potential. This noncontrollability is attributable to several phenomena such as degraded polarization of the ferroelectric layer and polarization shielding by the mobile charges of electrons, holes, and ions under the depolarization field. We have identified four plausible origins of the degradation of the retention characteristics (Fig. 1.5); (1) insufficient polarization, (2) polarization relaxation of the ferroelectric layer, (3) ion drift, and (4) leakage current, which carries interface charges shielding the polarization. Origin (1) can be avoided by applying a sufficiently large voltage to a high-quality ferroelectric thin film. To achieve this objective, thicknesses of the ferroelectric and insulator films must be adequate to

Fig. 1.5 Origins of degradation of memory retention in MFIS structure

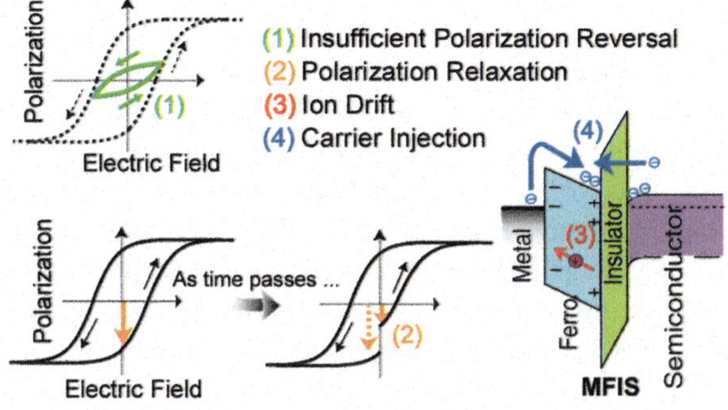

fully polarize the ferroelectric. Origin (2) depends on whether a switched polarization can be thermally relaxed under the depolarization field. The remanent polarization of the annealed SBT film is relatively robust to degradation, so relaxation does not seriously contribute to the retention problem. Regarding origin (3), ion drift is estimated to have short responses (several tens of seconds) [21], so is largely irrelevant to long-term retention. Therefore, origins (1)–(3) are assumed to be noncritical in this problem. At the present, charge injection from the top metal electrode and from the semiconductor into the insulator–ferroelectric interface is considered to significantly degrade the MFIS structure. Therefore, the most likely sources of memorized state degradation in MFIS structures are leakage current and charge storage at this interface.

1.2.2 Theoretical Analysis of the Band Profile and Retention Degradation of MFIS Capacitors

Charges at the ferroelectric–insulator interface are altered by currents passing across the metal–ferroelectric Schottky junction and across the semiconductor–insulator junction. This section investigates how currents in the ferroelectric and insulator layers affect the retention characteristics of MFIS structures [21–25].

The retention degradation of an MFIS capacitor was simulated by obtaining the temporal variations in its band profile. At each time step, the simulations solved several simultaneous equations involving terms such as current continuity and Gauss' law. The updates were terminated when the field distribution had sufficiently converged [23]. The currents through the ferroelectric and insulator layers were assumed to depend solely on the fields applied to these layers.

In the presence of an electric field E_f, the current density $J_f(E_f, t)$ flowing in the ferroelectric layer decays over time as [21]:

$$J_f(E_f, t) = J_{f0}(E_f) \cdot t^{-\beta}, \tag{1.1}$$

where t is the elapsed time, $J_{f0}(E_f)$ is the current density, and $\beta = 0.52$ was experimentally determined in a real Al/SBT/Pt capacitor [23]. The SBT thin film of this capacitor was prepared by the PLD method with a stoichiometric SBT target. $J_f(E_f, t)$ is the sum of two components: Schottky (J_f^S) and Fowler–Nordheim (J_f^F) emission current densities. The Schottky current dominates in low fields and the Fowler–Nordheim current dominates in high fields. The current density $J_i(E_i)$ flowing in the insulator layer under an electric field E_i is time-independent, and is similarly assumed to be the sum of Schottky (J_i^S) and Fowler–Nordheim (J_i^F) emission current densities. J_f^S, J_f^F, J_i^S and J_i^F depend on the voltages applied to the ferroelectric (V_f) and the insulator (V_i) layers and on the Schottky barrier height (ϕ_B); that is [26],

$$J_{f,i}^S \propto T^2 \exp\left\{ \left(a\sqrt{V_{f,i}} - q\phi_{Bf,i} \right)/kT \right\} \tag{1.2}$$

and

$$J_{f,i}^F \propto V_{f,i}^2 \exp\left\{ -b\left(q\phi_{Bf,i} \right)^{3/2}/V_{f,i} \right\}, \tag{1.3}$$

where q is the elementary charge; a and b are constants independent of V_f, V_i, and ϕ_B; T is the absolute temperature, and k is the Boltzmann constant. The polarization is assumed to be analytically described by Miller's equations [27].

1.2.3 Calculated Time Dependences of Band Profile and Capacitance of the MFIS Structure

The time-dependent potential distribution and electrical characteristics of the MFIS structure were numerically obtained by iterating the above-mentioned equations. The thicknesses of the ferroelectric and insulator layers were 400 and 10 nm, respectively; the flat band voltage was 0 V; the dielectric constants of the ferroelectric and insulator were 50 and 3.9, respectively; and T was 300 K. The silicon substrate was p-type with an acceptor concentration of 10^{15} cm^{-3}, the spontaneous polarization P_s was 1.2 μC/cm^2, the remanent polarization P_r was 1.0 μC/cm^2, and the coercive field E_c was 50 kV/cm. The MFIS structure was memorized by applying a +10 V or −10 V pulse during the write cycle and holding the gate voltage at zero. The current density through the ferroelectric layer in the MFIS structure was expected to be as large as 2×10^{-9} A/cm^2 under a field of 10 kV/cm, with a time dependence of $t^{-0.52}$. The assumed current through the insulator layer was approximately 1×10^{-15} A/cm^2 under a field of 500 kV/cm, as obtained in studies on high-quality SiO$_2$ thin films [28]. The degradation process was modeled by characterizing the time dependences of several basic parameters [22]. In the initial state, the ferroelectric layer in the MFIS structure exhibited a remanent polarization at zero voltage in the defined hysteresis loop. When the MFIS structure was aged, the absolute value of E_f decreased, with consequent decrease in the injected current J_f. Finally, the electric fields became heavily reduced in the ferroelectric, insulator, and semiconductor layers of the MFIS structure. The time dependence of the total capacitance of the MFIS was calculated from the obtained band profiles. Figure 1.6 plots the calculated time dependences of the capacitance and the band profiles for $V_g = 0$ V, after writing data at $V_g = \pm 10$ V. In Fig. 1.6, the labels (I), (II), and (III) refer to a fully memorized state, an intermediate state (depletion state), and an aged state, respectively.

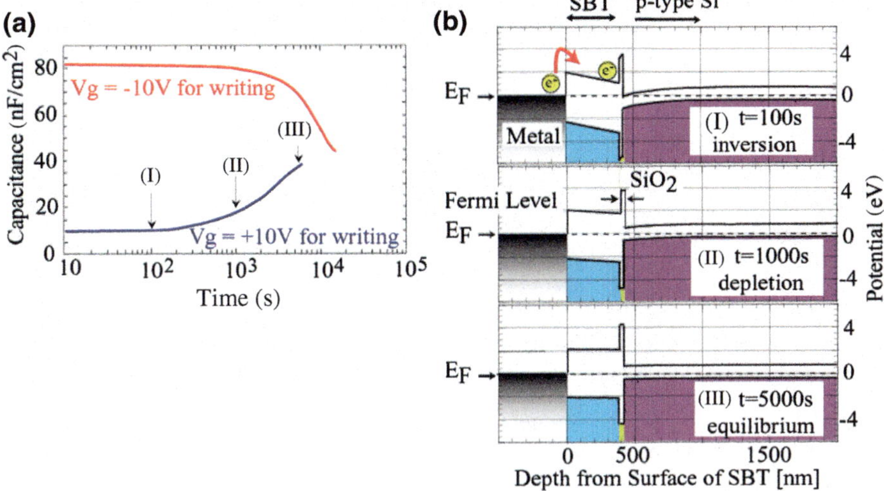

Fig. 1.6 **a** Capacitance retention characteristics and **b** band profiles for MFIS with SBT/SiO$_2$/p-Si structure: (*I*) inversion state, (*II*) depletion state, and (*III*) aged state

1.2.4 Effects of Currents Through the Ferroelectric and Insulator Layers on the Retention Characteristics of the MFIS Structure

Next, the effects of $J_i(E_i)$ and $J_f(E_f, t)$ on the retention characteristics of the MFIS structure were investigated. The retention time was defined as the time at which the capacitance difference between the ON and OFF states of the memorized MFIS was reduced to half of its initial value.

The variation of $J_i(E_i)$ with elapsed time in the MFIS was investigated with all other parameters fixed. Let us assume that $J_i(E_i)$ increases as $AJ_i(E_i)$. This assumption is plausible as the current can vary under real interface conditions such as changes in the barrier height and inhomogeneity of the Schottky junction. Figure 1.7a plots the capacitance versus elapsed time for various A. The retention time decreases rapidly as A increases beyond 10^4, whereas for $A < 10^4$, its maximum remains almost steady at 5×10^3 s. This result implies that when A is below $\sim 10^4$, the retention time is relatively robust, but if A exceeds $\sim 10^4$, the retention characteristics are seriously degraded.

The variation of $J_f(E_f, t)$ with elapsed time in the MFIS was investigated with all other parameters fixed. Here, $J_f(E_f, t)$ was assumed to vary as $BJ_f(E_i, t)$. $J_f(E_f, t)$ at a certain time was experimentally determined in [23]. When $B = 1$, meaning that $J_{f0}(E_f = 35 \text{ kV/cm}) \cong 10^{-8}$ A/cm^2, the time-varying capacitance of the calculated MFIS well agrees with the experimental results for Al/SBT/SiO$_2$/Si [23]. As shown in Fig. 1.7b, the retention time decreases rapidly as B increases above 10^{-3}, but its maximum remains almost constant (at 3×10^8 s) for $B < 10^{-3}$. This result implies

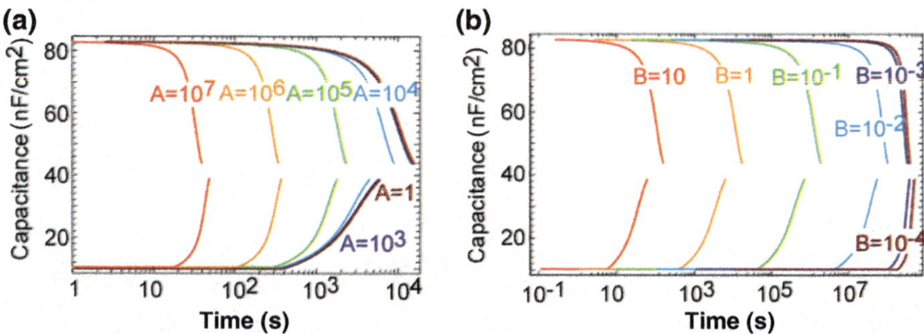

Fig. 1.7 Capacitance retention characteristics for different proportionality factors A and B of currents **a** across the semiconductor–insulator junction and **b** through the metal-ferroelectric layers

that, under the above assumption, the retention time can be extended by decreasing B to below 10^{-3}.

Comparing panels (a) and (b) of Fig. 1.7, we observe that the retention characteristics are improved more by reducing $J_f(E_f, t)$ than by reducing $J_i(E_i)$. For a favorable outcome in this calculation, the current through the ferroelectric should also rapidly decrease with time. Therefore, it is desired that the current through the ferroelectric is minimized as far as possible.

1.2.5 Methods for Suppressing Leakage Current Through the MFIS Structure

To enhance the memory retention time, several mechanisms for reducing the currents across the MFIS junctions were considered in this study. The main improvements can be made at the metal–ferroelectric Schottky barrier, ferroelectric layer, and insulator–semiconductor junction, as shown in Fig. 1.8.

Figure 1.8 shows five ways of suppressing the leakage current across the metal–ferroelectric Schottky junction: ① increasing the Schottky barrier height by decreasing/increasing the metal work function, ② inserting an ultrathin insulator layer between the metal and ferroelectric layers, ③ improving the quality of the surface region of the ferroelectric film, ④ reducing the number of trap states for decreasing the current through the ferroelectric layer, and ⑤ suppressing the Schottky and Fowler–Nordhaim currents by using a high-k insulator film.

The leakage current across the Schottky junction can be reduced by enlarging the barrier height, as shown in Eqs. (1.2) and (1.3). By selecting an appropriate metal material that increases the work function, the barrier height increase ($\Delta\phi_B$) might reach 0.12 eV, reducing the leakage current by a factor of 10^{-2}, and thereby enlarging the retention time by a factor of 10^4. However, the barrier height might not be increased by judicious choice of the metal electrode, because the Fermi level

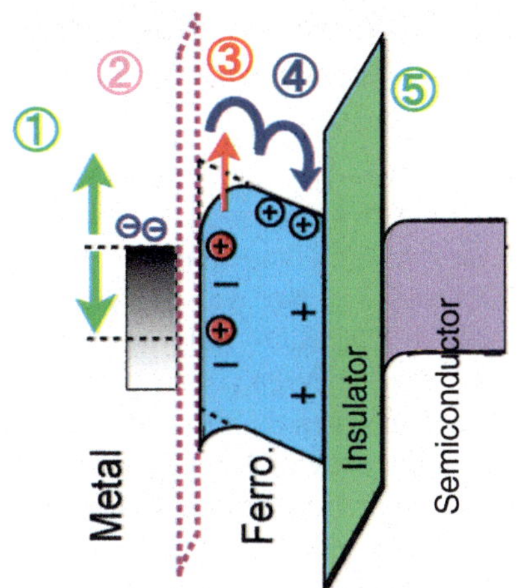

Fig. 1.8 Various mechanisms for reducing leakage current through MFIS structure: ① Decrease of metal work function. ② Insertion of Insulator film. ③ Surface improvement of ferroelectric film. ④ Decrease of trap density. ⑤ High-k insulator film

of a ferroelectric thin film tends to be pinned regardless of the metal's work function. Therefore, suppressing the current by selecting a metal that enhances the work function is considered unrealistic for improving the retention characteristics.

Alternatively, the current across the Schottky junction should be suppressed by inserting an ultrathin insulator layer between the metal and ferroelectric layers. The inserted layer should be as thin as possible, because a thick insulator would increase the depolarization field in the ferroelectric layer as well as the applied gate voltage for writing the states. If a 1-nm-thick SiO_2 film is inserted between the metal and the ferroelectric layers, the calculated retention times can extend over 10 years. However, a MgO film deposited on a SBT ferroelectric film only slightly improved the retention time, because current continued to penetrate the rough surface of the ferroelectric thin film. Thus, the insertion of an ultrathin insulator is not expected to dramatically improve the retention time.

A current passing through the ferroelectric layer encounters many traps that are produced during deposition, heat treatment, and device fabrication. Trapping and detrapping processes alter the charges, and charge-assisted carrier transport induces a space-charge-limited current. Reducing the trap density is required to suppress the leakage current.

Polarization of the ferroelectric layer induces a large electric field even if no voltage is applied to the MFIS structure. This electric field can be reduced by using a film of high dielectric constant (k) as the insulator layer. Many high-k dielectric thin films have been investigated in the fabrication of ULSI devices, and typical materials are HfO_2 (k = 20–25), La_2O_3 (k = 27), and Pr_2O_3 (k = 31). If the dielectric constant is increased from 3.9 (that of SiO_2) to 20–30, the retention time is expected to increase very much. MFIS FETs using HfO_2 high-k film yield excellent retention characteristics.

1.2.6 Retention Improvement by Heat and Radical Treatments

As discussed in the previous section, high-quality MFIS structures have been fabricated by inserting good insulator layers and by reforming the SBT ferroelectric layers through annealing and radical treatments [29–31]. A SiO_2 insulator film of thickness 7.5 nm was grown on an n-type Si wafer by the thermal oxidation method. The SiO_2 film was treated by a nitrogen radical beam produced by a radio-frequency gun. This beam consisted mainly of excited molecular and atomic neutral nitrogens with small quantities of N_2 molecules and N ions (the latter are repelled by the high field normal to the ion beam). X-ray photoelectron (XPS) and photoyield spectroscopies clarified that the surface of the SiO_2 film was densified by nitridation. The enhanced dielectric constant of the nitrided SiO_2 film and improvement of the insulator–Si interface improvement were confirmed by the enhanced accumulation capacitance and the steep accumulation–depletion transition in the C–V curve of the MOS structure. Thus, besides providing a high-k insulator, the nitride layer might also inhibit the thermal diffusions of the elements in the ferroelectric material. An SBT thin film was prepared on the nitride oxide/Si structure by chemical-solution deposition. The coated SBT film was rapidly thermally annealed in N_2 gas, and was again annealed at 750 °C for 120 min. After deposition and annealing, the SBT/insulator/Si was additionally treated by nitrogen radicals. The radical treatment smoothed the surface morphology of the SBT, removing some of the projected parts that induce large electric fields. Thus, the radical treatment was expected to improve the insulating property of the SBT film. Indeed, the Pt/SBT/insulator/Si MFIS structure exhibited very low leakage current, as shown in Fig. 1.9. As confirmed by ultraviolet-ray photoyield spectroscopy and XPS, the barrier height of the metal–SBT Schottky junction was enhanced by the nitridation. Figure 1.10 plots the capacitance retention characteristics of the

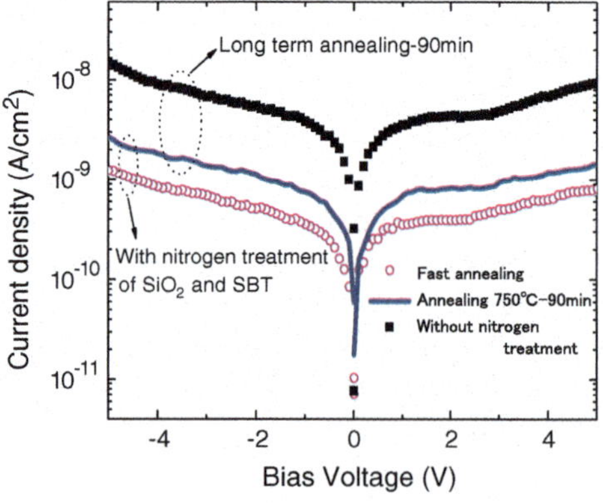

Fig. 1.9 Leakage current density-voltage characteristics of MFIS structures with and without nitrogen radical treatment

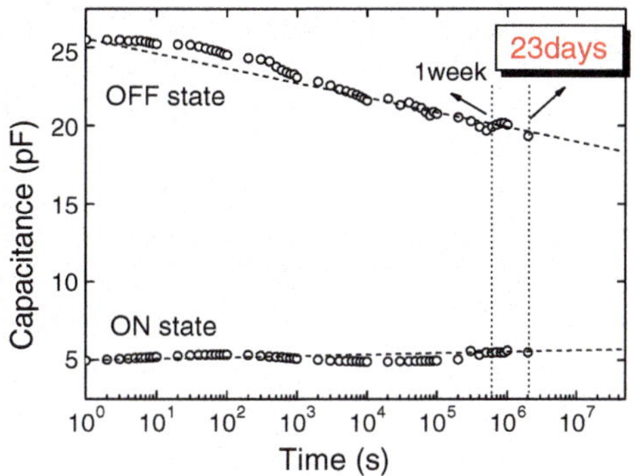

Fig. 1.10 Retention characteristics of capacitance of MFIS structures with nitrogen radical treatment

Pt/SBT/nitride oxide/Si structure. The MFIS structure fabricated by nitrogen radical irradiation and fast annealing shows excellent performance, with satisfactory retention over 23 days.

1.3 Improvement of Ferroelectric Gate FETs

High-performance memories have been fabricated from various kinds of ferro-electric gate FETs. Several of the successfully fabricated devices are described below.

1.3.1 1T2C-Type FET

The charge induced on a SiO_2 thin film over the current channel region has been reduced by connecting two same-sized ferroelectric capacitors to a metal gate electrode. This memory device, called the 1T2C-type FET [32, 33], is schematized in Fig. 1.11. Because the two capacitors are oppositely polarized in write mode, the charge on the metal electrode on the SiO_2 side is much suppressed, as the positive

Fig. 1.11 Structure of 1T2C type FET and an example of its operation voltages

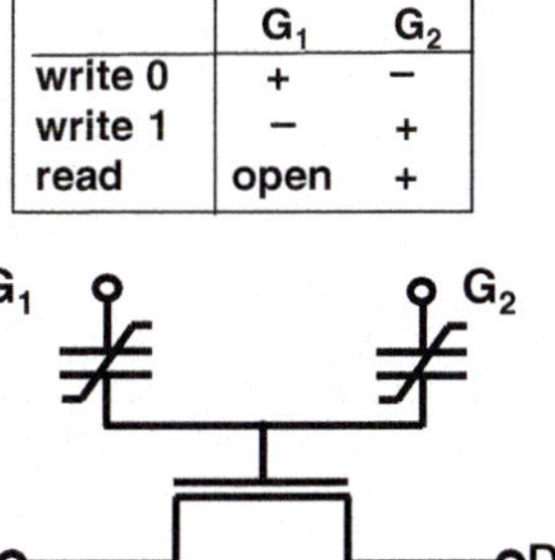

	G_1	G_2
write 0	+	−
write 1	−	+
read	open	+

and negative charges are induced and compensated. Consequently, almost no depolarization field is generated in the ferroelectric films. When the memorized state is read out, a voltage is applied to one capacitor, while the other capacitor is opened. In this device, the memory retention of the ON/OFF currents was prolonged up to 10^5 s. An integrated 1T2C device with small cell size has been fabricated on a silicon-on-insulator substrate.

1.3.2 MFMIS FET

A floating metal electrode inserted between the ferroelectric and insulator layers provides a highly ferroelectric thin film without damaging the SiO_2 gate layer. This gate stack is called an MFMIS structure. Multilayer films of $Ir/IrO_2/PZT/Ir$ have been successively stacked onto the poly-Si/SiO_2 gate of a conventional MOSFET by sputtering and sol-gel methods, as shown in Fig. 1.12 [34–36]. The IrO_2 thin film between the PZT/Ir and poly-Si/SiO_2 layers effectively suppresses element diffusion to the SiO_2/Si gate interface during PZT thin-film deposition. As mentioned in Sect. 1.1.2, the PZT film with Ir/IrO_2 electrode exhibits strong ferroelectric properties, resisting fatigue for up to 10^{12} polarization-reversal cycles [35]. The polarization and thickness of the ferroelectric as well as the thicknesses of the SiO_2 films should be adjusted to allow for sufficient voltage application to the SiO_2/Si structure. The drain-current–gate voltage characteristics of the fabricated MFMIS FET reveal significant hysteresis, corresponding to the polarization hysteresis of PZT film.

To improve the device characteristics of the MFMIS FET, researchers have optimized the area ratio of the MIS capacitor to the ferroelectric MFM capacitor [37, 38]. If a BIT thin film with a remanent polarization of approximately 13 $\mu C/cm^2$ is directly applied to the SiO_2/Si structure and is fully polarized, the additional electric field in the SiO_2 film is 38 MV/cm, almost four times larger than the breakdown field of high-quality SiO_2 film. In other words, when applying a small voltage to the MFIS structure to avoid breakdown in the SiO_2 film, the electric field in the ferroelectric reverses only the partial polarization, yielding insufficient polarization hysteresis. In the MFMIS structure, the polarization can be fully reversed by reducing the polarization charge, thus improving the characteristics

1.3.3 High-k Insulating Layer

MFIS FETs have been fabricated with high-k insulating films rather than with SiO_2 films. In these devices, HfO_2 or Hf–Al–O is used as the high-k insulating layer, and an SBT or BLT thin film forms the ferroelectric layer [20, 39–43]. When the MFIS structure is biased at 0 V, the depolarization field is much lower in the high-k film of the metal/ferroelectric/HfO_2/Si structure than in the SiO_2 film of the

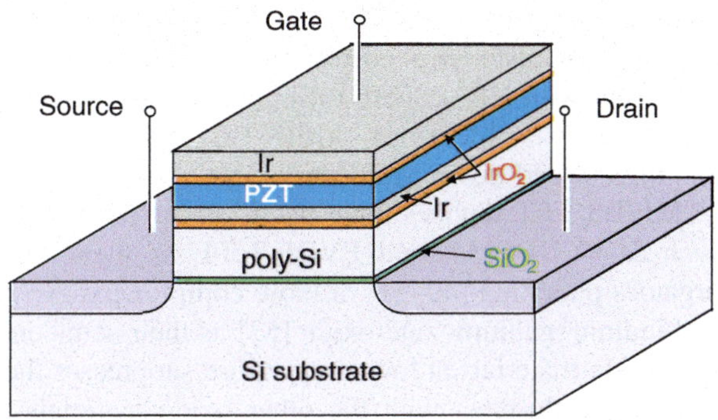

Fig. 1.12 Structure of MFMIS FET

metal/ferroelectric/SiO$_2$/Si structure for the same polarization of the ferroelectric film. Consequently, the depolarization field is reduced, current flows across the metal–ferroelectric Schottky and high-k film–Si junctions are suppressed, and the memory retention time is extended. The drain current–gate voltage (I_D–V_G) characteristics are rendered strongly hysteretic by polarization hysteresis (Fig. 1.13a). The heavily reduced gate current greatly extends the memory retention. Figure 1.13b plots typical data retention characteristics of FETs with SBT/H$_f$AlO gate structures. The drain currents of the ON and OFF states are largely unchanged after 37.0 and 33.5 days, respectively. Extrapolating the obtained retention curves of the ON and OFF states, the drain current ON/OFF ratio should exceed 10^4 at 10 years after writing the data.

1.3.4 New Materials

Recently, doping HfO$_2$ thin film with small amounts of elements has been shown to produce the good polarization hysteresis structure of the film [44]. MFIS FETs with

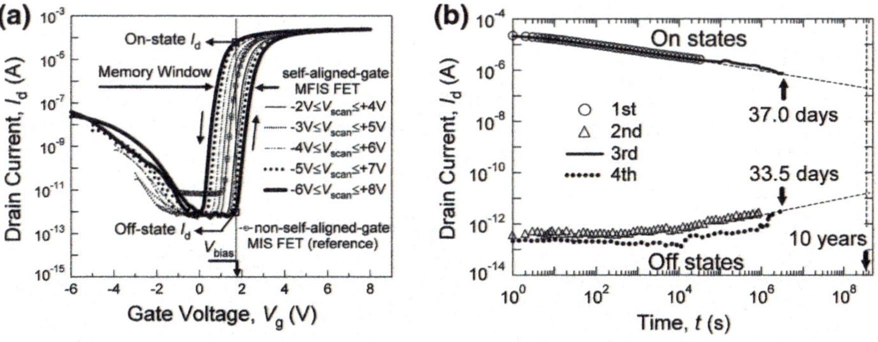

Fig. 1.13 a Drain current–gate voltage characteristics and **b** retention characteristics of Pt/SBT/HfAlO/Si MFIS FET. Copyright 2005 The Japan Society of Applied Physics

HfO$_2$:Si ferroelectric and HfO$_2$ high-k thin films exhibit good memory effects such as I$_D$–V$_G$ characteristics corresponding to ferroelectric hysteresis [45]. MFIS FETs fabricated with HfO$_2$:Si/HfO$_2$/Si gates display good memory retention and endurance behaviors. Moreover, MFIS FET performance is preserved when the gate length is scaled down to 28 nm.

MF(I)S FET devices have been fabricated with organic ferroelectric materials such as PVDF-TrFE and PVDF-TeFE as their ferroelectric layers [46–48] and organics plus ZnO [49–51] or more complex oxides such as indium–tin–oxide [52] and indium–gallium–zinc-oxide [53] as their semiconductor layers. The deposition of organic materials at low temperature suppresses thermal diffusion, and the oxide materials little react with the other oxide materials in combined oxide semiconductor–oxide ferroelectric materials. Therefore, in many instances, these materials require no insulator layer between the semiconductor and ferroelectric layers.

1.4 Conclusion

Feature, principle and improvement of nonvolatile ferroelectric-gate field effect transistors (FeFET) are overviewed. Switching and memory characteristics of the FeFET have been improved very much by adopting various structures, good ferroelectric and insulating thin films, and optimum fabrication processes. Especially, the memory retention has been largely extended to be more than thirty days which is available for the practical applications. It is expected very much that integrated memories of the FeFETs would be manufactured, and used to wide applications requiring nonvolatality, better scalability, higher read-write speeds, lower dissipation powers, higher tamper resistances and higher radioactivity tolerance.

Acknowledgments The author would like to thank Dr. Mitsue Takahashi and Profs. Minoru Noda and Takeshi Kanashima for helping this work, and Enago (www.enago.jp) for the English language review.

References

1. J.L. Moll, Y. Tarui, IEEE Trans. Electron Devices **ED-10**, 338 (1963)
2. R. Zuleeg, H.H. Wiede, Solid State Electr. **9**, 657 (1966)
3. S.S. Perlman, K.H. Ludewig, IEEE Trans. Electron Devices **ED-14**, 816 (1967)
4. J.H. McCuster, S.S. Perlman, IEEE Trans. Electron Devices **ED-15**, 182 (1968)
5. G.G. Teather, L. Young, Solid State Electr. **9**, 527 (1968)
6. J.C. Crawford, F.L. English, IEEE Trans. Electron Devices **ED-16**, 525 (1969)
7. S.-Y. Wu, IEEE Trans. Electron Devices **ED-21**, 499 (1974)
8. K. Sugibuchi, Y. Kurogi, N. Endo, J. Appl. Phys. **46**, 2877 (1975)
9. Y. Higuma, Y. Matsui, M. Okuyama, T. Nakagawa, Y. Hamakawa, in Proceedings of the 9th Conference on Solid State Devices, Tokyo (1977)

10. Y. Hamakawa, Y. Matsui, Y. Higuma, Y. Hamakawa, in Proceedings of IEEE IEDM Conference, Washington D.C. (1977)
11. Y. Matsui, Y.Higuma, M. Okuyama, T. Nakagawa, Y. Hamakawa, in Proceedings of the 1st Conference on Ferroelectric Material Applications, Kyoto (1977)
12. W.I. Kinney, W. Sheoherd, W. Miller, J. Evans, R. Womack, in Technical Digest of IEEE (IEDM, Washington, D.C., USA, Dec., 1987), p. 850
13. S.S. Eaton, D.B. Butler, M. Parris, D. Wilson, H. McNeillie, in Digest of Technical IEEE Papers of International Solid State Circuit Conference, San Francisco, USA, 31 Feb 1988, p. 130
14. J.F. Scott, *Ferroelectric Memories*. Springer Series on Advanced Microelectronics, vol. 3 (Springer, Berlin, 2000)
15. H. Ishiwara, M. okuyama, Y. Arimoto, *Ferroelectric Random Access Memories*. Topics in Applied Physics 93 (Springer, Berlin, 2004)
16. M. okuyama, Y. Ishibashi: *Ferroelectric Thin Films*. Topics in Applied Physics 98 (Springer, Berlin, 2005)
17. C.A.-P. Araujo, J.D. Cuchiaro, L.D. McMillan, M.C. Scott, J.F. Scott, Nature **374**, 627 (1995)
18. B.H. Park, B.S. Kang, S.D. Bu, T.W. Noh, J. Lee, W. Jo, Nature **401**, 682 (1999)
19. T. Nakamura, Y. Nakao, A. Kamisawa, H. Takasu, Jpn. J. Appl. Phys. **34**, 5184 (1995)
20. K. Takahashi, K. Aizawa, B.-E. Park, H. Ishiwara, Jpn. J. Appl. Phys. **44**, 6218 (2005)
21. M. Okuyama, M. Noda, *Topics in Applied Physics,* vol. 98, eds. by M. Okuyama, Y. Ishibashi (Springer, Berlin, 2005), p. 219
22. M. Okuyama, M. Takahashi, H. Sugiyama, T. Nakaiso, K. Kodama, M. Noda, in Proceedings of the 12th IEEE International Symposium on Applications of Ferroelectrics (2000) pp. 337–340
23. M. Okuyama, H. Sugiyama, T. Nakaiso, M. Noda, Integr. Ferroelectr. **34**, 37 (2000)
24. M. Okuyama, M. Takahashi, K. Kodama, T. Nakaiso, M. Noda, Mat. Res. Soc. Symp. Proc. **655**, cc13.10.1 (2000)
25. M. Takahashi, H. Sugiyama, T. Nakaiso, K. Kodama, M. Noda, M. Okuyama, Jpn. J. Appl. Phys. **40**, 2923 (2001)
26. S.M. Sze, *Physics of Semiconductor Devices*, 2nd ed, Chap. 7 (A Wiley-Interscience Publication, New York, 1981), p. 403
27. S.L. Miller, P.J. McWhorter, J. Appl. Phys. **72**, 5999 (1992)
28. T. Ohmi, M. Morita, A. Teramoto, K. Makihara, K.S. Tseng, Appl. Phys. Lett. **60**, 2126 (1992)
29. L. Van Hai, T. Kanashima, M. Okuyama, *INTECH, Ferroelectric Materials—Material Aspects*, ed. by M. Lallart, Chap. 7 (2011), p. 129
30. L. Van Hai, T. Kanashima, M. Okuyam, Integr. Ferroelectr. **96**, 27 (2008)
31. L. Van Hai, T. Kanashima, M. Okuyama, Integr. Ferroelectr. **84**, 179 (2006)
32. S.-M. Yoon, H. Ishiwara, IEEE Trans. Electron Devices **48**, 2002 (2001)
33. S. Yamamoto, H.-S. Kim, H. Ishiwara, Jpn. J. Appl. Phys. **42**, 2059 (2003)
34. Y. Nakao, T. Nakamura, A. Kamisawa, H. Takasu, Integr. Ferroelectr. **6**, 23 (1995)
35. T. Nakamura, Y. Nakao, A. Kamisawa, H. Takasu, Integr. Ferroelectr. **9**, 179 (1995)
36. Y. Fujimori, N. Izumi, T. Nakamura, A. Kamisawa, Jpn. J. Appl. Phys. **37**, 5207 (1998)
37. E. Tokumitsu, G. Fuji, H. Ishiwara, Jpn. J. Appl. Phys. **39**, 2125 (2000)
38. T. Suzuki, E. Tokumitsu, Jpn. J. Appl. Phys. **41**, 6886 (2002)
39. S. Sakai, R. Ilangovan, IEEE Electron Device Lett. **25**, 369 (2004)
40. K. Aizawa, B.-E. Park, Y. Kawashima, K. Takahashi, H. Ishiwara, Appl. Phys. Lett. **85**, 3199 (2004)
41. S. Sakai, R. Ilangovan, M. Takahashi, Jpn. J. Appl. Phys. **43**, 7876 (2004)
42. M. Takahashi, S. Sakai, Jpn. J. Appl. Phys. **44**, L800 (2005)
43. X. Zhang, K. Takeuchi, M. Takahashi, S. Sakai, Jpn. J. Appl. Phys. **51**, 04DD01 (2012)
44. U. Schroeder, S. Mueller, J. Mueller, E. Yurchuk, D. Martin, C. Adelmann, T. Schloesser, R. van Bentum, T. Mikolajick, ECS J. Solid State Sci. Technol. **2**, N69 (2013)

45. J. Muller, T.S. Boscke, U. Schroeder, R. Hoffman, T. Mikolajick, IEEE Electron. Device Lett. **33**, 185 (2012)
46. R.C.G. Naber, C. Tanase, P.W.M. Blom, G.H. Gelinck, A.W. Marsman, F.J. Touwslager, S. Setayesh, D.M. De Leeuw, Nat. Mater. **4**, 243 (2005)
47. G.G. Lee, B.E. Park, J. Kor. Phys. Soc. **56**, 1484 (2010)
48. T. Watanabe, H. Miyashita, T. Kanashima, M. Okuyama, Jpn. J. Appl. Phys. **49**, 04DD14 (2010)
49. S.-H. Noh, W. Choi, M.S. Oh, S, Jang, E. Kim, Appl. Phys. Lett. **90**(25), 253504 (2007)
50. Y. Kato, Y. Kaneko, H. Tanaka, Y. Shimada, Jpn. J. Appl. Phys. **47**, 2719 (2008)
51. T. Fukushima, T. Yoshimura, K. Masuko, A. Ashida, N. Fujimura, Jpn. J. Appl. Phys. **47**, 8874 (2008)
52. E. Tokumitsu, M. Senoo, T. Miyasako, Microelectr. Eng. **80**(Suppl.), 305 (2005)
53. S.-M. Yoon, S.-H. Yang, S.-W. Jung, E. Tokumitsu, H. Ishiwara, Appl. Phys. Lett. **96**, 232903 (2010)

Part II

Practical Characteristics of Inorganic Ferroelectric-Gate FETs: Si-Based Ferroelectric-Gate Field Effect Transistors

Chapter 2
Development of High-Endurance and Long-Retention FeFETs of Pt/Ca$_y$Sr$_{1-y}$Bi$_2$Ta$_2$O$_9$/(HfO$_2$)$_x$(Al$_2$O$_3$)$_{1-x}$/Si Gate Stacks

Mitsue Takahashi and Shigeki Sakai

Abstract Studies of our Pt/Ca$_y$Sr$_{1-y}$Bi$_2$Ta$_2$O$_9$(CSBT(y))/(HfO$_2$)$_x$(Al$_2$O$_3$)$_{1-x}$(HAO (x))/Si MFIS FeFETs were reviewed which were originated from the Pt/SrBi$_2$Ta$_2$O$_9$(SBT)/HAO(x = 0.75)/Si FeFET invented in 2002. Electrical properties of the fist FeFET were introduced which were 10^6 s-long retention, 10^{12} cycles-high endurance and 4×10^{-8} s-demonstrated writing speed. Stable I_d–V_g curves and I_d-retentions were measured up to 85 °C using p-channel FeFETs. Individual requirements to the M, F, I and IL layers as the components of MFIS were discussed using a band profile of the Pt/SBT/HAO(x = 0.75)/Si. Experimental studies for improving the HAO(x) and IL layer qualities were introduced. The composition ratio x in HAO(x) was optimized using single HAO(x) films and the MIS characters which all underwent a standard 800 °C annealing for SBT poly-crystallization. The ratio $x \geqq 0.75$ was found to be suitable for the I layer in the MFIS. As an ambient gas in depositing HAO(x = 0.75) by PLD, O$_2$ and N$_2$ were compared. In the Pt/SBT/HAO(x = 0.75)/Si FeFET, the HAO worked as a material-diffusion barrier only when it was deposited in N$_2$. Effect of increasing the ambient N$_2$ pressure was studied using the FeFETs. The pressure should be less than 40 Pa for keeping a clear interface between the SBT and HAO. Direct nitriding Si was studied for enlarging the memory window of Pt/SBT/HAO(x = 0.75)/Si FeFET. Oxinitriding Si was also demonstrated as a modified way to decrease the subthreshold-voltage swing of the FeFET. Experimental works to use CSBT(y) instead of the SBT was also introduced. The Pt/CSBT(y = 0.1, 0.2)/HAO(x = 0.75)/Si FeFETs showed wider pulse-memory window V_{plsW} than the reference Pt/SBT/HAO(x = 0.75)/Si FeFET at the common measurement conditions. When $(V_E, V_P) = (-5, 7\ \mathrm{V})$ and $t_{pls} = 1\ \mu s$, the Pt/CSBT(y = 0.1, 0.2)/HAO(x = 0.75)/Si

M. Takahashi (✉) · S. Sakai
National Institute of Advanced Industrial Science and Technology,
Tsukuba, Ibaraki 305-8568, Japan
e-mail: mitsue-takahashi@aist.go.jp

S. Sakai
e-mail: shigeki.sakai@aist.go.jp

© Springer Science+Business Media Dordrecht 2016
B.-E. Park et al. (eds.), *Ferroelectric-Gate Field Effect Transistor Memories*,
Topics in Applied Physics 131, DOI 10.1007/978-94-024-0841-6_2

FeFETs showed $V_{\text{plsW}} = 0.35$ V which was 13 % larger V_{plsW} than the reference FeFET.

2.1 Introduction

In developing semiconductor memories, downsizing has been a general trend common to all memories for long time. The most successful semiconductor memory today is a flash memory which has advantages of not only data non-volatility with low-power consumption but also high scalability. Recently the flash memory has the feature size under 20 nm and is coming close to the limit of downsizing worth developing and manufacturing at a huge cost [1]. Alternative or supplemental nonvolatile memories, so-called emerging memories, have been intensively developed which have some additional values of faster access speed, higher endurance and lower power consumption than the conventional flash memory. The emerging memories are using functional materials such as resistive switching, phase change, magnetic and ferroelectric [2].

Ferroelectric-gate field effect transistor (FeFET) is a nonvolatile memory transistor using a ferroelectric material in the gate insulator [3]. The FeFET is a voltage-driven one-transistor (1T) memory as well as a flash memory cell. The 1T memories have advantages of non-destructive-read operation and potential high scalability in integrating memory cells by $4F^2$. The F is a feature size which is the minimum size manufacturable in every semiconductor-process generation. An FeFET is using ferroelectric-polarization switching for programming and erasing the data. Hence it has the intrinsic low-power consumption and high endurance against program-and-erase cycles. Actually we reported that FeFETs of metal/ferroelectric/insulator/semiconductor (MFIS) gate stacks had about 1/3 as small program voltage and 10^4 times as high endurance as a flash-memory cell [4, 5]. As far, we have investigated the MFIS FeFET with two directions. One is finding a good fabrication process for downsizing the single FeFETs which leads to the future manufacturing process. The other is demonstrating operations of multiple-FeFET integrated circuits introduced later in Chap. 13. This chapter is a part of the former research directions in which we discuss materials and the fabrication process of MFIS FeFETs to increase the memory windows in preparation for the prospective reduction of the F layer thicknesses. The thinning of the Flayer is important for downsizing an FeFET because the F layer is occupying the most volume of the gate stack. Since we reported the first self-aligned-gate FeFET with at least 33.5 days-long retention in 2005 [6], gate length of FeFET was decreased from the early 2 μm to the recent 100 nm with our technological progress in lithography, etching and ion implantation. On the basis of the studies reviews in this chapter, the 100 nm-gate size FeFETs with 4×10^5 s long retentions and 10^8 cycles endurances were achieved [7, 8].

The first long-retention FeFETs were invented in 2002 by Sakai. The structures and production method were patent-applied [9] and the electrical properties were press-released [10]. Later the characterization of the 1×10^6 s-long retention, the 10^{12} cycles-high endurance and writing speed was reported by Sakai and Ilangovan in an academic paper [11]. The FeFETs had MFIS gate stacks of Pt/SrBi$_2$Ta$_2$O$_9$(SBT)/(HfO$_2$)$_x$(Al$_2$O$_3$)$_{1-x}$(HAO)/Si [9–11] or Pt/SBT/HfO$_2$/Si [9]. The SBT is a kind of bismuth-layered perovskite ferroelectric materials. Weak fatigues and small leakage currents of metal/ferroelectric/metal (MFM) capacitors using the SBT were reported [12]. The ferroelectric SBT is often made from a precursor oxide Sr-Bi–Ta-O by a poly-crystallization annealing. The annealing temperature at least 650 °C is necessary and that around 800 °C is sufficient for exhibiting the ferroelectric performance effectively [13]. Otherwise, if the annealing temperature is 550 °C for example, the raw material will grow into cubic crystals with paraelectric natures [14]. Since 2002, we have developed the MFIS FeFET according to the device fabrication policy as follows. The top priority was using bismuth-layered perovskite ferroelectrics like the SBT and CSBT for the F layer in the expectation of producing high-endurance FeFETs derived from the fatigue-free natures of the ferroelectric materials [12]. The second was annealing the MFIS in high temperature at about 800 °C for maximizing the ferroelectric quality. Materials for the M and I were selected from heat resistant materials. We tested their compatibility with Si and the bismuth-layered perovskite by stacking them in MFIS and annealing at about 800 °C altogether. The MFIS of Pt/SBT/HAO/Si was the first successful solution to realize an excellent FeFET with long retention and high endurance [9–11]. Since then, single FeFETs have been developed on a basis of advancing the Pt/SBT/HAO/Si as follows.

2.2 Basic Fabrication Process and Characterization of Pt/SBT/HAO/Si FeFETs

2.2.1 Fabrication Process

We introduce our basic fabrication process of n-channel non-self-aligned-gate FeFETs consisting of Pt/SBT/HAO/Si stack. Schematic cross-section of an n-channel non-self-aligned-gate MFIS FeFET is shown in Fig. 2.1. First, p-type Si wafers were prepared on which n^+ source-and-drain and p^+ substrate regions for probing were formed in advance. After removing a sacrificial SiO$_2$ on the Si by

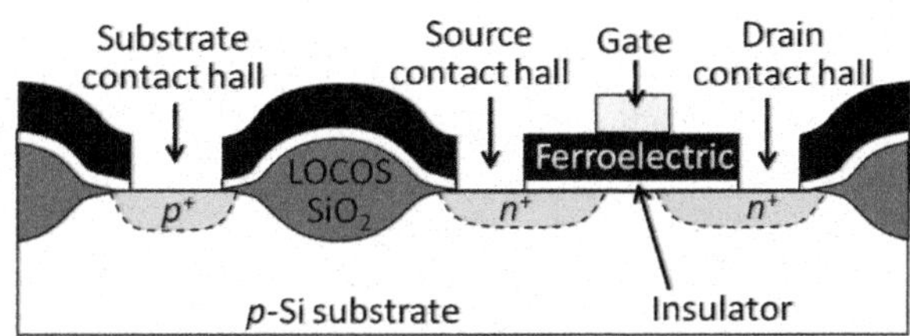

Fig. 2.1 Schematic cross-section of non-selfaligned gate FeFET

buffered hydrofluoric acid (BHF), HAO($x = 0.75$) was deposited on the Si by KrF-excimer pulsed-laser-deposition (PLD) in 13 Pa N_2 ambient at 200 °C substrate temperature. A ceramic target of Hf–Al-O with the molar ratio corresponding to HfO_2:$Al_2O_3 = 3$:1 was used for depositing the HAO. Then $SrBi_2Ta_2O_9$ (SBT) was deposited on the HAO/Si by the PLD in 13 Pa O_2 ambient at 400 °C substrate temperature. A ceramic target of Sr–Bi–Ta–O with the element ratio Sr: Bi:Ta = 1:3:2 was used for depositing the SBT. On the SBT/HAO/Si, Pt was deposited in vacuum by 50 kV electron-beam (EB) evaporator. The thicknesses of the Pt, SBT and HAO layers were controlled by the individual deposition-time lengths. For example, the thicknesses we often used were about 200 nm for the Pt, 400 nm for the SBT and 13 nm for the HAO. The Pt thickness was verified by a stylus profiler. The SBT and HAO thicknesses were verified by an ellipsometer. After preparing the Pt/SBT/HAO/Si, the stack was once annealed in 1 atm O_2 at 400 °C for 30 min in a furnace. Photo-resist masks of the FeFET gates were patterned on the Pt/SBT/HAO/Si by photolithography using a g-line stepper. The next process was an etching by Ar^+ ion milling with the beam voltage 500 V and current 20 mA in 1.3×10^{-4} torr Ar ambient. The Pt gate electrodes were formed by the Ar^+ milling. The typical gate length (L) of a non-self-aligned gate FeFET was $L = 10$ μm. The gate widths (W) were $W = 10$, 20, 40, 50, 80, 100, 150 and 200 μm which were the typical pattern sizes in a photo-mask set of ours. The Pt/SBT/HAO/Si was annealed in O_2 at 800 °C for the SBT poly-crystallization. Then the photolithography was used again for making photo-resist patterns to open source-and-drain and substrate contact holes on the Si by Ar^+ ion milling. Consequently, we completed fabrication of an FeFET. The FeFET was a kind of metal-oxide-semiconductor (MOS) FETs having four terminals of gate, drain, source and substrate. The FeFETs were characterized by measurements of drain current versus gate voltage (I_d–V_g) curves, data retention, endurance, and writing speed as follows.

2.2.2 Static Memory Window

Static memory window is basic information of an FeFET quality as a memory device. It is characterized by I_d–V_g curves measured using a semiconductor parameter analyzer. The I_d–V_g curve is drawn by measuring I_d values of an FeFET with static scanning V_g from $V_{base} - V_{SA}$ to $V_{base} + V_{SA}$ and back to $V_{base} - V_{SA}$ or in a simplified description $V_g = V_{base} \pm V_{SA}$. The V_{base} is a base V_g and the V_{SA} is V_g scan amplitude.

Figure 2.2a showed an I_d–V_g of a Pt/SBT/HAO($x = 0.75$)/Si FeFET measured at $V_g = \pm6$ V which meant $V_{base} = 0$ V and $V_{SA} = 6$ V [11]. The drain voltage (V_d), source voltage (V_s), substrate voltage (V_{sub}) were kept at $V_d = 0.1$ V and $V_s = V_{sub} = 0$ V. The n-channel FeFET shows the counterclockwise I_d–V_g loop with the right upward- and left downward-paths as indicated in Fig. 2.2a. A p-channel FeFET shows the opposite clockwise I_d–V_g loop as later shown in

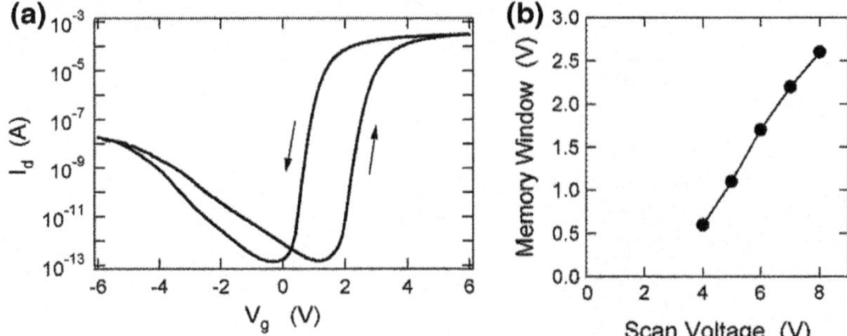

Fig. 2.2 **a** I_d–V_g curve of an n-channel Pt/SBT/HAO(x = 0.75)/Si FeFET with L = 10 μm and W = 200 μm. Thicknesses in the gate stack were 200 nm Pt, 400 nm SBT and 13 nm HAO. **b** V_w − V_{SA} extracted from the I_d–V_g curves for various V_{SA}. Modified from [11]

Fig. 2.3 x dependence of static memory window at V_g = ±6 V of n-channel Pt/SBT/HAO(x)/Si FeFETs. Thicknesses in the gate stack were 200 nm Pt, 400 nm SBT and 14 nm HAO(x). The gate area sizes were L = 10 μm and W = 200 μm. Modified from [15]

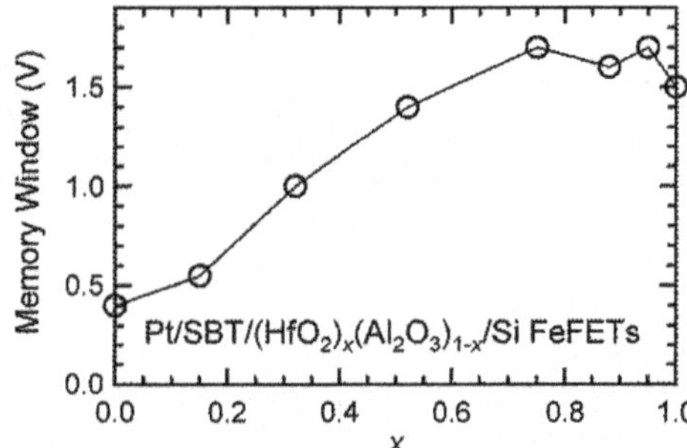

Fig. 2.8. Static memory window (V_w) of an FeFET was defined as a difference between two threshold voltage (V_{th}) values in the I_d–V_g loop which were regarded as the V_g values at I_d = 10^{-6} A for simplification in this work. The reason of using x = 0.75 in HAO was the largest static memory window of the Pt/SBT/HAO (x = 0.75)/Si FeFET among the Pt/SBT/HAO(x)/Si FeFET with the x ranged from 0 to 1.0 [9, 15] as indicated in Fig. 2.3. The memory windows at V_g = 2 ± 6 V were extracted from the I_d–V_g curves of the FeFETs which were annealed at 800 °C in O_2 for 1 h. Thicknesses were 200 nm for the Pt, 14 nm for the HAO and 400 nm for the SBT.

I_d–V_g curves of the Pt/SBT/HAO(x = 0.75)/Si FeFET were precisely measured at various V_g ranged from V_g = ±4 V to ±8 V. The V_w extracted from the I_d–V_g curves exhibited a monotonic increase as the V_{SA} was raised from 4 to 8 V as shown in Fig. 2.2b [11]. The increasing V_W in Fig. 2.2b indicated that ferroelectric polarization of the SBT in the FeFET was not saturated even at the large V_{SA} of 8 V. The reason of the unsaturated SBT polarization was a growth of an interfacial layer (IL) between the HAO and Si during the SBT crystallization annealing at 800 °C [15–17]. As shown in Fig. 2.4, cross-sectional transmission electron microscopy (TEM) and the point analyses by energy dispersive X-ray spectroscopy

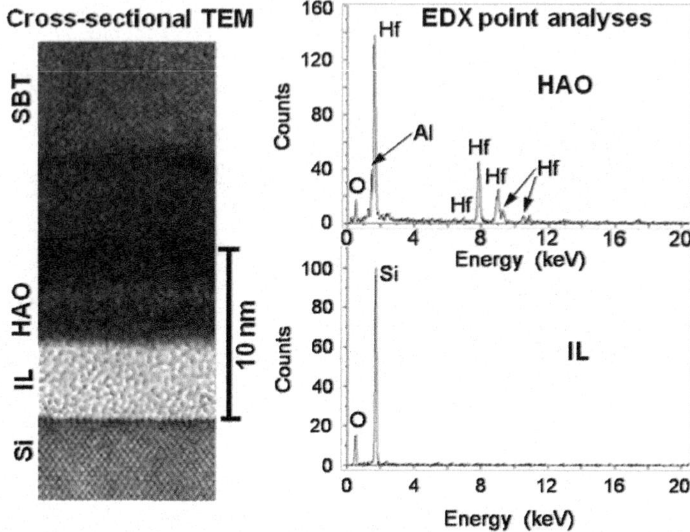

Fig. 2.4 Cross-sectional TEM picture of an SBT/HAO($x = 0.75$)/Si sample and the EDX point analyses at the HAO and the IL. The sample underwent an O_2 annealing at 800 °C for 1 h. The deposited initial thicknesses were 400 nm SBT and 14 nm HAO. Modified from [15]

(EDX) suggested that the IL was SiO_2 with a low dielectric constant of $\varepsilon_{IL} = 3.9$. The total V_g across the MFIS was divided and shared among the layers of F, I and S, in proportion to the inverse of the individual capacitances connected in series. Under a given V_g, the voltage across the IL-SiO_2 was fairly large because of the low ε_{IL}. As a result, the voltage across the SBT became so small that the FeFET shows the unsaturated polarization as indicated in Fig. 2.2b. From the different points of view, however, the growth of the IL also gives a benefit of protecting the Si interface by the flat SiO_2 film with the uniform quality as later discussed in Sect. 2.3. The good Si interface was suggested by the steep I_d–V_g curves with a small subthreshold-voltage swing (S). The S in our FeFET works was basically about $S = 100$ mV/decade as we later described in Sect. 2.3.2.

2.2.3 Retention

Retention of an FeFET is hold-time dependence of either I_d or V_{th} binary state. It is the evidence of data nonvolatility of the FeFET. In this study, we measured I_d retentions which were the time-dependences of two I_d values for on- and off-states, respectively. The on-state I_d versus time (I_d–t) was measured after a programming gate voltage $V_g = V_P$ was applied. The off-state I_d–t was measured after an erasing gate voltage $V_g = V_E$ was applied. Each I_d–t curve was measured with keeping a common hold gate voltage $V_g = V_{hold}$. As shown in Fig. 2.5, a Pt/SBT/HAO/Si

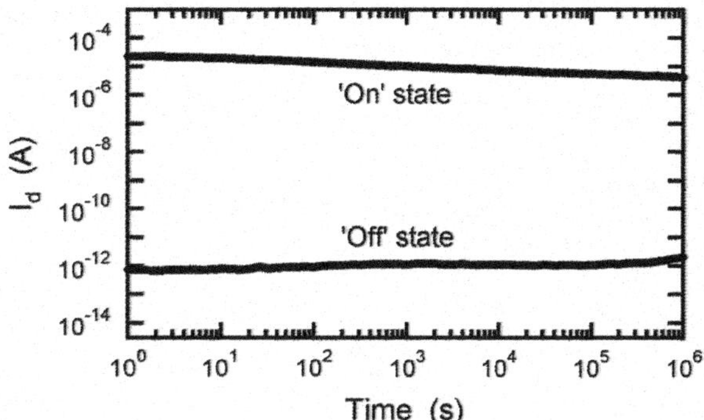

Fig. 2.5 I_d-retention of an n-channel FeFET measured at $V_{hold} = 1.7$ V after the poling by $V_g = \pm 6$ V. The FeFET had Pt/SBT/HAO($x = 0.75$)/ Si. Thicknesses in the gate stack were 200 nm Pt, 400 nm SBT and 13 nm HAO. The gate area size was $L = 10$ µm and $W = 200$ µm. Modified from [11]

FeFET exhibited the I_d retentions with no significant degradation for at least 1×10^6 s or 11.6 days [11]. They were measured at room temperature. For the on-state I_d measurement, $V_g = V_P = 6$ V was applied for poling then reduced to $V_g = V_{hold} = 1.7$ V. The V_g was applied by a dc voltage source connected to the gate and substrate of the FeFET. During the V_g swing, we kept V_d, V_s and V_{sub} at $V_d = V_s = V_{sub} = 0$ V. The on-state I_d measurement was immediately started and was continued for 1×10^6 s with keeping $V_d = 0.1$ V and $V_s = V_{sub} = 0$ V. The I_d–t was measured by a time-sampling mode of a semiconductor parameter analyzer connected to the drain and source of the FeFET. For the off-state I_d measurement, $V_g = V_E = -6$ V was applied for poling then raised to $V_g = V_{hold} = 1.7$ V by a dc voltage source with keeping $V_d = V_s = V_{sub} = 0$ V. The off-state I_d measurement was immediately started and was continued for 1×10^6 s with keeping $V_d = 0.1$ V and $V_s = V_{sub} = 0$ V by the semiconductor parameter analyzer in the time-sampling mode. In this study, we empirically determined the V_{hold} to maximize the ratio of the on-state $\log(I_d)$ to the off-state $\log(I_d)$. There was a strong positive correlation between the V_{hold} and the flat-band voltage of the MFIS, therefore, we could adjust the V_{hold} to $V_{hold} = 0$ V by optimizing the ion-implantation condition of the Si [18]. The other option for reducing the V_{hold} to 0 V may be using another heat-resistive metal for the M layer which has a smaller work function than Pt [19].

2.2.4 Endurance

Endurance of an FeFET shows how many times of program and erase operations are accepted before incorrect writing begins due to the FeFET degradation. The endurance of a Pt/SBT/HAO/Si FeFET was investigated by measuring static I_d–V_g curves after many cycles of endurance pulses were imposed. In this study, the endurance pulses were V_g pulses of alternate $V_P = 8$ V and $V_E = -8$ V with 1 µs period as shown in the inset of Fig. 2.6 [11]. The endurance V_g pulses applied on the gate were outputted from a pulse generator and transmitted through a 50 Ω coaxial cable terminated with a 50 Ω resistance. Accuracy of the pulse-wave forms

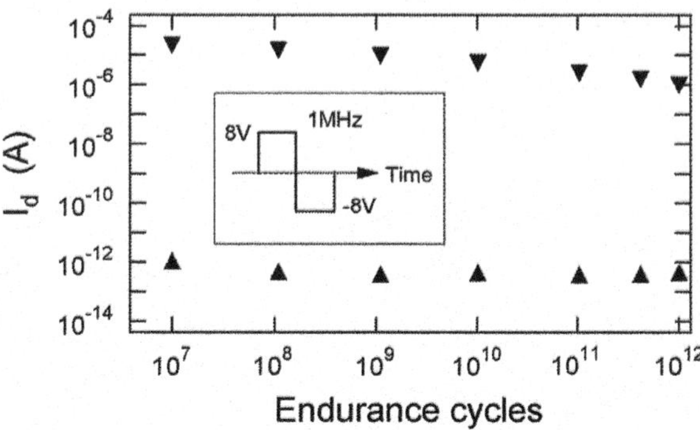

Fig. 2.6 Endurance of an n-channel Pt/SBT/HAO ($x = 0.75$)/Si FeFET. The inset shows a V_g endurance-pulse cycle. Thicknesses in the gate stack were 200 nm Pt, 400 nm SBT and 13 nm HAO. The gate area size was $L = 10$ μm and $W = 200$ μm. Modified from [11]

were checked by an oscilloscope in advance. We fixed V_d, V_s and V_{sub} at $V_d = V_s = V_{sub} = 0$ V during the endurance pulses were imposed. Every time after counting the accumulated numbers of endurance cycles to 10^7, 10^8, 10^9, 10^{10}, 10^{11}, 5×10^{11} and 10^{12}, the pulse application was interrupted and the V_g connection was changed to a semiconductor parameter analyzer. Then a static I_d–V_g curve was drawn by scanning V_g from -6 to 6 V and back to -6 V, or $V_g = \pm 6$ V in a simplified description, with keeping $V_d = 0.1$ V and $V_s = V_{sub} = 0$ V. On- and off-state I_d values were extracted from the I_d–V_g loop at a common $V_g = 2$ V and plotted in Fig. 2.6. After drawing the static I_d–V_g curve by $V_g = \pm 6$ V, the V_g connection was changed back to the pulse generator and the endurance-pulse application was resumed toward the next accumulated number of endurance cycles. As shown in Fig. 2.6, the FeFET maintained more than 6 order difference between the on- and off-state I_d values even after 10^{12} endurance cycles. Note that V_{th} endurance was recently investigated as discussed in Sects. 2.5 and 2.6 instead of the I_d endurance introduced in this study.

2.2.5 Writing Speed

Pulse writing is a substantial writing technique rather than the static writing for the practical FeFET use. Writing speed of an FeFET is estimated by measuring V_g-pulse-width (t_{pls}) dependence of either I_d or V_{th} binary state. To be exact, the writing is programming and erasing. As mentioned in Sect. 2.2.2, ferroelectric polarization in FeFETs is usually unsaturated. According to a study of ferroelectric switching kinetics [20], unsaturated polarizations tend to have slower switching speeds than saturated polarizations. The study suggests that FeFETs have significant t_{pls} dependence of the memory windows due to the unsaturated ferroelectric polarizations. As t_{pls} increases, memory windows measured by the pulsed V_g applications become close to those measured by static V_g applications.

In this work, we introduced the t_{pls} dependence of I_d in programming and erasing a Pt/SBT/HAO($x = 0.75$)/Si FeFET as shown in Fig. 2.7 [11]. The gate and

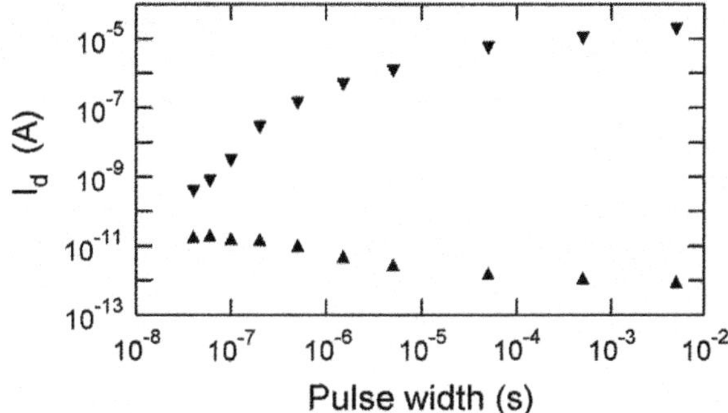

Fig. 2.7 Pulse width t_{pls} dependence of on- and off-state I_d value of an n-channel Pt/SBT/HAO ($x = 0.75$)/Si FeFET. Thicknesses in the gate stack were 200 nm Pt, 400 nm SBT and 13 nm HAO. The gate area size was $L = 10$ μm and $W = 200$ μm. Modified from [11]. Unpublished data were added

substrate of the FeFET were connected to a pulse generator. The drain and source of the FeFET were connected to a semiconductor parameter analyzer. The single V_g pulses were outputted from the pulse generator and transmitted through a 50 Ω coaxial cable terminated with a 50 Ω resistance. Accuracy of the pulse shape was confirmed using $(2 \times t_{pls})$-period ± 8 V V_g wave by an oscilloscope in advance. The pair of $I_d - t_{pls}$ characteristics for 8 and -8 V V_g-pulse applications was measured with stepping up the t_{pls} from 4×10^{-8} to 5×10^{-3} s. For measuring an on-state-I_d at a t_{pls}, a single programming V_g pulse of a fixed height V_P and the time width t_{pls} was applied by the pulse generator. During the programming V_g pulse application, we fixed V_d, V_s and V_{sub} at $V_d = V_s = V_{sub} = 0$ V. In the programming V_g pulse, V_g started from $V_{base} = 2$ V, increased to $V_P = 8$ V, kept at the V_P for t_{pls} and back to the V_{base}. With holding the V_{base} on the gate, the on-state-I_d measurement immediately began using the semiconductor parameter analyzer in a time-sampling mode at $V_d = 0.1$ V and $V_s = V_{sub} = 0$ V. Next, for measuring an off-state-I_d at the t_{pls}, a single erasing V_g pulse of a fixed height V_E and the time width t_{pls} was applied by the pulse generator. During the erasing V_g pulse application, we fixed V_d, V_s and V_{sub} at $V_d = V_s = V_{sub} = 0$ V. In the erasing V_g pulse, V_g started from $V_{base} = 2$ V, decreased to $V_E = -8$ V, kept at the V_E for t_{pls} and back to the V_{base}. With holding the V_{base} on the gate, the off-state-I_d measurement immediately began using the semiconductor parameter analyzer in the time-sampling mode at $V_d = 0.1$ V and $V_s = V_{sub} = 0$ V. As shown in Fig. 2.7, the FeFET still remained the on/off-I_d ratio of more than one order, exactly 26, at a fastest writing speed of $t_{pls} = 4 \times 10^{-8}$ s.

2.2.6 I_d–V_g and Retention at Elevated Temperatures

Finally, we introduce electrical properties of p-channel Pt/SBT/HAO/Si FeFETs [21]. The difference from the n-channel one was using n-type Si wafers with p^+ source-and-drain and n^+ substrate regions for probing. Thicknesses of the Pt, SBT and HAO layers were about 200, 600 and 7 nm, respectively. Figure 2.8 shows

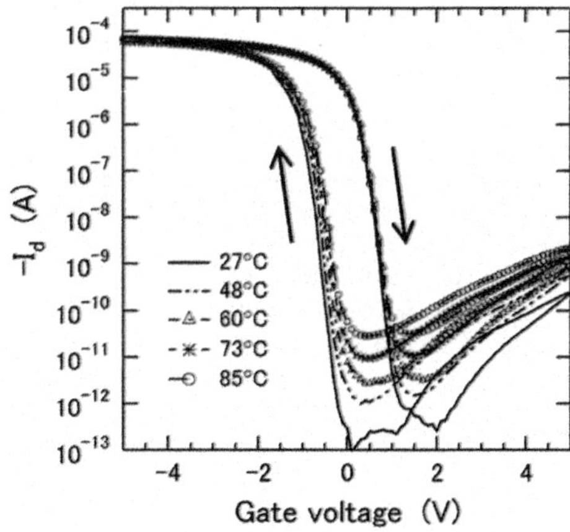

Fig. 2.8 static I_d–V_g curves of a p-channel Pt/SBT/HAO ($x = 0.75$)/Si FeFET measured at elevated temperatures from 27 to 85 °C. Thicknesses in the gate stack were 200 nm Pt, 600 nm SBT and 7 nm HAO. The gate area size was $L = 10$ μm and $W = 200$ μm. Modified from [21]

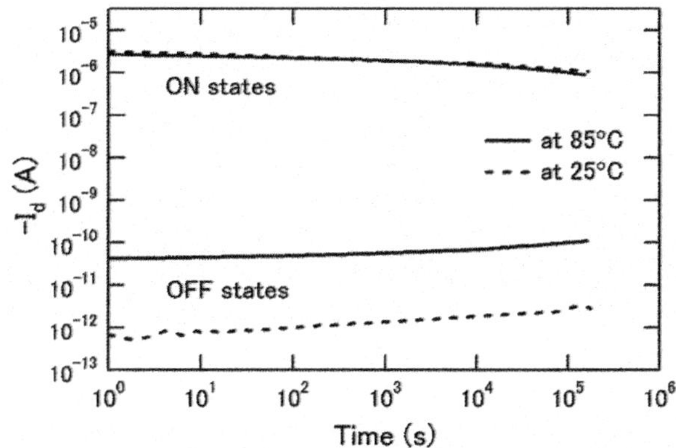

Fig. 2.9 I_d-retentions of a p-channel Pt/SBT/HAO ($x = 0.75$)/Si FeFET measured at 27 and 85 °C. Thicknesses in the gate stack were 200 nm Pt, 600 nm SBT and 7 nm HAO. The gate area size was $L = 10$ μm and $W = 200$ μm. Modified from [21]

static I_d–V_g curves of the FeFET measured at elevated temperatures from 27 to 85 °C [21]. The temperature was the sample-stage temperature of a manual prober. The I_d–V_g loops of the p-channel FeFET were drawn in clockwise directions by scanning V_g from −5 to 5 V and back to −5 V with keeping $V_d = -0.1$ V and $V_s = V_{sub} = 0$ V. As shown in Fig. 2.9, the I_d-retentions measured at 27 and 85 °C showed stable curves for more than 10^5 s [21]. The on-state I_d was measured at $V_g = V_{hold} = 0$ V, $V_d = -0.1$ V and $V_s = V_{sub} = 0$ V, after the poling by $V_g = V_P = -5$ V and $V_d = V_s = V_{sub} = 0$ V. The off-state I_d was measured also at $V_g = V_{hold} = 0$ V, $V_d = -0.1$ V and $V_s = V_{sub} = 0$ V, after the poling by $V_g = V_E = 5$ V and $V_d = V_s = V_{sub} = 0$ V. The V_{hold} of 0 V during the retention measurements indicated that the p-channel FeFET had the appropriate impurity concentration in the Si channel as mentioned in Sect. 2.2.3.

2.3 Requirements to the Layers in MFIS

2.3.1 Requirements to the Layers M, F, I

We will review the requirements to the M, F, and I layers. All the material of the M, F and I layers must be heatproof enough to show material stabilities even through an annealing process of the MFIS stack all at once. The material stability means that they do not have either significant reactions or element diffusions at the interfaces among the layers of the M, F, I and S. The annealing temperature is determined by crystallization temperature of the F-layer material for securing the ferroelectric performance. The temperature is about 800 °C in using SBT for the F material as we mentioned in Sect. 2.2.1. In this chapter, we introduced the experimental works only using Pt as the M layer and Si as the S substrate. With regard to the M layer, there may be an option of choosing another material with smaller work function [19] than the Pt in expectation of adjusting the V_{th} of the FeFET for $V_{hold} = 0$ V.

Requirements to the F and the I layers are very much related with each other on the points of equivalent oxide thickness (EOT) and band alignment. Figure 2.10a shows the band profile of the Pt/SBT/HAO($x = 0.75$)/Si FeFET at $V_g = 5.95$ V for programming [22]. The assumed thicknesses were 200 nm Pt, 200 nm SBT, 7 nm HAO and 3.5 nm IL. The IL was a thermally grown SiO_2 as indicated in Fig. 2.4. The band profile was drawn on the assumption that the F layer shows the unsaturated ferroelectric polarization along the minor loop of polarization *versus* electric field (*P–E*) as indicated in Fig. 2.10b [23]. As we discussed in Sect. 2.2.2, the total V_g across the MFIS was divided and shared among the layers of F, I and S, in proportion to the inverse of the individual capacitances connected in series. The F

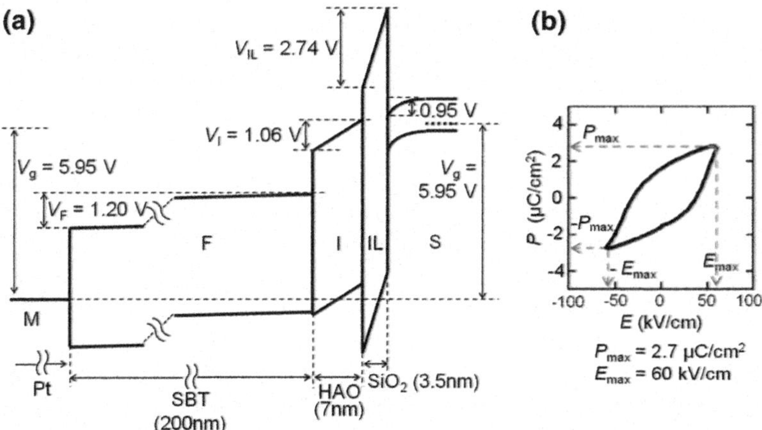

Fig. 2.10 **a** Band profile of the Pt/SBT/HAO($x = 0.75$)/Si FeFET at $V_g = 5.95$ V for programming. **b** Assumed P–E curve of the SBT in the FeFET. The curve was measured using an MFM capacitor of SBT. Modified from [22]

layer had a small V_g share of $V_F = 1.2$ V because the IL took much voltage of $V_{IL} = 2.74$ V as indicated in Fig. 2.10a. The ways to increase the memory window V_w of the FeFET from the view point of EOT, the F layer should have a relatively large EOT(F), and the I-and-IL total layers should have a relatively small EOT (I, IL) in comparison with their present states. Increasing the EOT(F) is difficult because the F material usually have a high dielectric constant ε_F and the F layer thickness d_F is to be reduced for the physical downsizing of the FeFET as we discussed layer in Sect. 2.6. Therefore the EOT(I, IL) must be decreased either by reducing the physical thicknesses d_I and d_{IL}, or by increasing the dielectric constants ε_I and ε_{IL} of the I-and-IL layers.

2.3.2 Requirements Especially to the I-and-IL Layers

Thinning the d_{IL} and increasing the ε_{IL} are required as discussed in Sect. 2.3.1. The d_{IL} can be decreased by much reducing the FeFET annealing temperature. It will be possible by drastic changing of the F material which is out of the scope in this chapter. Our work of increasing the ε_{IL} in later introduced in Sect. 2.5. The existence of the IL has some benefits to the FeFET. As indicated in Fig. 2.10a, the IL has a large band gap which aligned in the MFIS band profile as a good electrical barrier against charge injection from the S toward the F. Moreover the IL has a role of preserving good Si interface with little state densities. The good Si interfaces were indicated by many steep I_d–V_g curves with small subthreshold-voltage swings (S) about $S = 100$ mV/decade as we demonstrated in many works reviewed in this chapter. The S value was defined as $S = \ln10 \cdot dV_g/d(\ln I_d)$ [24].

With regard to the I layer, thinning the d_I and increasing the ε_L is required in addition to the material stability at high temperature about as discussed in Sect. 2.3.1. A thin-deposited high-k material will satisfy the requirements. From the viewpoint of band alignment in the MFIS, there is another requirement of the I layer. It is a higher barrier of the I layer than that of the F layer both against electrons and holes. Otherwise, the I layer forms a potential well between the SBT and a high-barrier IL which causes charge trapping. The uncontrollably trapped charges degrade the FeFET V_{th} stability. To summarize the requirements to the I layer, high-temperature-proof, thin-deposited high-k material with higher barriers than the F layer against electrons and holes is wanted.

A good candidate material of the I layer is $(HfO_2)_x(Al_2O_3)_{1-x}$ (HAO). It is 800 °C-resistive as we demonstrated in Fig. 2.4 and high-k of ε_I ranged from 9 for $x = 0$ to 25 for $x = 1.0$ [25]. The HAO also has a larger energy gap than the SBT without forming a potential well between the SBT the SiO_2 [26]. Thus we investigated the HAO as the I layer of the Pt/SBT/HAO/Si FeFETs.

2.4 Preparation of HAO for Pt/SBT/HAO/Si Gate Stack

2.4.1 Single HAO(x) and the MIS Characters at Various Composition Ratios

We investigated the composition ratio x of $(HfO_2)_x(Al_2O_3)_{1-x}$ ($HAO(x)$) to be amorphous in Pt/SBT/HAO(x)/Si MFIS FeFETs even after an annealing for the SBT polycrystallization [15]. X-ray diffraction (XRD) of 50 nm-thick HAO (x) films deposited on Si substrates were studied. The x was varied from 0 to 1.0. All the samples for the XRD underwent an annealing at 800 °C in O_2 for 1 h. As shown in Fig. 2.11, only the HAO(x) with $x = 1.0$ and 0.95 exhibited a peak corresponding to HfO_2 crystallization. Cross-sectional TEM images of Pt/HAO(x)/ Si MIS FETs suggested the HfO_2 crystallization in the HAO(x) with $x = 1.0$ and 0.95 [15]. In the TEM pictures, there were some spots of stripe patterns derived from the HfO_2 crystal lattice which were not found in Al_2O_3-rich HAO cross sections. The MIS FETs were fabricated by the same as described in Sect. 2.2.1, except for varying the x and no depositing the SBT. The gate-area sizes were $L = 10$ μm and $W = 200$ μm formed by photolithography and Ar^+ milling. Thicknesses were 200 nm for the Pt and 14 nm for the HAO(x). All of the MIS FETs were annealed at 800 °C in O_2 for 1 h in the same way as our basic Pt/SBT/HAO(x)/Si MFIS FETs underwent.

We measured static I_d–V_g and I_g–V_g curves of the MIS FETs for all the x [15]. There seemed negligibly small hystereses indicating no significant trapped charges in the I_d–V_g. As shown in Fig. 2.12, the MIS FETs had small gate-leakage current in the all x range. Precisely speaking, the I_g of the MIS FETs tended to be small as x was reduced. The cross-sectional TEM pictures indicated that the IL between the HAO(x) and Si in the MIS FETs became thick as the x was reduced toward

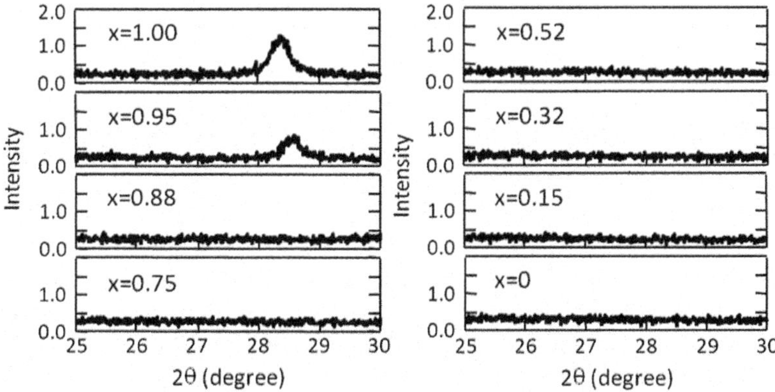

Fig. 2.11 XRD analyses of 50 nm-thick HAO(x) films deposited on Si substrates which all underwent annealing at 800 °C in O_2 for 1 h. The x was varied from 0 to 1.0. Modified from [15]

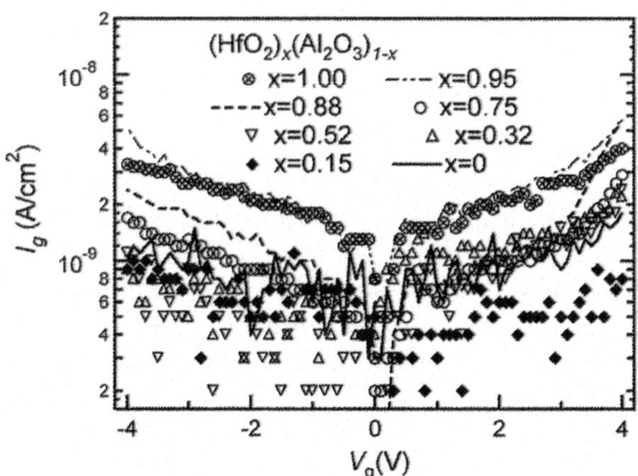

Fig. 2.12 Gate leakage currents of Pt/HAO(x)/Si MIS FETs with various x from 0 to 1.0. The all MIS FETs underwent annealing at 800 °C in O$_2$ for 1 h. They had the gate lengths of L = 10 μm and the gate widths of W = 80 and 20 μm. Deposited thicknesses were 200 nm for the Pt and 14 nm for the HAO common to the all MIS FETs. Modified from [15]

Al$_2$O$_3$-rich side [15]. The thick grown IL might be the reason of the I_g decrease as the x was reduced as indicated in Fig. 2.12.

In the MIS stack, EOT of the gate oxide was investigated by measuring capacitance vs V_g (C–V) curves [15]. The gate oxide was the double layer of HAO (x) and IL. Pt/HAO(x)/p-Si MIS diodes for various x from x = 0 to 1.0 were prepared. The gate area size was 200 μm × 200 μm formed by photolithography and Ar$^+$ milling. All of the MIS diodes were annealed at 800 °C in O$_2$ for 1 h. Hence they had IL grown on the Si. The MIS diode had two terminals: gate and substrate. The C–V curves were measured at 10 kHz by an LCR meter. The V_g was scanned and the V_{sub} was fixed to 0 V. There were no significant hystereses observed in the all C–V as shown in Fig. 2.13a. The capacitances at V_g = −2 V in the accumulation region were extracted from the C–V curves and plotted in Fig. 2.13b. The maximum C with the minimum EOT was observed at around x = 0.95. As a result of the works in this section, x = 0.75 was often used when we wanted amorphous and small-leakage HAO(x). We also used x = 1.0 for the largest ε_I = 25 [25] when suppressing the total EOT of HAO(x) and IL was prioritized.

2.4.2 Comparison of O$_2$ and N$_2$ Ambient in Depositing HAO

We verified which gas was better, O$_2$ or N$_2$, for the ambient during the HAO deposition by PLD [27]. Two types of Pt/SBT/HAO/p-Si diodes were prepared. One had the HAO named HAO(O) which was deposited to the physical thickness 15 nm in 13 Pa O$_2$ at 200 °C substrate temperature. The other had the HAO named HAO(N) which was deposited to the physical thickness 15 nm in 13 Pa N$_2$ at 200 °C substrate temperature. Fabrication process except for using O$_2$ ambient in deposition the HAO(O) was the same as described in Sect. 2.2.1. Thicknesses were 130–150 nm for the Pt, 400 nm for the SBT and 15 nm for both the HAO(N) and

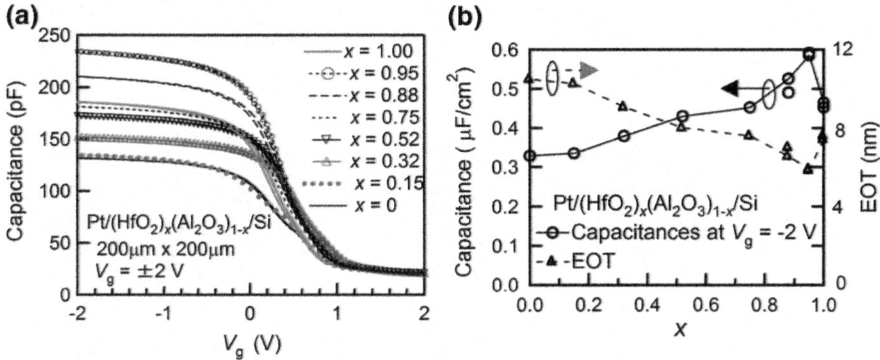

Fig. 2.13 **a** C–V curves of Pt/HAO(x)/p-Si MIS diodes for various x from $x = 0$ to 1.0. **b** x dependence of capacitors and the estimated EOT values of Pt/HAO(x)/p-Si MIS diodes. Deposited thicknesses were 200 nm for the Pt and 14 nm for the HAO common to the all diodes. The gate-electrode areas were 200 μm × 200 μm. Modified from [15]

HAO(O). An ellipsometer was used for understanding the precise deposition speeds of the HAO(N) and HAO(O) which significantly depended on the ambient gas kinds. Shapes of the Pt electrodes were 1.5 mm-diameter dots deposited using a metal hard mask. All the capacitors of Pt/SBT/HAO(N)/Si and Pt/SBT/HAO(O)/Si underwent an annealing process in 1 atm O_2 at 800 °C for 1 h. Backsides of the Si substrates were mechanically ground and covered with Al deposited by the EB evaporator. C–V characteristics of the Pt/SBT/HAO(N)/Si and Pt/SBT/HAO(O)/Si were measured at 10 kHz by an LCR meter. As shown in Fig. 2.14a, b, the C–V hysteresis curves were drawn in clock-wise directions which indicated ferroelectric-polarization switching on the p-type semiconductor. At every scan range of V_g, the Pt/SBT/HAO(N)/Si in Fig. 2.14a showed a larger V_w than the Pt/SBT/HAO(O)/Si in Fig. 2.14b. For example at $V_g = \pm6$ V, the Pt/SBT/HAO (N)/Si showed $V_w = 0.9$ V while the Pt/SBT/HAO(O)/Si showed $V_w = 0.3$ V as indicated in Fig. 2.15. The reason of the small V_w of the Pt/SBT/HAO(O)/Si was a thick IL grown between the HAO(O) and Si. The IL thicknesses were confirmed by a cross-sectional scanning transmission electron microscope (STEM) [27]. The IL thickness in the Pt/SBT/HAO(O)/Si was 7.4 nm which was much larger than that in the Pt/SBT/HAO(N)/Si 3.4 nm. The IL has a dielectric constant $\varepsilon_{IL} = 3.9$ [15] which is the lowest among those of the SBT, HAO(O) and HAO(N). Hence much decrease of the V_w was caused by the 4 nm increase in the IL thickness.

We found that the IL growth on the Si was promoted by the HAO(O) because the HAO(O) did not work sufficiently as a material-diffusion barrier. Backside secondary-ion-mass-spectrometry (SIMS) gave us useful information about the quality as the material-diffusion barrier of the HAO(O) and HAO(N) as shown in Fig. 2.16. The elements Sr, Bi and Ta of the SBT in Fig. 2.16c–e had the intensity peaks at the depth of the HAO(O) while the same elements did not show penetrations into the HAO(N) location. As a conclusion of this work, O_2 was not a

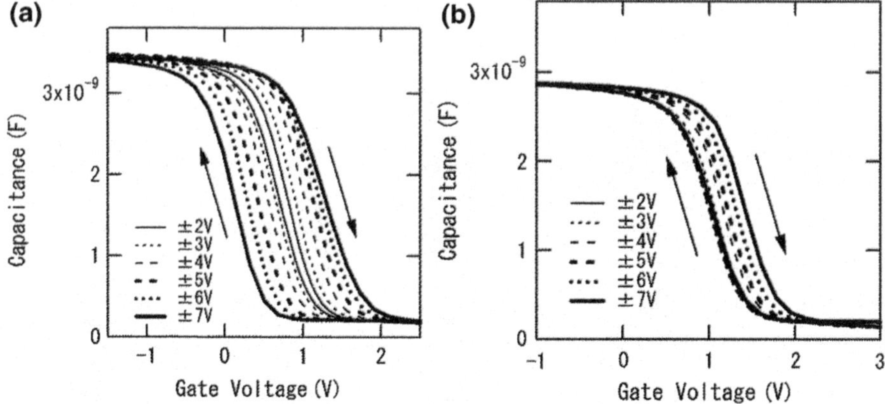

Fig. 2.14 *C–V* hysteresis curves of **a** Pt/SBT/HAO(N)/*p*-Si and **b** Pt/SBT/HAO(O)/*p*-Si MFIS diodes. Thicknesses were 130–150 nm Pt, 400 nm SBT and 15 nm for the HAO(N) and HAO(O). The gate-electrode areas were 1.5 mm-diameter dots. Modified from [27]

Fig. 2.15 Memory windows of Pt/SBT/HAO(N)/*p*-Si and Pt/SBT/HAO(O)/*p*-Si MFIS diodes extracted from *C–V* curves for various scan ranges of V_g in Fig. 2.14a, b. Modified from [27]

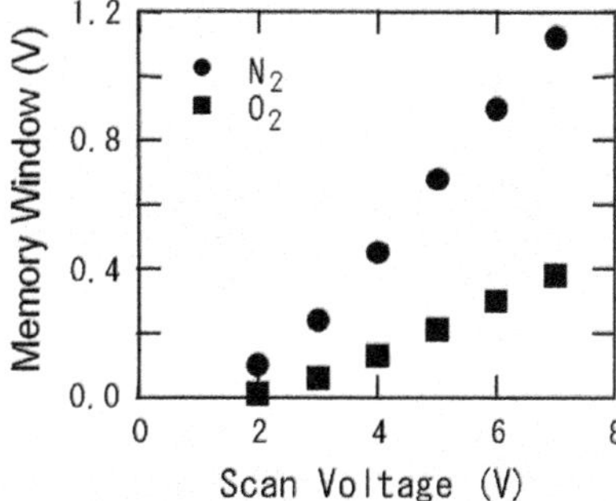

suitable ambient gas in depositing the HAO by PLD for preparing the Pt/SBT/HAO/Si stack. Therefore we selected N_2 ambient in depositing the HAO by the PLD.

2.4.3 Effect of N_2 Ambient Pressure Increase in Depositing HAO

We investigated effects of increasing ambient N_2 pressure during HAO deposition (P_{HAO}) in Pt/SBT/HAO/*p*-Si [28]. Fabrication process except for varying the P_{HAO} was the same as described in Sect. 2.2.1. The P_{HAO} was varied from 7 to 158 Pa. The all FeFETs had the gate area of $L = 10$ μm and $W = 200$ μm formed by photolithography and Ar^+ milling. Thicknesses were 250 nm for the Pt, 450 nm for the SBT and 10 nm for the HAO. I_d–V_g hysteresis curves of the FeFETs of

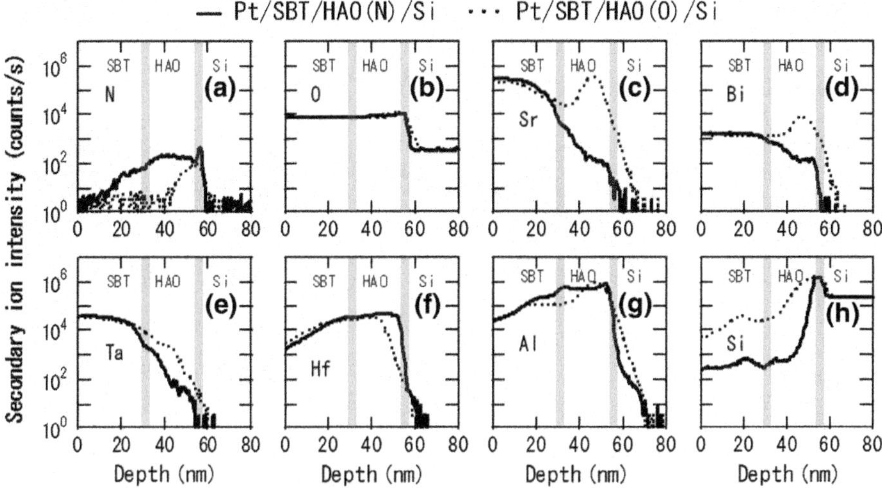

Fig. 2.16 SIMS profiles of Pt/SBT/HAO(N)/p-Si and Pt/SBT/HAO(O)/p-Si MFIS diodes. *Solid lines* were for Pt/SBT/HAO(N)/p-Si. *Dashed lines* were for Pt/SBT/HAO(O)/p-Si. Elements of **a** N, **b** O, **c** Sr, **d** Bi, **e** Ta, **f** Hf, **g** Al and **h** Si were studied. Modified from [27]

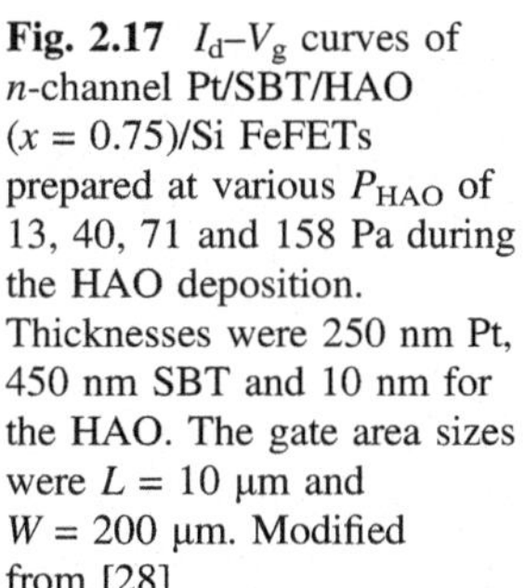

Fig. 2.17 I_d–V_g curves of n-channel Pt/SBT/HAO ($x = 0.75$)/Si FeFETs prepared at various P_{HAO} of 13, 40, 71 and 158 Pa during the HAO deposition. Thicknesses were 250 nm Pt, 450 nm SBT and 10 nm for the HAO. The gate area sizes were $L = 10$ μm and $W = 200$ μm. Modified from [28]

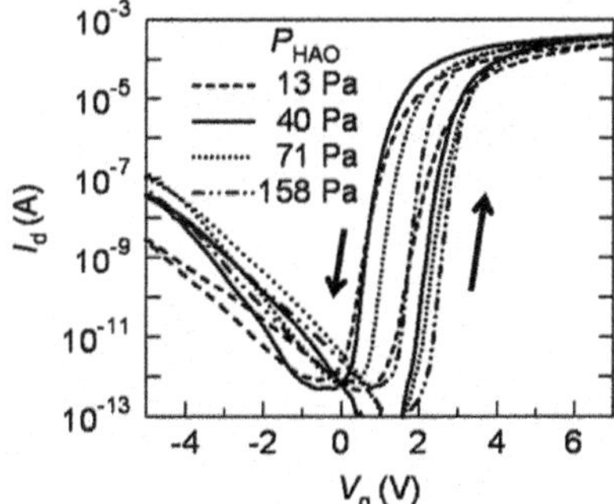

P_{HAO} = 13, 40, 71 and 158 Pa were measured as shown in Fig. 2.17. The curves were drawn in counterclockwise directions which indicated ferroelectric polarization switching on p-Si substrates. Static memory windows (V_w) extracted from the I_d–V_g curves had broad peaks around $P_{HAO} = 40$ Pa. The largest V_w was about $V_w = 1.5$ V at $V_g = 1 \pm 6$ V. Relatively small V_w at $P_{HAO} > 40$ Pa indicated that such large P_{HAO} on the contrary increased the total EOT of the HAO and IL. As shown in Fig. 2.18, cross-sectional STEM images of the FeFETs indicated the increasing tendency of the IL thickness and the HAO roughness as P_{HAO} was raised from 40 Pa. The rough HAO interface suggested that inter-diffusion of materials

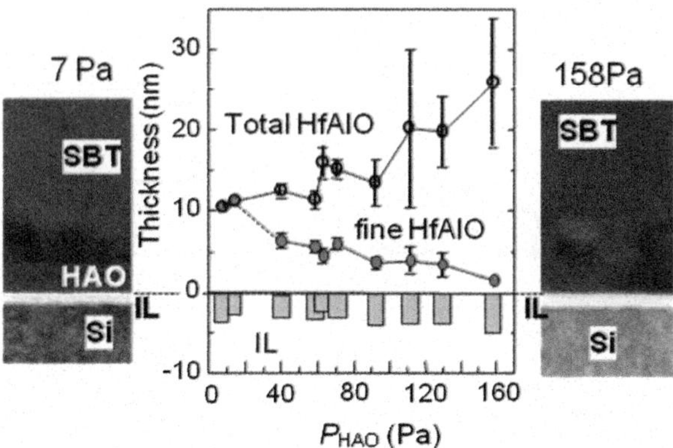

Fig. 2.18 Physical thicknesses of HAO and IL-SiO$_2$ measured by cross-sectional STEM images of n-channel Pt/SBT/HAO ($x = 0.75$)/Si FeFETs. They were prepared at various P_{HAO} ranged from 7 to 158 Pa during the HAO deposition. As-deposited HAO was 10 nm thick each. Modified from [28]

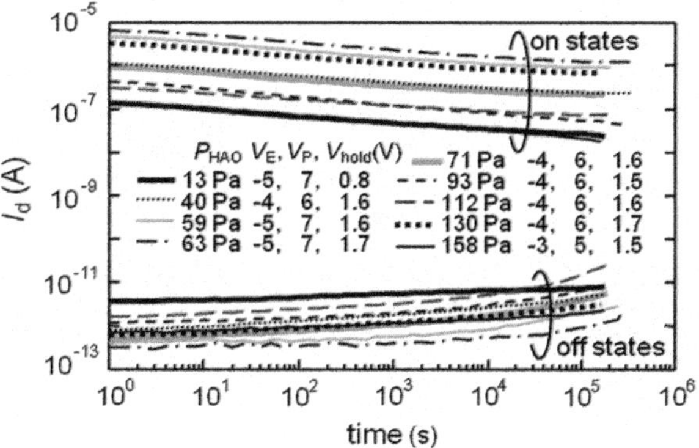

Fig. 2.19 I_d retentions of Pt/SBT/HAO/Si FeFETs in which the HAO layers were deposited in N$_2$ at P_{HAO} ranged from 13 to 158 Pa. Deposited thicknesses were 250 nm Pt, 450 nm SBT and 10 nm HAO common to all the FeFETs. The gate area sizes were $L = 10$ μm and $W = 200$ μm. Modified from [28]. Unpublished data were added

between the HAO and the SBT became significant as the P_{HAO} was raised. At $P_{HAO} = 13$ Pa, the boundary between the SBT and the HAO was clear with showing the HAO thickness of 10 nm which was as same as the designed as-deposited thickness. As P_{HAO} was raised, however, the HAO thickness seemed increasing with unclear boundary between the SBT and the HAO. The IL also grew thick as the P_{HAO} was raised. As shown in Fig. 2.19, stable I_d retentions until at least 2×10^5 s were indicated for all the FeFETs made by P_{HAO} ranged from 13 to 158 Pa in spite of the thick IL and the rough HAO observed at $P_{HAO} > 40$ Pa as shown in Fig. 2.18. The I_d retentions were measured by the same way discussed in Sect. 2.2.3. The V_P, V_E and V_{hold} we used for investigating the I_d-retentions were described in Fig. 2.19. Thanks to the thick SBT, all the FeFETs even at $P_{HAO} > 40$ Pa seemed to have stable retentions despite the thick-grown HAO with rough morphology suggested in Fig. 2.18. However, $P_{HAO} \ll 40$ Pa during the PLD deposition would be appropriate for suppressing total EOT of the HAO and IL.

2.5 Nitriding and Oxinitriding Si of MFIS FeFET

2.5.1 Direct Nitriding Si for Large Memory Window of FeFET

As mentioned in Sect. 2.3, we investigated EOT reduction of the I-and-IL layers. In this study, for the purpose of increasing the ε_{IL}, thin silicon nitride (Si–N) was grown on the Si surface in advance before depositing the MFI stack [29]. Pt/SBT/HfO$_2$/Si-N/p-Si FeFETs were prepared and characterized. A rapid-thermal-nitridation (RTN) machine was custom-designed. Si–N was formed on the Si by lamp-annealing in NH$_3$ ambient. All the initial Si substrates were transferred into vacuum immediately after sacrificial SiO$_2$ layers on the Si surfaces were removed by BHF. Fabrication process except for introducing the RTN was the same as described in Sect. 2.2.1 except for introducing the RTN. The Si-N layers were formed on the Si in NH$_3$ at temperatures T_{RTN} ranged from 800 to 1190 °C. T_{RTN} dependence of the Si–N thickness was estimated by investigating cross-sectional STEM pictures of Si–N grown on Si at T_{RTN} = 800, 1020 and 1190 °C as shown in Fig. 2.20a–d. The T_{RTN} was calibrated in advance using a thermocouple connected on a Si wafer. The ambient NH$_3$ pressure during the RTN was fixed at 532 Pa. Times for increasing, keeping and decreasing the RTN temperatures were controlled as 50, 10 and 30 s below 400 °C, respectively. Finally the Pt/SBT/HfO$_2$ stacks were deposited on the Si–N/Si by the same process as introduced in Sect. 2.2.1. Thicknesses were 200 nm for the Pt, 450 nm for the SBT and 6 nm for the HfO$_2$. The gate areas were $L = 10$ μm and $W = 200$ μm formed by photolithography and Ar$^+$ milling. Reference FeFETs of Pt/SBT/HfO$_2$/Si were also prepared by the conventional process without using the RTN. In the reference FeFETs, thicknesses of the Pt, SBT and HfO$_2$ were designed as the same as those in the RTN-processed Pt/SBT/HfO$_2$/Si–N/Si FeFETs.

Figure 2.21a–f show I_d–V_g hysteresis curves of the Pt/SBT/HfO$_2$/SiN/Si FeFETs. The Si–N layers were grown at T_{RTN} = 880, 940, 1000, 1020, 1080 and 1160 °C. All the curves were drawn in counterclockwise directions. The T_{RTN} very much affected static memory windows V_w. The memory windows V_W at

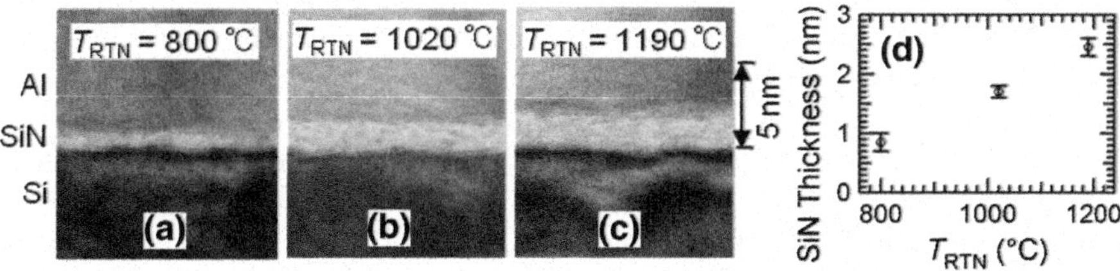

Fig. 2.20 Cross-sectional STEM pictures of Si–N/Si reference samples. The Si–N was grown on Si at **a** T_{RTN} = 800 °C, **b** 1020 °C and **c** 1190 °C. **d** T_{RTN} dependence of the Si–N thickness. Al was deposited for clear observations. Modified from [29]

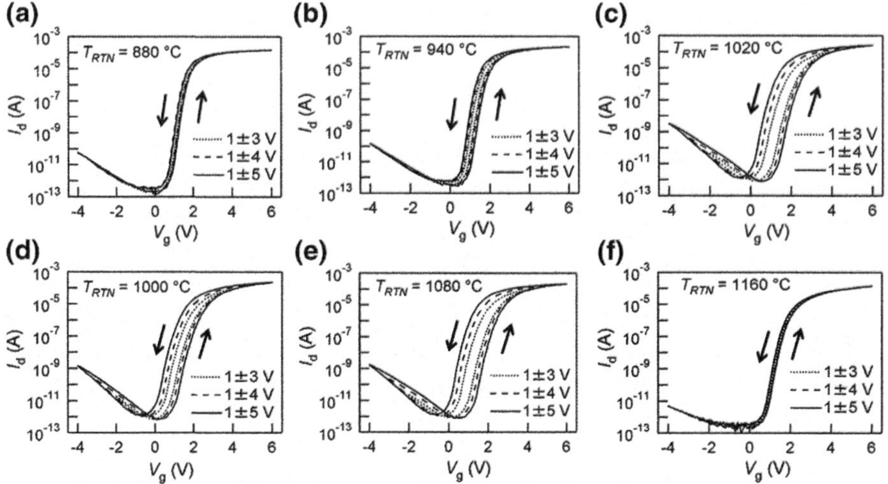

Fig. 2.21 I_d–V_g curves of Pt/SBT/HfO$_2$/Si–N/Si FeFETs in which the Si–N layers were prepared by RTN at various T_{RTN} of **a** 880 °C, **b** 940 °C, **c** 1000 °C, **d** 10,200 °C, **e** 1080 °C and **f** 1160 °C. Thicknesses were 200 nm Pt, 450 nm SBT and 6 nm HfO$_2$. The gate area sizes were $L = 10$ µm and $W = 200$ µm. Modified from [29]

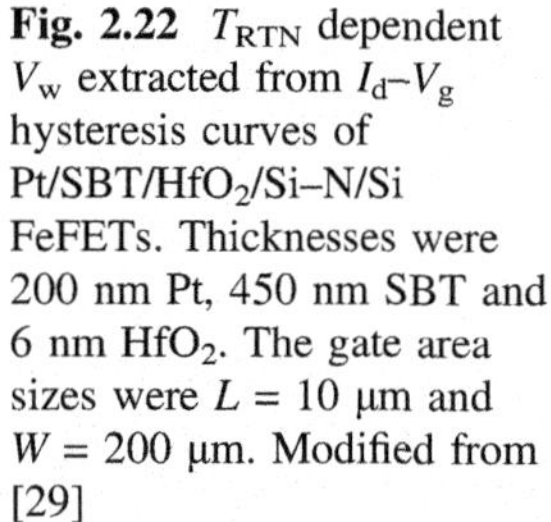

Fig. 2.22 T_{RTN} dependent V_w extracted from I_d–V_g hysteresis curves of Pt/SBT/HfO$_2$/Si–N/Si FeFETs. Thicknesses were 200 nm Pt, 450 nm SBT and 6 nm HfO$_2$. The gate area sizes were $L = 10$ µm and $W = 200$ µm. Modified from [29]

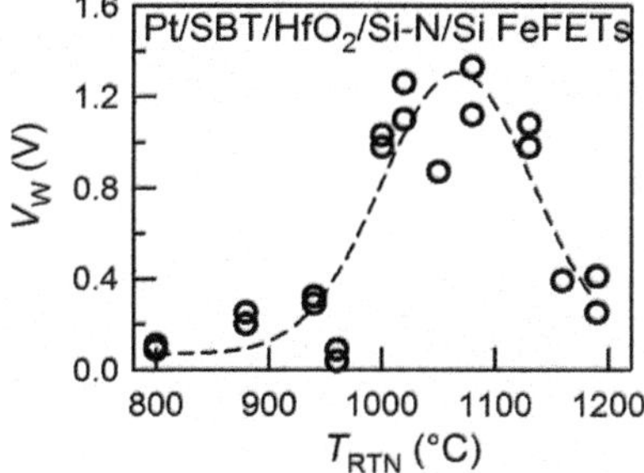

$V_g = 1 \pm 5$ V were extracted from the I_d–V_g of the FeFETs at various T_{RTN} and plotted in Fig. 2.22. The T_{RTN} dependent V_w showed a peak at the T_{RTN} ranged from 1020 to 1130 °C. The maximum V_w at $V_g = 1 \pm 5$ V was $V_w = 1.36$ V when $T_{RTN} = 1080$ °C. It was about 10 % larger than of 1.24 V which was the averaged V_w of 16 reference Pt/SBT/HfO$_2$/Si FeFETs measured at $V_g = 1 \pm 5$ V. In other words, introducing the RTN into the FeFET process increased the V_w by 10 %. The Pt/SBT/HfO$_2$/SiN/Si FeFET at $T_{RTN} = 1080$ °C showed stable on- and off-state I_d retained until at least 1.7×10^5 s or 2 days each as shown in Fig. 2.23. The I_d retentions were measured by the same way discussed in Sect. 2.2.3. The writing voltages were $V_P = 5.5$ V for the on state and $V_E = -4.5$ V for the off state. The I_d retention was measured with keeping $V_g = V_{hold} = 1.3$ V, $V_d = 0.1$ V and $V_s = V_{sub} = 0$ V.

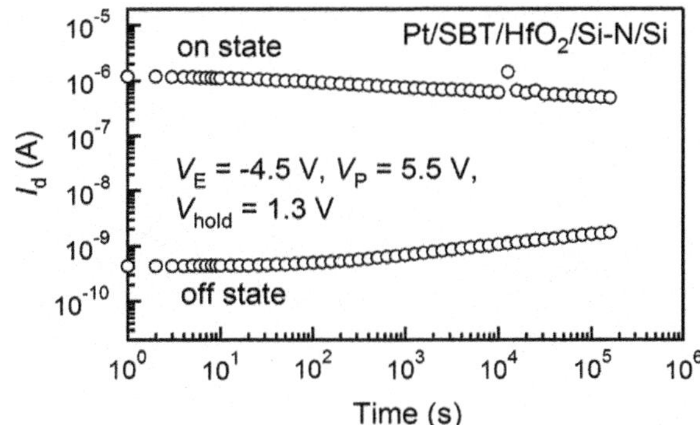

Fig. 2.23 T_{RTN} dependent V_w extracted from I_d–V_g hystereses of Pt/SBT/HfO$_2$/Si–N/Si FeFETs. Thicknesses were 200 nm Pt, 450 nm SBT and 6 nm HfO$_2$. The gate area sizes were L = 10 μm and W = 200 μm. Modified from [29]

We investigated cross-sectional TEM to know the effect of the RTN process on the IL thickness. The cross-sectional TEM images indicated the same physical thickness of 3.5 nm for the two IL layers: one in the Pt/SBT/HfO$_2$/SiN/Si at T_{RTN} = 1020 °C and the other in the reference Pt/SBT/HfO$_2$/Si without using the RTN [29]. The initial Si–N thickness of 1.7 nm at T_{RTN} = 1020 °C was estimated as indicated in Fig. 2.20d. As shown in Fig. 2.24, backside SIMS analysis showed the element N had an intensity peak at the IL depth. The IL also included oxygen. The reason of the O inclusion in the IL was that the FeFET underwent the 800 °C annealing in O$_2$ for the SBT poly-crystallization. Judging from the cross-sectional TEM and the backside SIMS studies, the RTN process did not change the IL thickness but increase the ε_{IL} due to the nitrogen inclusion.

Gate leakage currents of the Pt/SBT/HfO$_2$/SiN/Si FeFETs at various T_{RTN} were also measured as shown in Fig. 2.25. Generally small I_g values were obtained at V_g = ±5 V. In detail, the FeFETs at around T_{RTN} = 800 and 1190 °C showed relatively low I_g. The small I_gs were due to abnormally thick IL of about 30 nm confirmed by cross-sectional STEM studies [29]. The STEM pictures also showed

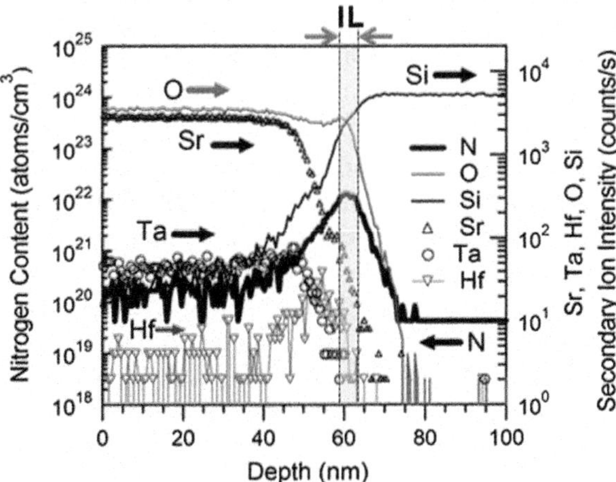

Fig. 2.24 Backside SIMS profile of a Pt/SBT/HfO$_2$/SiN/Si FeFET in which the Si–N grown at T_{RTN} = 1080 °C. Signal of Bi was less than the detection limit. Modified from [29]

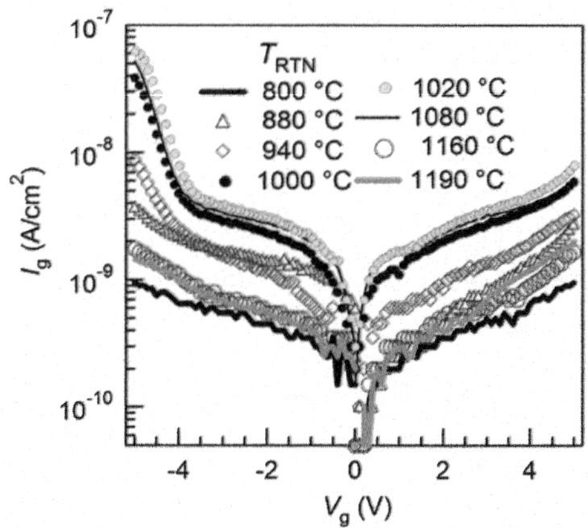

Fig. 2.25 Gate leakage currents of Pt/SBT/HfO$_2$/Si–N/Si FeFETs at various T_{RTN} ranged from 800 to 1190 °C. Modified from [29]

jagged boundaries between the IL and the Si of the FeFETs at T_{RTN} = 800 and 1190 °C. The Si surfaces might be irregularly eroded by the RTN and the subsequent poly-crystallization annealing for SBT at 800 °C. Figure 2.21a, f also supported that the very thick IL layers much reduced the V_w of the FeFETs at T_{RTN} = 880 and 1160 °C. Therefore the optimum T_{RTN} was around 1080 °C. The other T_{RTN} far from the optimum resulted in thick IL caused by irregular Si erosions.

2.5.2 Oxinitriding Si for Improving the Si Interface of FeFET

In Sect. 2.5.1, we introduced the RTN as a process to prepare Pt/SBT/HfO$_2$/Si-N/Si FeFETs which had larger V_w than reference Pt/SBT/HfO$_2$/Si FeFETs. On the other hand, however, subthreshold-voltage swing of $S = \ln 10 \cdot dV_g/d(\ln I_d)$ [24] was increased by the RTN. The reference Pt/SBT/HfO$_2$/Si FeFET usually had about S = 100 mV/decade while the Pt/SBT/HfO$_2$/Si-N/Si FeFET at T_{RTN} = 1080 °C showed S = 173 mV/decade. The large S might suggest an increase of Si surface-state density [24] caused by the direct nitriding the Si. In this work, we prepared and compared four kinds of FeFETs which had differently treated Si substrates such as reference-Si, Si-N/Si, SiO$_2$/Si and Si-O-N/Si [30].

The four kinds of substrates were prepared by the process using the RTN and rapid thermal oxidation (RTO) as follows. In every case, the initial Si substrate was transferred into vacuum immediately after sacrificial SiO$_2$ layer on the Si surface was removed by BHF. First, one out of the four which was named as the reference Si directly proceeded to the HfO$_2$ deposition. The reference Si underwent neither RTN nor RTO. Second, the Si–N/Si was prepared by RTN of the Si in 532 Pa NH$_3$ at T_{RTN} = 1080 °C. Third, the SiO$_2$/Si was made by RTO of the Si in 532 Pa O$_2$ at the RTO temperature T_{RTO} = 1000 °C. Fourth, the Si–O–N/Si was prepared by a two-step heating, first by the RTO of the Si in 532 Pa O_2 at a fixed T_{RTO} = 1000 °C, and second by the RTN in 532 Pa NH_3 at various T_{RTN} ranged from 920 to 1150 °C. After the first RTO, the substrate for the Si–O–N/Si was once cooled below 400 °C and held in the process chamber during exchanging the ambient gas from O_2 to NH_3 via vacuum. In both the RTN and the RTO, the temperature was increased in 50 s to the target value, kept for 10 s at the target and decreased in 30 s below 400 °C. On the four kinds of substrates, Pt/SBT/HfO$_2$ stacks were commonly deposited by the same way as described in Sect. 2.2.1. Thicknesses were 200 nm for the Pt, 450 nm for the SBT and 6 nm for the HfO$_2$. The gate area sizes were L = 10 μm and W = 200 μm or W = 100 μm which were formed by photolithography and Ar$^+$ milling.

We optimized the T_{RTN} in preparing the Si–O–N/Si for maximizing V_w and minimizing S of the I_d–V_g curves. Figure 2.26 shows T_{RTN} dependences of V_w extracted from the I_d–V_g curves of the Pt/SBT/HfO$_2$/Si–O–N/Si FeFETs. The I_d–V_g curves were measured at V_g = 1 ± 3 V, 1 ± 4 V and 1 ± 5 V. The V_w had a broad peak at around T_{RTN} = 1050 °C as shown in Fig. 2.27. The largest V_w at V_g = 1 ± 4 V was V_w = 1.02 V. The S values of the I_d–V_d curves at V_g = 1 ± 4 V were plotted in Fig. 2.28. The S was the smallest S = 100 mV/decade around T_{RTN} = 1020 °C. Therefore in this work, T_{RTN} = 1050 °C was determined as the

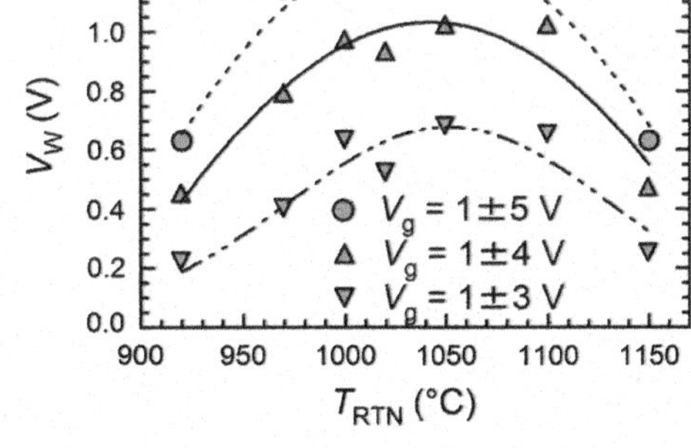

Fig. 2.26 T_{RTN} dependences of V_w extracted from the I_d–V_g curves of the Pt/SBT/HfO$_2$/Si–O–N/Si FeFETs measured at V_g = 1 ± 3 V, 1 ± 4 V and 1 ± 5 V. Modified from [30]

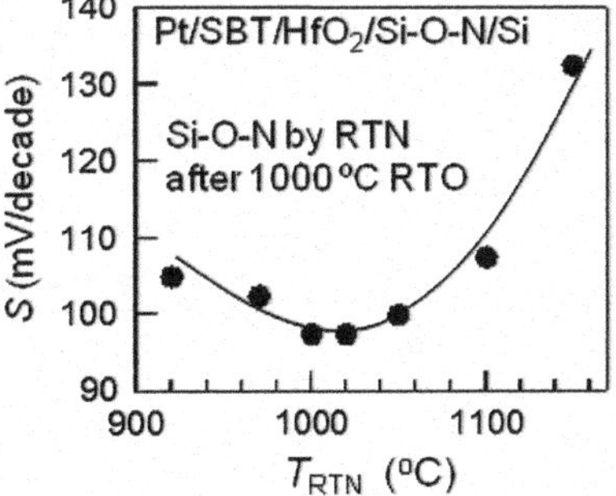

Fig. 2.27 T_{RTN} dependences of S extracted from the I_d–V_g curves of the Pt/SBT/HfO$_2$/Si–O–N/Si FeFETs measured at V_g = 1 ± 4 V. Modified from [30]

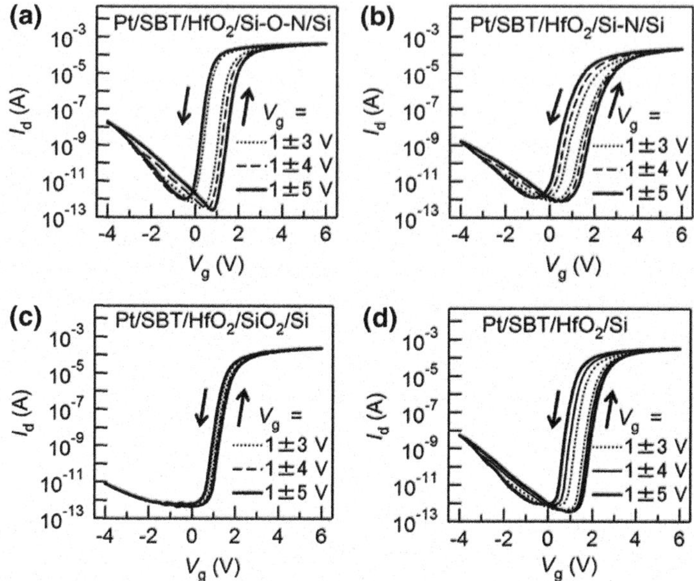

Fig. 2.28 I_d–V_g curves of FeFETs having the MFIS gate stacks such as **a** Pt/SBT/HfO$_2$/Si–O–N/Si, **b** Pt/SBT/HfO$_2$/Si–N/Si FeFETs, **c** Pt/SBT/HfO$_2$/SiO$_2$/Si and **d** Pt/SBT/HfO$_2$/(IL)/Si. Thicknesses were 200 nm Pt, 450 nm SBT and 6 nm HfO$_2$. The gate area sizes were $L = 10$ µm and $W = 200$ µm. Modified from [30]

best for obtaining the largest V_w and the smallest S of Pt/SBT/HfO$_2$/Si-O-N/Si FeFETs.

Static I_d–V_g curves were measured and compared among the four kinds of FeFETs having the following gate stacks: reference-Pt/SBT/HfO$_2$/Si, Pt/SBT/HfO$_2$/Si–N/Si, Pt/SBT/HfO$_2$/SiO$_2$/Si and Pt/SBT/HfO$_2$/Si–O–N/Si. As shown in Fig. 2.28, only the Pt/SBT/HfO$_2$/SiO$_2$/Si FeFET showed very small V_w among the four FeFETs. The small V_w suggested that the SiO$_2$ intentionally grown by the RTO induced the further thick SiO$_2$ grown on the Si finally. Particularly small I_g of the Pt/SBT/HfO$_2$/SiO$_2$/Si FeFET also implied the thick SiO$_2$ in the FeFET [30]. With regard to the S value, the Pt/SBT/HfO$_2$/Si–O–N/Si FeFET showed $S = 100$ mV/decade which was much smaller than $S = 173$ mV/decade of the Pt/SBT/HfO$_2$/Si–N/Si FeFET. The two-step Si annealing by first RTO and then RTN demonstrated in this study was found to be effective for preserving the good Si interface.

The Pt/SBT/HfO$_2$/Si–O–N/Si FeFET tended to show a large V_w especially at small V_g scan. As indicated in Fig. 2.28a, the FeFET had $V_w = 0.64$ V at $V_g = 1 \pm 3$ V, which was the largest among the four FeFETs. The other three FeFETs in Fig. 2.28 showed (b) $V_w = 0.51$ V, (c) $V_w = 0.09$ V and (d) $V_w = 0.42$ V, at $V_g = 1 \pm 3$ V. The large V_w at the small V_g scan as shown in

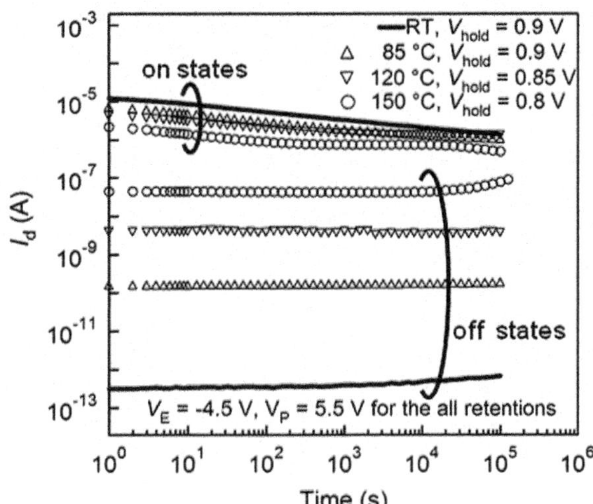

Fig. 2.29 I_d retentions of a Pt/SBT/HfO$_2$/Si–O–N/Si FeFET measured at room temperature, 85, 120 and 150 °C. The FeFET underwent T_{RTN} = 1050 °C. Thicknesses were 200 nm Pt, 450 nm SBT and 6 nm HfO$_2$. The gate area sizes were L = 10 μm and W = 100 μm. Modified from [30]. Unpublished data were added

Fig. 2.28a suggested that the Si–O–N reduced the EOT of the IL and relatively increased V_F across the SBT.

I_d retentions of the Pt/SBT/HfO$_2$/Si–O–N/Si FeFET were measured by the same way described in Sect. 2.2.3. The V_P, V_E and V_{hold} were described in Fig. 2.29. The retentions were measured at room temperature (RT), 85, 120 and 150 °C. The gate area sizes were L = 10 μm and W = 100 μm. The I_d retentions at the all temperatures remained stable until at least 1.0×10^5 s. As a conclusion of the section, forming the Si–O–N at the optimum T_{RTO} and T_{RTN} indicated a good potential to produce high quality FeFETs. By introducing the Si–O–N into the FeFET fabrication process, the operational voltages would be decreased with the S values kept small.

2.6 Using CSBT Instead of SBT in FeFET

We introduced the works for decreasing the I layer EOT in Sect. 2.4 and for decreasing the IL layer EOT in Sect. 2.5. In addition to the studies, we modified the F material for directly increasing the V_W of the FeFET as introduced in this section. We focused on Ca$_y$Sr$_{1-y}$Bi$_2$Ta$_2$O$_9$ (CSBT(y)) among ferroelectric materials of bismuth-layered perovskite in expectation of the high-endurance nature as well as the SrBi$_2$Ta$_2$O$_9$ (SBT). Since some reports of CSBT indicated that MFM capacitors using the CSBT had larger coercive field (E_c) than the SBT [31–33], we began to investigate Pt/CSBT/HAO/Si FeFETs [34].

We fabricated and characterized Pt/CSBT(y)/HAO(x = 0.75)/Si FeFETs for demonstrating a larger V_w than the conventional Pt/SBT/HAO(x = 0.75)/Si FeFETs. We had a strong motivation to change the F material from the SBT to the

CSBT because we recently set the F thickness $d_F \leqq 200$ nm which was less than half of our early-stage $d_F = 400$ nm in 2002 [9–11]. The purpose of the reducing d_F was enabling the etching of the high-aspect gate stacks for prospective downsizing FeFETs [7, 8]. We had three expectations in applying the CSBT to the FeFETs. One was that the CSBT would have a similar nature to the SBT and would not show significant material diffusion with Si across the HAO. Second was that MFIS FeFETs using the CSBT might show a larger V_w than that using the SBT because a larger E_c of the CSBT than that of the SBT was reported in major P-E loops of the MFM capacitors [31–33]. The last was that the crystallization temperature of the CSBT was expected to be lower than that of the SBT because good ferroelectric properties of $CaBi_2Ta_2O_9$ (CBT or CSBT($y = 1.0$)) capacitances were reported which were annealed at 700 and 750 °C [35].

In order to optimize the Ca composition ratio y and the annealing temperature, we prepared Pt/CSBT(y)/HAO($x = 0.75$)/Si FeFETs where the y was varied as $y = 0, 0.1, 0.2, 0.5$ and 1.0. Fabrication process was basically the same as described in 2.1 except for using a custom-designed large-area PLD [36], replacing the SBT with the CSBT(y), thinning the F to 200 nm, and searching for the best annealing temperature T_{an} for the CSBT(y) poly-crystallization. The HAO($x = 0.75$) was deposited by the PLD in 15 Pa N_2 at the substrate temperature 220 °C. A ceramic target of $(HfO_2)_{0.75}(Al_2O_3)_{0.25}$ was used. The CSBT(y) was deposited by the PLD in 7 Pa O_2 at 415 °C using Ca–Sr–Bi–Ta–O ceramic targets with the element ratio Ca:Sr:Bi:Ta = y:1 − y:3:2. All the FeFETs were annealed in 1 atm O_2 for 30 min at various T_{an} from 748 to 833 °C for the CSBT(y) poly-crystallization. Gate areas of all the FeFETs had a fixed length $L = 10$ μm and a various widths of $W = 200$, 150, 100, 80, 50, 40, 20 and 10 μm formed by photolithography and Ar^+ milling. Thicknesses were 200 nm for the Pt, 200 nm for the CSBT(y) and 7 nm for the HAO($x = 0.75$).

We prepared numerous FeFETs of Pt/CSBT(y)/HAO($x = 0.75$)/Si with $y = 0$, 0.1, 0.2, 0.5 and 1.0. Figure 2.30 show the T_{an} dependence of the V_w values which were extracted from static I_d-V_g curves measured the FeFETs at $V_g = 1 \pm 5$ V. In each V_w-T_{an} curve for y, the individual Pt/CSBT(y)/HAO($x = 0.75$)/Si FeFETs underwent annealing at various T_{an}. The FeFETs with $y = 0$ using the SBT had the maximum $V_w = 0.75$ V at $T_{an} = 813$ °C. The FeFETs with $y = 0.1$ had the maximum $V_w = 0.89$ V at $T_{an} = 800$ °C. Those with $y = 0.2$ had the maximum $V_w = 0.84$ V also at $T_{an} = 800$ °C. When $y = 1.0$ using the CBT, however, the FeFETs showed almost zero V_w which was unexpected results from the MFM studies using the CBT with large E_c [35, 37, 38]. This case is a good example that MFM could not fully predict the behaviors of the F in MFIS. As shown in Fig. 2.31a–c, cross-sectional TEM pictures indicated that IL thickness of the Pt/CSBT($y = 0.1$)/HAO($x = 0.75$)/Si FeFETs became thick as the T_{an} was raised from 748 to 833 °C. The interfaces between the IL and the HAO($x = 0.75$) seemed clear in the all Pt/CSBT($y = 0.1$)/HAO($x = 0.75$)/Si FeFETs prepared at $T_{an} = 748$, 788 and 833 °C. However, the Pt/CBT(or CSBT($y = 1.0$))/HAO($x = 0.75$)/Si FeFET showed about 7 nm-thick IL exhibiting no clear interface across the IL, HAO($x = 0.75$) and CBT as indicated in Fig. 2.31d. The Si surface of the

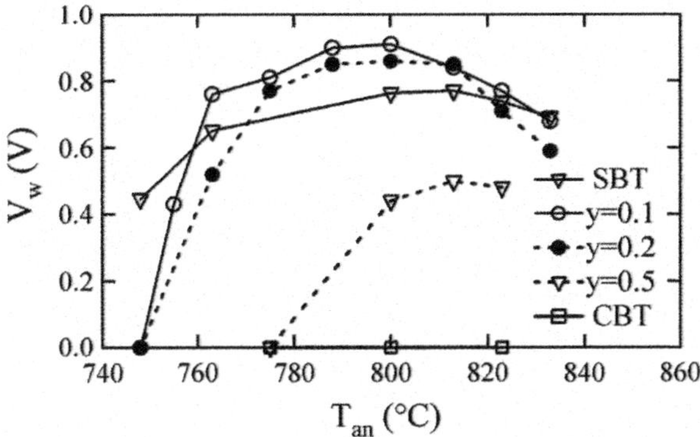

Fig. 2.30 Annealing-temperature dependence of static memory windows extracted from I_d–V_g curves of Pt/CSBT(y)/HAO(x = 0.75)/Si FeFETs. The Ca composition ratio y was varied as y = 0, 0.1, 0.2, 0.5 and 1.0. The CSBT(y = 1.0) was the CBT. The I_d–V_g curves were measured at V_g = 1 ± 5 V. Thicknesses were 200 nm Pt, 200 nm SBT and 7 nm HAO. The gate lengths were L = 10 µm. Modified from [34]

Pt/CBT/HAO(x = 0.75)/Si FeFET showed a wavy profile. The cross-sectional TEM study suggested that the pure CBT had material diffusions among the HAO (x = 0.75) and Si. Owing to the thick grown IL, the gate-leakage current of the Pt/CSBT(y = 1.0)/HAO(x = 0.75)/Si FeFET were as small as about 1/10 of the other FeFETs with y = 0, 0.1, 0.2 and 0.5 [34].

Electrical properties of Pt/CSBT(y)/HAO(x = 0.75)/Si FeFETs with y = 0, 0.1 and 0.2 were investigated. They were (1) I_d retentions, (2) endurances, and (3) pulse memory windows measured by a system updated from our conventional one described in Sect. 2.2. The three kinds of FeFETs with y = 0, 0.1 and 0.2 were annealed at each optimum T_{an} for maximizing V_w which were T_{an} = 813 °C for y = 0, and T_{an} = 800 °C for both y = 0.1 and 0.2 as indicated in Fig. 2.30.

1. I_d retentions of the FeFETs with y = 0, 0.1 and 0.2 were measured using a semiconductor parameter analyzer controlled by a Labview program. All of the on-and off-state I_d–t curves were stable for at least 4×10^5 s as shown in Fig. 2.32. The hold gate voltage values were V_{hold} = 1.1, 1.2 and 1.3 V for the FeFETs with y = 0, 0.1 and 0.2, respectively. The measurement procedure was basically the same as described in Sect. 2.2.3 except for fixing the writing or poling time at t_{pls} = 0.1 s. For investigating the on-state I_d–t, a programming V_g pulse was applied first. During the programming V_g pulse, we fixed V_d, V_s and V_{sub} at V_d = V_s = V_{sub} = 0 V. In the programming pulse, the V_g began from 0 V, raised to V_P = 6 V and kept for t_{pls} = 0.1 s, then finally stepped down to V_{hold}. Then on-state I_d retentions started to be measured till 4×10^5 s. At the time of every marker in the on-state I_d–t curve in Fig. 2.32, V_d was raised from 0 to 0.1 V and the on-state I_d was read out with keeping V_g = V_{hold} and V_s = V_{sub} = 0 V. In the other time, the FeFETs were held with V_g = V_{hold} and V_d = V_s = V_{sub} = 0 V. Similarly, for investigating the off-state I_d–t, erasing a V_g

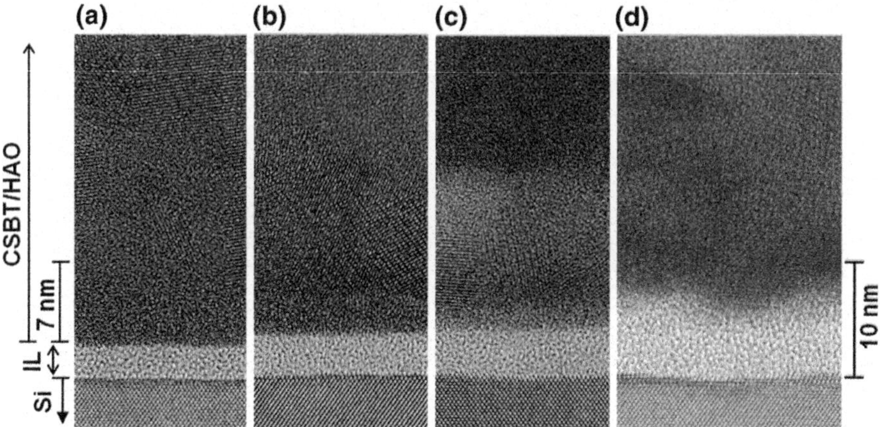

Fig. 2.31 Cross-sectional TEM pictures of Pt/CSBT($y = 0.1$)/HAO($x = 0.75$)/Si FeFETs prepared by annealing at **a** $T_{an} = 748$ °C, **b** $T_{an} = 788$ °C and **c** $T_{an} = 833$ °C. **d** Cross-sectional TEM picture of Pt/CSBT($y = 1.0$)/HAO($x = 0.75$)/Si FeFET prepared by annealing at $T_{an} = 800$ °C. The CSBT($y = 1.0$) was the CBT. Thicknesses were 200 nm Pt, 200 nm SBT and 7 nm HAO. Modified from [34]

Fig. 2.32 I_d retentions of Pt/CSBT(y)/HAO($x = 0.75$)/Si FeFETs with $y = 0$, 0.1 and 0.2. The FeFETs with $y = 0$, 0.1 and 0.2 were annealed at $T_{an} = 813$, 800 and 800 °C, respectively. Thicknesses were 200 nm Pt, 200 nm SBT and 7 nm HAO. The gate lengths were $L = 10$ μm. Modified from [34]

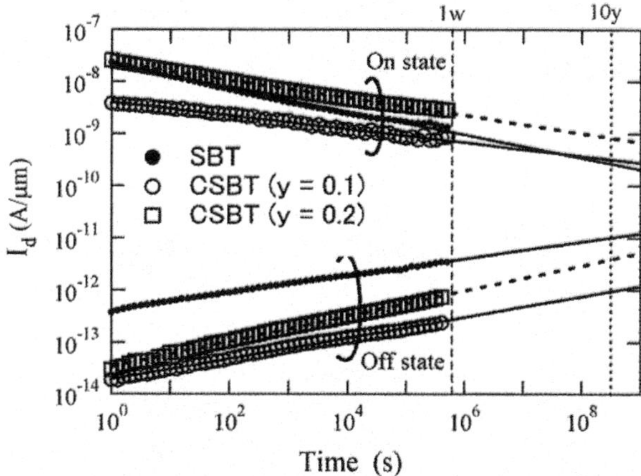

pulse was applied first. During the erasing V_g pulse, we fixed V_d, V_s and V_{sub} at $V_d = V_s = V_{sub} = 0$ V. In the erasing pulse, the V_g began from 0 V, reduced to $V_E = -4$ V and kept for $t_{pls} = 0.1$ s, then finally stepped down to V_{hold}. Then off-state I_d retentions started to be measured till 4×10^5 s. At the time of every marker in the off-state I_d–t curve in Fig. 2.32, V_d was raised from 0 to 0.1 V and the off-state I_d was read out with keeping $V_g = V_{hold}$ and $V_s = V_{sub} = 0$ V. In the other time, the FeFETs were held with $V_g = V_{hold}$ and $V_d = V_s = V_{sub} = 0$ V.

2. The V_{th} endurances of the FeFETs with $y = 0$, 0.1 and 0.2 were measured using a semiconductor parameter analyzer, a pulse generator and a dc voltage source

which were conducted by a Labview program. As shown in Fig. 2.33a–d, the
FeFETs with $y = 0, 0.1$ and 0.2 showed no significant V_{th} shifts until at least 10^8
endurance cycles of 20 µs period 1 ± 5 V. The measurement procedure was
basically the same as described in Sect. 2.2.4 except for reading not I_d but V_{th}
values from the static I_d–V_g curves which were drawn every time after the
accumulated numbers of endurance cycles were counted to 10^n with the n in-
cremented from 0 to 8. In this work, endurance pulses were 20 µs period V_g
pulses of alternately applied $V_P = 6$ V and $V_E = -4$ V which were outputted
from a pulse generator. During the endurance pulses were applied, we fixed V_d,
V_s and V_{sub} at $V_d = V_s = V_{sub} = 0$ V. After the accumulated numbers of
endurance cycles were counted to 10^n at every marker in Fig. 2.33d, the pulse
application was interrupted and V_d was raised from 0 to 0.1 V. Then a static
I_d–V_g curve was drawn by scanning V_g from -4 to 6 V and back to -4 V (or
$V_g = 1 \pm 5$ V), with keeping $V_s = V_{sub} = 0$ V. The I_d was normalized by the
gate width W and was expressed in units of A/µm. Two V_{th} values corresponding to the programmed and erased states were extracted from the
I_d–V_g loop where the V_{th} values were defined as the gate voltages which gave
$I_d = 10^{-8}$ A/µm in the I_d–V_g loop. After recording the static I_d–V_g curve

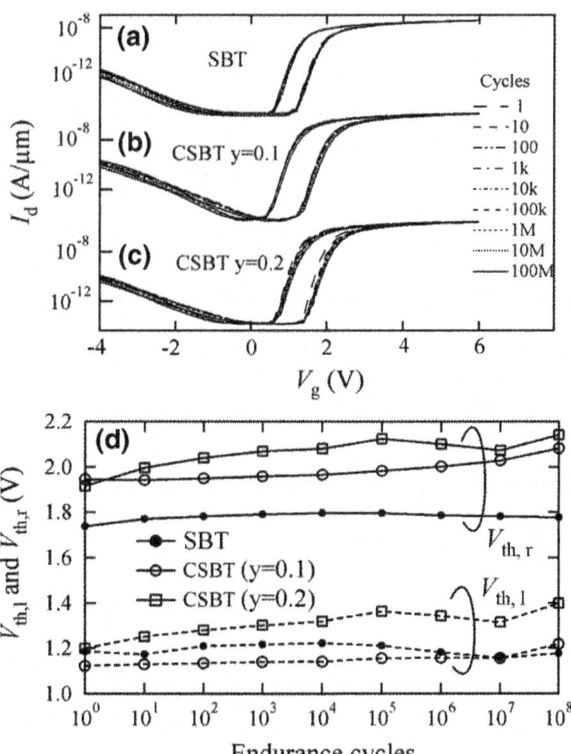

Fig. 2.33 V_{th} endurances of Pt/CSBT(y)/HAO($x = 0.75$)/ Si FeFETs with $y = 0, 0.1$ and 0.2. The FeFETs with $y = 0$, 0.1 and 0.2 were annealed at $T_{an} = 813$, 800 and 800 °C, respectively. Thicknesses were 200 nm Pt, 200 nm SBT and 7 nm HAO. The gate lengths were $L = 10$ µm. Modified from [34]

corresponding to the 10^n cycles, V_d was reduced back to 0 V. Then the V_g endurance-pulse application was resumed.

3. The pulse memory windows of the FeFETs with $y = 0$, 0.1 and 0.2 were investigated. The pulse-memory-window measurement has essentially the same meaning as the writing speed measurement as discussed in Sect. 2.2.5. In order to discuss the practical ability of an FeFET, the memory window should be investigated by writing with a pulsed V_g not a static V_g. The pulse memory window was defined as a V_{th} difference (V_{plsW}) of $V_{plsW} = V_{thE} - V_{thP}$. The V_{thE} and V_{thP} were the V_{th} values of an FeFET at the erased and programmed states. The V_{thE} and V_{thP} were defined as the V_g values at $I_d/W = 10^{-8}$ A/μm. The V_{thE} was the V_{th} after a negative V_g pulse was applied for erasing the FeFET. The negative V_g pulse had a height V_E and a width t_{pls}. Similarly, The V_{thP} was the V_{th} after a positive V_g pulse was applied for programming the FeFET. The positive V_g pulse had a height V_P and the width t_{pls}. The measurement procedure was indicated in Fig. 2.34. A pulse generator and a dc voltage source were used for outputting the V_g wave in Fig. 2.34. The role of the dc voltage source was to give a trigger signal to the pulse generator. I_d values in read operation were measured by a semiconductor parameter analyzer. All the measurement systems were conducted by a Labview program. The V_{thE} and V_{thP} were measured in a cycle in Fig. 2.34 with fixing (V_E, V_P) and at increasing t_{pls} cycle by cycle.

For erasing the FeFET, a single negative V_g pulse was applied which started from a base V_{base}, reduced to V_E kept for t_{pls} and stepped back to the V_{base}. During the negative V_g pulse application, we fixed $V_d = V_s = V_{sub} = 0$ V. After the negative V_g pulse application, the V_d was raised to $V_d = 0.1$ V for reading. Then I_d was measured by static scanning V_g from one end of the read voltage (V_{readA}) to the other end (V_{readB}) with keeping $V_s = V_{sub} = 0$ V. The V_{thE} was determined as the V_g at $I_d = 10^{-8}$ A/μm in the I_d–V_g curve. In this work, we set as $V_{base} = 1.2$ V for the FeFETs with $y = 0.1$ and $y = 0$. The V_{base} was $V_{base} = 1.5$ V for the FeFET with $y = 0.2$. We used $V_{readA} = 2$ V and $V_{readB} = 1$ V.

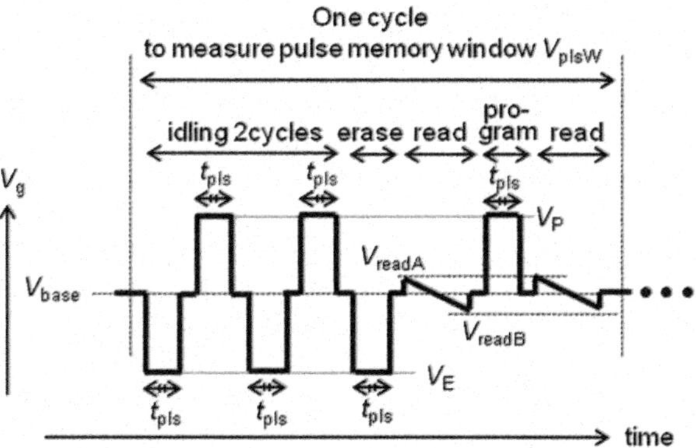

Fig. 2.34 Schematic view of the V_g wave form used for investigating V_{plsW} of FeFETs. The V_{plsW} was measured at t_{pls} elevated with (V_E, V_P) fixed

For programming the FeFET, a single positive V_g pulse was applied which started from the V_{base}, raised to V_P kept for t_{pls} and stepped back to the V_{base}. During the positive V_g pulse application, we fixed $V_d = V_s = V_{sub} = 0$ V. After the positive V_g pulse application, the V_d was raised to $V_d = 0.1$ V for reading. Then I_d was measured by static scanning V_g from one end of the read voltage (V_{readA}) to the other end (V_{readB}) with keeping $V_s = V_{sub} = 0$ V. The V_{thP} was determined as the V_g at $I_d = 10^{-8}$ A/μm in the I_d–V_g curve. Note that the read V_g range from V_{readA} to V_{readB} was common to the read after the erase and the read after the program. The small range was selected so that the read operation did not significantly affect the V_{thP} and V_{thE}.

The V_{thE} and V_{thP} obtained at various height combinations (V_E, V_P) with increasing t_{pls} were shown in Fig. 2.35a–c for the individual FeFETs with $y = 0$, 0.1 and 0.2. In all of Fig. 2.35a–c, the broken lines were for (V_E, V_P) = (−6, 8 V), the dotted lines were for (V_E, V_P) = (−5, 7 V) and the solid lines were for (V_E, V_P) = (−4, 6 V). In each (V_E, V_P), the V_{th} difference of $V_{thE} - V_{thP}$ became large as the pulse width t_{pls} was increased. At every t_{pls}, the larger the $V_P - V_E$ was, the wider the $V_{thE} - V_{thP}$ was. The increasing tendency of the $V_{thE} - V_{thP}$ as the writing-V_g-pulse width and amplitude indicated unsaturated-ferroelectric polarization in the FeFET as we have discussed since the early stage [17].

For all the FeFETs with $y = 0$, 0.1 and 0.2, pulse memory windows $V_{plsW} = V_{thE} - V_{thP}$ were plotted as a function of t_{pls} at every (V_E, V_P) in Fig. 2.36. As shown in Fig. 2.36, the FeFETs with $y = 0.1$ and 0.2 tended to show larger pulse memory windows than the FeFET with $y = 0$ at a common t_{pls} in every

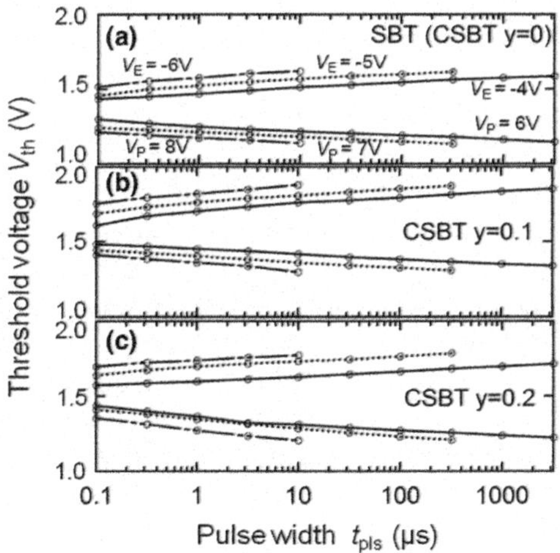

Fig. 2.35 Pulse width t_{pls} dependent V_{thE} and V_{thP} of three kinds of Pt/CSBT(y)/HAO($x = 0.75$)/ Si FeFETs with **a** $y = 0$, **b** $y = 0.1$ and **c** $y = 0.2$. The *broken lines* were for (V_E, V_P) = (−6, 8 V), the *dotted lines* were for (V_E, V_P) = (−5, 7 V) and the *solid lines* were for (V_E, V_P) = (−4, 6 V). The FeFETs with $y = 0$, 0.1 and 0.2 were annealed at $T_{an} = 813$, 800 and 800 °C, respectively. Thicknesses were 200 nm Pt, 200 nm SBT and 7 nm HAO. The gate lengths were $L = 10$ μm. Modified from [34]

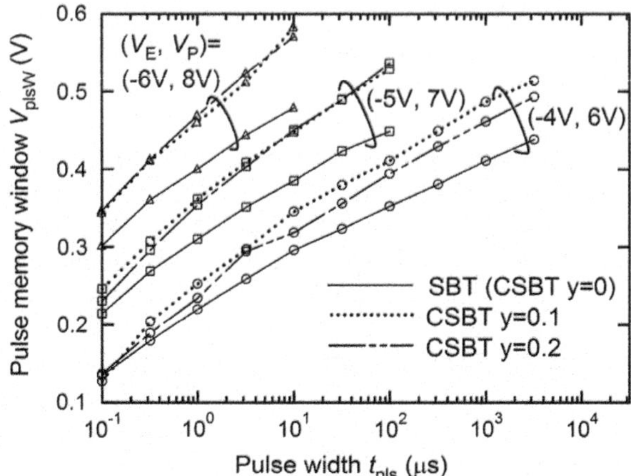

Fig. 2.36 Pulse width t_{pls} dependences of the pulse memory windows $V_{plsW} = V_{thE} - V_{thP}$ at $(V_E, V_P) = (-4, 6$ V$), (-5, 7$ V$)$ and $(-6, 8$ V$)$. The V_{plsW} versus t_{pls} were extracted from the $V_{thE} - t_{pls}$ and $V_{thP} - t_{pls}$ curves in Fig. 2.35a–c. The *solid lines* were for y = 0, the *dotted lines* were for y = 0.1 and the *broken lines* were for y = 0.2. The FeFETs with y = 0, 0.1 and 0.2 were annealed at T_{an} = 813, 800 and 800 °C, respectively. Thicknesses were 200 nm Pt, 200 nm SBT and 7 nm HAO. The gate lengths were L = 10 μm. Modified from [34]

(V_E, V_P). From the other view point, the t_{pls} of the FeFETs with y = 0.1 and 0.2 were shorter than that of the FeFET with y = 0 to show the same pulse memory window. For example when $(V_E, V_P) = (-5, 7$ V$)$ and $t_{pls} = 1$ μs were fixed, the reference FeFET with y = 0 had $V_{plsW} = 0.31$ V while the FeFET with y = 0.1, 0.2 showed $V_{plsW} = 0.35$ V which was 13 % larger V_{plsW} than the reference FeFET. On focusing on the t_{pls} necessary to reach $V_{plsW} = 0.4$ V at $(V_E, V_P) = (-5, 7$ V$)$, $t_{pls} = 3$ μs was for the FeFET with y = 0.2 and $t_{pls} = 15$ μs was for the reference FeFET with y = 0.

In conclusion of this section, the Pt/CSBT(y)/HAO(x = 0.75)/Si FeFETs with y = 0.1 and 0.2 showed good I_d-retentions and endurances equivalent to those of the reference FeFET with y = 0. The Pt/CSBT(y)/HAO(x = 0.75)/Si FeFETs with y = 0.1 and 0.2 had larger static memory windows and pulse memory windows than the reference FeFET with y = 0. Therefore replacing the SBT with the CSBT will be a good solution to preserve the MFIS FeFET qualities with reducing the F thickness. Reducing the F thickness lead to decreasing the gate-stack aspect ratio which was a strong requirement for downsizing the FeFETs.

2.7 Summary

Studies of Pt/CSBT(y)/HAO(x)/Si MFIS FeFETs were reviewed as follows.

In Sect. 2.2, the first Pt/SBT/HAO(x = 0.75)/Si FeFET was characterized by measuring 10^6 s-long retention, 10^{12} cycles-high endurance and 4×10^{-8} s writing speed. Detailed methods of the measurements were also explained. I_d–V_g curves and I_d-retentions measured at most 85 °C were introduced using p-channel FeFETs.

In Sect. 2.3, requirements to the M, F, and I layers in MFIS were discussed. Especially, the requirements to the I layer were high-temperature-proof, thin-deposited high-k material with higher barriers than the F layer against electrons and holes. The good candidate was the HAO(x).

In Sect. 2.4, the optimum preparation conditions of the HAO(x) were investigated. For securing amorphous and small-leakage HAO(x), the preferable x was smaller than 0.95. We often used x $\geq$ 0.75 by expecting both the small leakage and large dielectric constant. The ambient gas kind O_2 and N_2 during the HAO (x = 0.75) deposition were compared. The Pt/SBT/HAO(N)/Si showed $V_w = 0.9$ V while the Pt/SBT/HAO(O)/Si showed $V_w = 0.3$ V at $V_g = \pm 6$ V. The IL thickness in the Pt/SBT/HAO(O)/Si was as thick as 7.4 nm because the HAO(O) did not work sufficiently as a material-diffusion barrier. We also introduced effects of increasing ambient N_2 pressure P_{HAO} during the HAO(x = 0.75) deposition in Pt/SBT/HAO/p-Si. In order to keep a clear interface between the SBT and HAO, $P_{HAO} \ll 40$ Pa was suitable.

In Sect. 2.5, direct nitriding Si was introduced as the way to increase the memory window of Pt/SBT/HAO(x = 0.75)/Si FeFET. The Pt/SBT/HAO (x = 0.75)/Si FeFET using the Si–N tended to show a large subthreshold-voltage swing S. Oxinitriding Si was demonstrated as an improved way to decrease the S value of the FeFET.

In Sect. 2.6, we changed the F material from the SBT to CSBT(y) for increasing the V_W of the FeFET. The Pt/CSBT(y = 0.1, 0.2)/HAO(x = 0.75)/Si FeFETs showed larger static-memory windows than the reference Pt/SBT/HAO(x = 0.75)/ Si FeFET. The Pt/CSBT(y = 0.1, 0.2)/HAO/Si also showed 13 % larger pulse memory windows than the reference. The Pt/CSBT(y = 0.1, 0.2)/HAO(x = 0.75)/Si FeFETs showed 10^8 cycles-high endurances and at least 4×10^5 s-long retentions which were equivalent to the reference Pt/SBT/HAO(x = 0.75)/Si FeFET.

Acknowledgments The authors thank all the researchers and the staffs who were sincerely engaged in this study.

References

1. International Technology Roadmap for Semiconductors, ITRS Tables' summaries, Emerging Research Devices, Table PIDS7a (2013)
2. International Technology Roadmap for Semiconductors, ITRS Tables' summaries, Emerging Research Devices, Tables ERD3, ERD4a and ERD4b (2013)
3. Y. Tarui, T. Hirai, K. Teramoto, H. Koike, K. Nagashima, Appl. Surf. Sci. **113–114**, 656 (1997)

4. S. Sakai, M. Takahashi, K. Takeuchi, Q.-H. Li, T. Horiuchi, S. Wang, K.-Y. Yun, M. Takamiya, T. Sakurai, in Proceedings of 23rd IEEE Non-Volatile Semiconductor Memory Workshop and 3rd International Conference on Memory Technology and Design (2008), p. 103
5. X. Zhang, M. Takahashi, K. Takeuchi, S. Sakai, Jpn. J. Appl. Phys. **51**, 04DD01 (2012)
6. M. Takahashi, S. Sakai, Jpn. J. Appl. Phys. **44**, L800 (2005)
7. L.V. Hai, M. Takahashi, W. Zhang, S. Sakai, Semicond. Sci. Technol. **30**, 015024 (2015)
8. L.V. Hai, M. Takahashi, W. Zhang, S. Sakai, Jpn. J. Appl. Phys. **54**, 088004 (2015)
9. S. Sakai, US Patent 7,226,795 (2005)
10. AIST press release on October 24, 2002
11. S. Sakai, R. Ilangovan, IEEE Electron Devices Lett. **25**, 369 (2004)
12. C.A. paz de Araujo, J.D. Cuchiaro, L.D. McMillan, M.C. Scott, J.F. Scott, Nature (London) **374**, 627 (1995)
13. I. Koiwa, T. Kanehara, J. Mita, T. Iwabuchi, T. Osaka, S. Ono, M. Maeda, Jpn. J. Appl. Phys. **35**, 4946 (1996)
14. S.J. Hyun, B.H. Park, S.D. Bu, J.H. Jung, T.W. Noh, Appl. Phys. Lett. **730**, 2518 (1998)
15. S. Sakai, M. Takahashi, R. Ilangovan, IEDM Tech. Dig. 915 (2004)
16. S. Sakai, R. Ilangovan, M. Takahashi, in Extended Abstracts of 2004 International Workshop on Dielectric Thin Films For Future ULSI Devices Science and Technology (IWDTF 2004), Tokyo (2004), pp. 55–56
17. S. Sakai, R. Ilangovan, M. Takahashi, Jpn. J. Appl. Phys. **43**, 7876 (2004)
18. Q.-H. Li, T. Horiuchi, S. Wang, M. Takahashi, S. Sakai, Semicond. Sci. Technol. **24**, 025012 (2009)
19. H.B. Michaelson, IBM J. Res. Dev. **22**, 72 (1978)
20. A.K. Tagantsev, I. Stolichnov, N. Setter, J.S. Cross, M. Tsukada, Phys. Rev. B **66**, 214109 (2002)
21. Q.-H. Li, S. Sakai, Appl. Phys. Lett. **89**, 222910 (2006)
22. S. Sakai, X. Zhang, L.V. Hai, W. Zhang, M. Takahashi, in Proceedings of 12th Non-volatile Memory Technology Symposium (NVMTS) (2012), pp. 55–59
23. K. Sakamaki, S. Sakai, unpublished
24. S.M. Sze, *Physics of Semiconductor Devices*, 2nd edn (Wiley, New York, 1981), p. 447
25. J. Robertson, Eur. Phys. J. Appl. Phys. **28**, 265 (2004)
26. H.Y. Yu, M.F. Li, B.J. Cho, C.C. Yeo, M.S. Joo, D.-L. Kwong, J.S. Pan, C.H. Ang, J.Z. Zheng, S. Ramanathan, Appl. Phys. Lett. **81**, 376 (2002)
27. T. Horiuchi, K. Ohhashi, M. Takahashi, S. Sakai, Funtai Oyobi Funmatsu Yakin **55**, 17 (2008) (in Japanese)
28. M. Takahashi, T. Horiuchi, S. Wang, Q.-H. Li, S. Sakai, J. Vac. Sci. Technol., B **26**, 1585 (2008)
29. T. Horiuchi, M. Takahashi, K. Ohhashi, S. Sakai, Semicond. Sci. Technol. **24**, 105026 (2009)
30. T. Horiuchi, M. Takahashi, Q.-H. Li, S. Wang, S. Sakai, Semicond. Sci. Technol. **25**, 055005 (2010)
31. Y. Noguchi, H. Shimizu, M. Miyayama, K. Oikawa, T. Kamiyama, Jpn. J. Appl. Phys. **40**, 5812 (2001)
32. R.R. Das, P. Bhattacharya, W. Perez, R.S. Katiyar, S.B. Desu, Appl. Phys. Lett. **80**, 637 (2002)
33. R.R. Das, P. Bhattacharya, W. Perez, R.S. Katiyar, Jpn. J. Appl. Phys. **42**, 163 (2003)
34. W. Zhang, M. Takahashi, S. Sakai, Semicond. Sci. Technol. **28**, 085003 (2013)
35. K. Kato, K. Suzuki, K. Nishizawa, T. Miki, J. Appl. Phys. **88**, 3779 (2000)
36. S. Sakai, M. Takahashi, K. Motohashi, Y. Yamaguchi, N. Yui, T. Kobayashi, J. Vac. Sci. Technol., A **25**, 903 (2007)
37. Y. Shimakawa, Y. Kubo, Y. Nakagawa, S. Goto, T. Kamiyama, H. Asano, F. Izumi, Phys. Rev. B **61**, 6559 (2000)
38. R.R. Das, P. Bhattacharya, W. Perez, R.S. Katiyar, Appl. Phys. Lett. **78**, 2925 (2001)

Chapter 3
Nonvolatile Field-Effect Transistors Using Ferroelectric Doped HfO$_2$ Films

Uwe Schroeder, Stefan Slesazeck and Thomas Mikolajick

Abstract Ferroelectrics are ideal for low power digital information storage since they can be switched purely field controlled with negligible current consumption and at the same time are nonvolatile. However, the incompatibility of classical ferroelectric materials with semiconductor technology has hindered the scaling of ferroelectric memory devices. Therefore, such devices are only used in niche applications today. In 2012, first reports indicated that hafnium oxide, which is a standard material in modern CMOS processes, can be transformed into a ferroelectric phase. Moreover, the specific properties such as much lower permittivity compared to classical perovskite based ferroelectrics and high coercive fields enable to realize scaled ferroelectric field effect transistors that show nonvolatile retention. Therefore, ferroelectric hafnium oxide can help to finally fully exploit the potential of ferroelectric memories. In this chapter, we will first show the basics of ferroelectric hafnium oxide. Then, the current understanding of the origin and technological parameters influencing the ferroelectricity in hafnium oxide are discussed. Finally, the current status and future prospects of a ferroelectric field effect transistor based memory technology are summarized and discussed.

3.1 Introduction

Ferroelectric materials show a remanent electrical polarization without applied electrical field. The direction of the remanent polarization can be switched by an external electrical field [1]. Therefore, ferroelectrics are a natural choice for the realization of nonvolatile memories [2]. The first attempts to realize a memory

U. Schroeder (✉) · S. Slesazeck · T. Mikolajick
Nanoelectronic Materials Laboratory—NaMLab gGmbH,
Noethnitzer Strasse 64, 01187 Dresden, Germany
e-mail: uwe.schroeder@namlab.com

U. Schroeder · S. Slesazeck · T. Mikolajick
IHM, Technische Universitaet Dresden, Noethnitzer Strasse 64,
01187 Dresden, Germany

© Springer Science+Business Media Dordrecht 2016
B.-E. Park et al. (eds.), *Ferroelectric-Gate Field Effect Transistor Memories*,
Topics in Applied Physics 131, DOI 10.1007/978-94-024-0841-6_3

Table 3.1 Comparison of three generations of ferroelectric memories

	Generation 1	Generation 2	Generation 3
Start of Development	early 1950s	late 1980s	late 2000s
Material	$BaTiO_3$	$Pb[Zr_xTi_{1-x}]O_3$ or $SrBi_2Ta_2O_9$	doped HfO_2
Cell architecture	1C	2T2C -> 1T1C	1T (possibly also 1T1C)
Scalability	large structures only	about 90 nm	below 20 nm
Status	development stopped	commercial products	research and development

based on a ferroelectric were already undertaken in the 1950s [3–5]. In the first generation of ferroelectric memories a barium titanate ($BaTiO_3$) ferroelectric single crystal was used. Metal electrodes were evaporated on both sides to form a cross point arrangement (see Table 3.1). This approach resulted in memory matrices that were not yet integrated with the support devices and had very low densities of 16 to 100 bits. Due to the non-ideal hysteresis curve of a typical ferroelectric like barium titanate, the disturb resulting from the half select that every cell sharing the same wordline or bitline like the addressed cell could not be solved. Therefore, the work on first generation ferroelectric memories was discontinued without reaching commercialization. Due to the success of magnetic core memories and later also semiconductor memories like static random access memory (SRAM) and dynamic random access memory (DRAM), the interest in developing ferroelectric memories rapidly declined in the following years.

In the late 1980s this situation changed. The semiconductor industry was searching for ways to increase the flexibility and decrease the cost of nonvolatile memories like electrically erasable programmable read-only memories (EEPROMs). The goal to realize a "universal memory" that would have the performance and cell size of a DRAM and the non-volatility of an EEPROM or flash memory revived the interest in ferroelectric memories. At that time, semiconductor technology was already available at an advanced level and SRAM, DRAM and EEPROM/Flash memories had become mainstream technologies. Therefore, it was clear that the ferroelectric material had to be integrated into the metal-oxide-semiconductor (MOS) process. From the former experience, it was also clear that the cross-point arrangement had to be modified using a cell architecture with an MOS transistor as a select element to protect unaddressed cells from disturbs. This resulted in a cell that was constructed of the combination of one transistor with one ferroelectric capacitor (1T1C cell see Table 3.1) very similar to the DRAM architecture proposed by Dennard [6]. In the first years, two of these cells were used in order to increase the available memory window resulting in a two transistor and two capacitor (2T2C) cell configuration. This development lead to the first commercial products in the early 1990s [7]. Polycrystalline lead zirconium titanate ($PZT = PbZr_{0.5}Ti_{0.5}O_3$) was the ferroelectric material of choice used in these devices. In the beginning, however, the endurance of this material was limited to

10^6–10^{10} cycles due to fatigue of the ferroelectric sandwiched between to noble metal electrodes. Strontium bismuth tantalate (SBT = $SrBi_2Ta_2O_9$) did not show this fatigue and therefore was tried as an alternative [8, 9]. However, the higher processing temperatures of this material resulted in even more severe integration challenges [10, 11] as was already the case for PZT. So, SBT was again replaced by PZT as soon as the reliability issues for PZT were resolved. The reliability problems were connected to the loss of oxygen in the vicinity of the electrodes and could consequently be resolved by implementing oxide electrodes like IrO_2 or strontium ruthenate (SRO) [12].

The reading signal generated by a ferroelectric capacitor is limited by the available polarization charge [13]. Therefore, with PZT, below 100 nm the technology would have to move to a 3-dimensional capacitor like it is used in DRAM for further scaling. This, however, has proven to be an extremely challenging task having such complex materials in mind [14]. As a result, the competitive position of ferroelectric memories in terms of cost per bit became even worse. Consequently the development work was deprioritized. Alternative materials like bismuth ferrite (BFO = $BiFeO_3$) with higher polarization charge cannot solve the issues for more than two more generations [13]. Therefore, with the 130 nm technology that is in production today [15], the second generation of ferroelectric memories is already close to its scaling limit and will stay limited to niche applications.

One alternative would be to integrate the ferroelectric into the gate stack of a metal-insulator-semiconductor (MIS) transistor. The polarization state could then be read using the drain current of the transistor rather than sensing the switched charge directly. This concept, called the ferroelectric field effect transistor (FeFET) dates back to the late 1950s [16, 17]. However, common ferroelectrics like PZT and SBT are incompatible with silicon. Consequently an interface layer is required between the ferroelectric and the silicon channel of the ferroelectric field effect transistor. This interface layer tends to have a significantly lower permittivity k than the ferroelectric. An internal depolarization field in the retention case is the result [18]. Nonvolatile retention could be shown by using hafnium oxide as the interface layer together with a very thick SBT layer [19], as discussed within this book in more detail.

In 2011, Boescke et al. [20] demonstrated for the first time that hafnium oxide doped with silicon can possess ferroelectric properties. Recently, it was confirmed that this is a ferroelectric effect based on the crystal structure obtained during specific processing conditions [21]. This finding may resolve the limited scalability of the second generation ferroelectric memory devices caused by the limited CMOS compatibility of PZT. In contrast to PZT, hafnium oxide is a standard material in sub 45 nm complementary metal-oxide-semiconductor (CMOS) fabrication. Therefore, well established CMOS proven fabrication and integration schemes are available. This discovery has revived the work on ferroelectric memories leading to its inclusion in the 2013 editions of the ITRS roadmap [22]. Table 3.1 compares the major properties of the three generations of ferroelectric memories. In the following the current status on the hafnium oxide based third generation ferroelectric memory is reviewed.

3.2　FeFET Integration

For a simple integration of a doped HfO_2 based ferroelectric material into a FeFET device a MFS transistor structure is used, which would be very similar to a standard high k metal gate device as introduced by Intel in 2007 for high performance applications [23]. The transistor stack is consisting of a $SiO_x(N_y)$ buffer layer (physical thickness ~ 1 nm), the ferroelectric HfO_2 layer (physical thickness ~ 10 nm), a TiN metal gate electrode followed by a highly doped poly Si layer. Here, the standard 2 nm HfO_2 dielectric was replaced by the ferroelectric HfO_2 layer. Due to the similarities in the dielectric layer properties no further process adjustments are necessary. Before the device is described in detail the material properties of doped HfO_2 are introduced.

3.2.1　Ferroelectric Doped HfO₂

Different polymorphs were reported for the HfO_2 bulk phase depending on the dopant content in HfO_2, temperature and pressure conditions [24]. In the CMOS processing temperature range below 1100 °C at atmospheric pressure typically a monoclinic phase (space group $P2_1/c$) is visible for a crystalline material. For higher dopant content a tetragonal (space group $P4_2/nmc$) or cubic phase (space group Fm3 m) is reported depending on the atomic radius of the dopant atom (see Sect. 3.2.3). For a ferroelectric layer thickness of 5–30 nm, as typically used in semiconductor processing, the ferroelectric and anti-ferroelectric properties were first discovered in 2007 by Boescke et al. at DRAM manufacturer Qimonda. Experiments continued at NaMLab, Fraunhofer CNT and RWTH Aachen and were first published in 2011 [20]. The newly discovered ferroelectric properties for thin HfO_2 layers are attributed to the formation of a non-centrosymmetric orthorhombic phase ($Pbc2_1$ or $Pca2_1$) which were recently confirmed by TEM results [21].

3.2.2　Si Doped HfO₂

First FeFET devices are fabricated by introducing Si doped HfO_2 as a ferroelectric material into high k metal gate transistor stacks. For this dopant the polarization hysteresis is evaluated on metal-ferroelectric-metal (MFM) structures in a dopant range from 0 to 12 mol% SiO_2 [25]. A maximum polarization could be verified for about 4 mol% SiO_2 (see Fig. 3.1). Lower and higher dopant amounts are resulting in a sharp drop of the remanent polarization. For a dopant range between 5 and 8 mol% SiO_2 a field induced ferroelectricity is proposed [20]. Applying an electric field beyond 1 MV/cm might lead to a field induced phase transformation of the tetragonal to a ferroelectric phase.

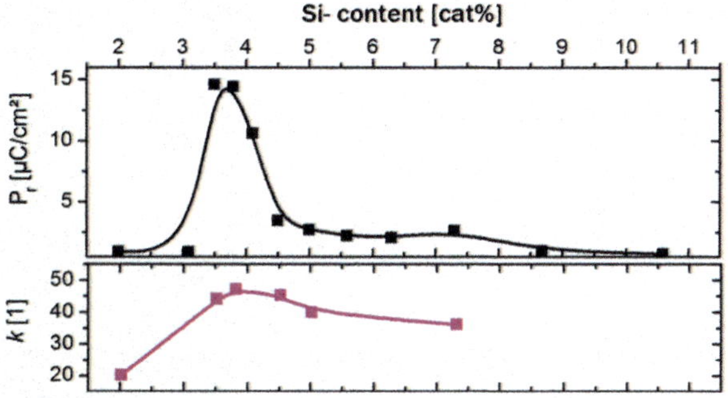

Fig. 3.1 *Top* Remanent polarization and *bottom* dielectric constant of 10 nm Si:HfO$_2$ for different Si dopant content in HfO$_2$ [25]

The ferroelectric polycrystalline HfO$_2$ layer within a FeFET device is consisting of grains with the thickness of the HfO$_2$ layer in horizontal direction and a $\sim$20–30 nm diameter in lateral dimension. TEM measurements are revealing only single grains with similar orientation [26]. The four Hf atoms form a distorted FCC lattice with the eight O atoms occupying the pseudo-tetrahedral interstitial sites. Polarization switching occurs by movement of four oxygen atoms between two local energy minima within the unit cell. Oxygen atoms are shifted between these two mirror symmetric positions along the c-axis thereby allowing spontaneous polarization in the Pbc2$_1$ cell.

Various positions of the c-axis are found to result in a distribution of local oriented grains and an according variation of the local polarization as verified by piezo force microscopy measurements [27]. On the macroscopic scale, large areas could be polarized in the two polarization directions indicating a good suitability of the material for introduction into ferroelectric semiconductor devices [27].

3.2.3 Other Doped HfO$_2$

Similar measurements as described for Si dopant are performed on MFM capacitors using different dopant atoms with a crystal radius larger than Si. Dopant atoms with different valences are evaluated. The different binding configurations resulting in different oxygen vacancy configurations had no effect on the general ability to form a ferroelectric HfO$_2$ phase. A pure field driven oxygen vacancy movement as the main driving root cause for the detected ferroelectric effects can be ruled out. After cycling the samples for at least 1000 wake-up cycles a de-pinching is visible and most samples showed clean hysteresis. The wake-up effect and fatigue is very similar to the Si doped HfO$_2$ sample with $\sim$4 mol% SiO$_2$. With increasing dopant concentrations for dopants materials larger than Hf, a reduction of the remanent polarization is detected.

Only a small dopant content range for Si and Al doped samples remained stable in the pinched hysteresis shape. The appearance of the stable pinched antiferroelectric-like hysteresis can be related to the phase transition determined for the characterized dopant materials according to Schroeder et al. [28]:

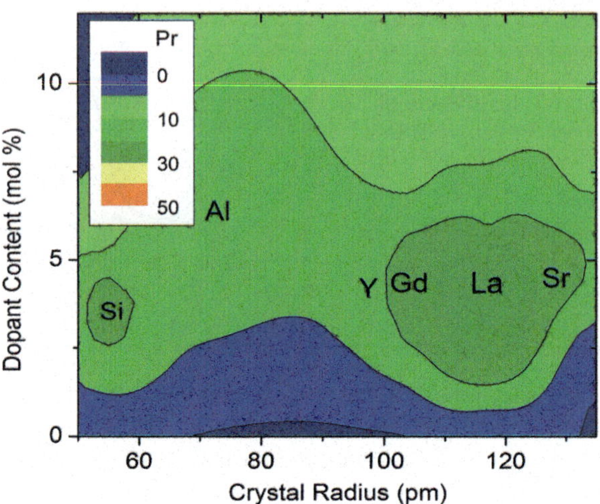

Fig. 3.2 Contour plot of the remanent polarization as a function of crystal radius and dopant content [28]

$$\text{Si, Al} \quad \text{monoclinic} \rightarrow \text{orthorhombic} \rightarrow \text{tetragonal} \rightarrow \text{cubic,}$$

in contrast to:

$$\text{Y, Gd, La, (Sr)} \quad \text{monoclinic} \rightarrow \text{orthorhombic} \rightarrow \text{cubic.}$$

An apparent correlation was visible: only for dopants with a crystal radius smaller than Hf a transition from a monoclinic to the tetragonal phase via the orthorhombic lattice is seen. Only these dopants indicated a stable pinched hysteresis. Reported values of the remanent polarization for HfO_2 with different dopants are summarized in Fig. 3.2. Simulation results predicted a maximum value of $P_r \sim 55\ \mu C/cm^2$ for the orthorhombic phase [29]. Since deposited layers within the MFM capacitor have a polycrystalline structure the polarization orientation can be random. Accordingly, lower polarization values are expected. A complete polarizability of the layer was confirmed by piezo force microscopy [27].

3.3 Memory Properties of Ferroelectric Hafnium Oxide

From the ferroelectric hysteresis loops, doped hafnium oxide looks promising for nonvolatile memory applications. However, to qualify for a real memory concept the material has to show [30]:

- fast switching with disturb free readout
- retention of 10 years at elevated temperatures
- endurance of at least 10^5 cycles (similar to flash memory, but the higher the better).

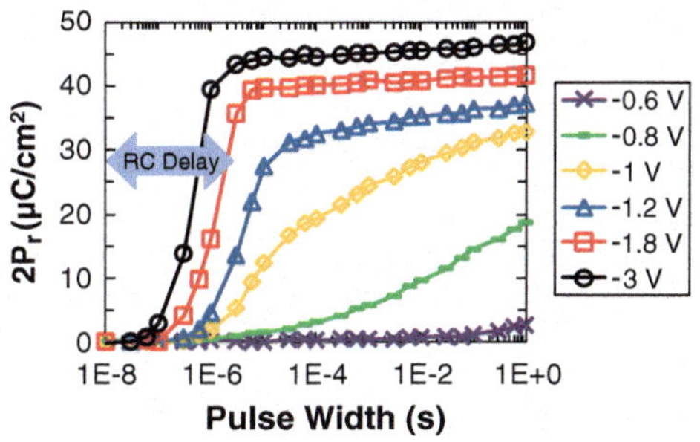

Fig. 3.3 Switched charge 2P$_r$ as a function of the pulse length for different pulse voltages. The measurement was performed using the PUND scheme

Figure 3.3 shows the programming characteristics of a 10 nm thick hafnium oxide film doped with 5 mol% SiO_2 sandwiched between TiN electrodes [31]. A PUND (positive up negative down) scheme is used for the measurement. It can be clearly seen that for a voltage of 3 V saturation is reached and that for very low voltages below 1 V only switching occurs if long program pulses are applied. Note that the measurement setup is limiting the speed of the measurement. It is established, that with voltages in the 4–5 V range switching in the few ns range can be achieved [28].

In general, retention is a key issue in nonvolatile memories and in ferroelectric memories in particular the opposite state retention [32] (degraded by the imprint effect) is of crucial importance. Figure 3.4 is showing the retention results of 10 nm thick $Si:HfO_2$ between TiN electrodes measured at 125 °C [32]. It can be seen, that both the same state as well as the other state of the capacitor is very stable even at elevated temperature of 125 °C.

The ability to repeatedly switch between the states is very important for non-volatile memories. Especially in the early days of the second generation of ferro-electric memories, when PZT was sandwiched between Pt electrodes, fatigue of the polarization with cycling was a severe issue. Due to the high coercive field in ferroelectric hafnium oxide, breakdown can be an issue if high stress is applied to the structure during cycling [32]. With more careful cycling, endurance in the 10^9–10^{10} cycle range can be demonstrated. Figure 3.5 shows an example where 10^9

Fig. 3.4 Retention of the same state (SS), the new same state (NSS) and of the opposite state (OS) of a capacitor having hafnium oxide doped with about 5 mol% of SiO_2 between TiN electrodes. The measurements were performed at a temperature of 125 °C

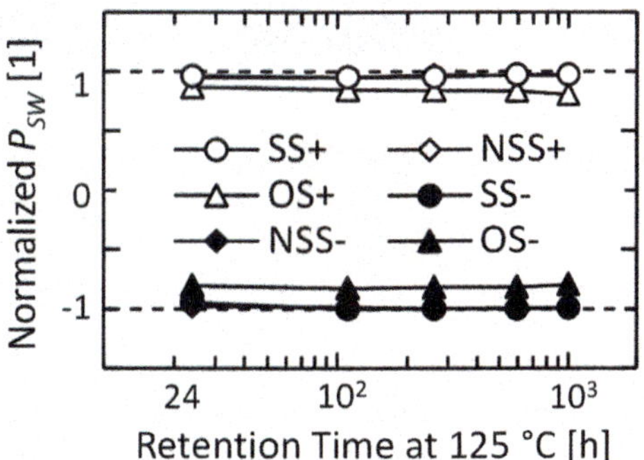

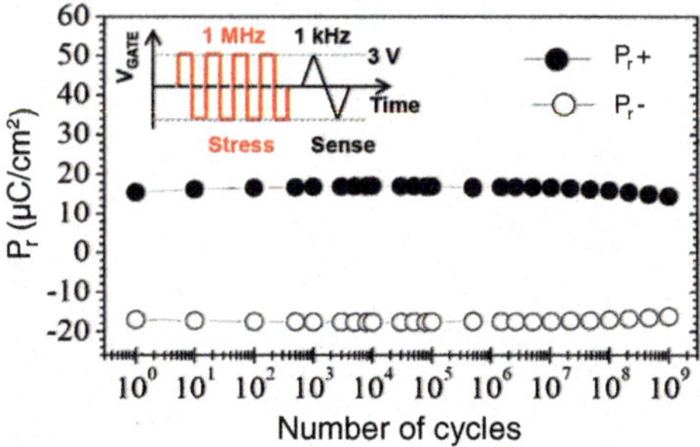

Fig. 3.5 Program/erase cycling characteristics of a capacitor with 10 nm Si:HfO_2 between TiN electrodes measured at a stress frequency of 1 MHz. The remanent polarization P_R and for both polarization states is shown as functions of the stress cycle count

cycles are reached [28]. The onset of fatigue can be observed after about 10^7 cycles. These are very encouraging results since no optimization of electrode interfaces etc., which are crucial to achieve high endurance in PZT [12], have been implemented yet.

Measurements are performed to understand the field cycling and fatigue effect in doped HfO_2 based MFM capacitors, here for 10 nm Sr:HfO_2. Figure 3.6a is displaying the transient current characteristics for the pristine (1), the 'after wake-up' (2), and the fatigue case. Two distinct current maxima are merging during field cycling and in the fatigue case a reduction of the peak maximum is coinciding with a broadening of the peak. First order reversal curve (FORC) measurements performed by Schenk et al. [33] are allowing a calculation of the internal bias field in

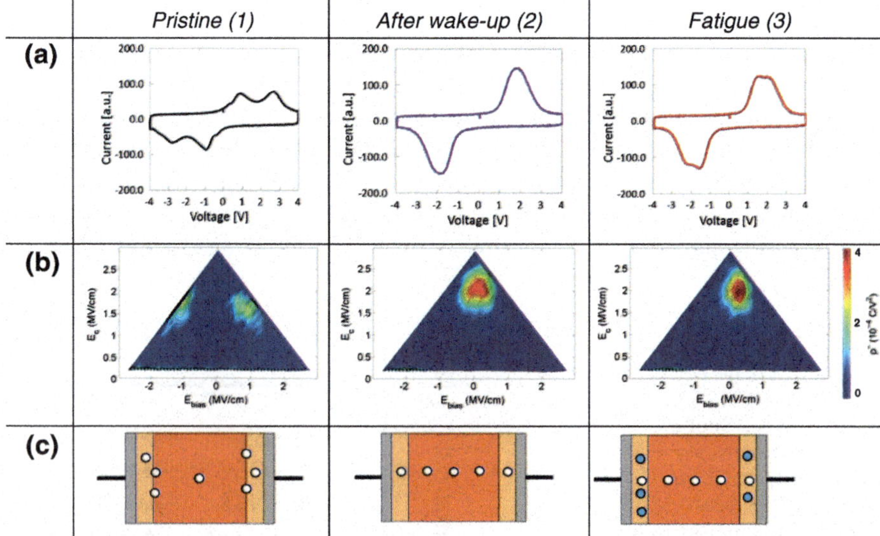

Fig. 3.6 **a** Transient current, **b** first order reversal curve measurements, and **c** trap distribution model for a 10 nm Sr:HfO_2 MFM capacitor in the pristine (*1*), 'after wake-up' (*2*), and fatigue (*3*) case. For the model: initial mobile traps are drawn in white and newly generated non-mobile static defects inside the interfacial TiO_xN_y layer are depicted in *blue*

addition to the coercive field (Fig. 3.6b) For the pristine case an internal bias field of ± 2 MV/cm is determined with only a minimal impact on the coercive field $E_c \sim 2$ MV/cm. After wake-up cycling the internal bias field diminished and a slight increase in the coercive field to $E_c \sim 2.2$ MV/cm is visible. During fatigue these values are staying constant, but the intensity is decreased. One hypothesis is that these experimental results could be correlated to a redistribution of trapped charges [34] as verified by simulation in Fig. 3.6c). Initially, most traps would be positioned at the interfacial TiO_xN_y layer boundary at the interface between the TiN electrode and the doped HfO_2 layer. During field cycling these mobile traps would uniformly distributed within the whole layer stack, resulting in a disappearance of the internal bias field of the MFM structure. During fatigue additional static defects would be generated mainly inside the interfacial layer resulting in a depolarization field at the ferroelectric layer and hence, in a reduction of the reduction of the remanent polarization. Ferroelectric domains with polarization direction almost parallel to the electrode might be prevented from switching. Simulated transient current curves are in good agreement with experimental data [34].

Dynamic hysteresis measurements one second after programming show a reduction of the relaxed remanent polarization value mainly during the first 100 cycles. The above mentioned internal bias field resulting in the depolarization field could be a cause for this effect.

3.4 Hafnium Oxide Based Ferroelectric Field Effect Transistor

3.4.1 Device Performance

Ferroelectric hafnium oxide seems to be an ideal candidate to realize a ferroelectric field effect transistor (FeFET) for two reasons. First, hafnium oxide has a much lower k value than classical inorganic ferroelectrics. Therefore, the depolarization field is much lower, even if an interface layer between the silicon and the ferroelectric hafnium oxide is still existing [18]. Second, hafnium oxide is the standard gate dielectric used in sub 45 nm CMOS processes [23] and a well-known material in a CMOS production environment. Therefore, the device architecture does not have to be changed to integrate the nonvolatile memory functionality.

For transistor fabrication, where HfO_2 is directly deposited on a Si surface, a $SiO_x(N_y)$ buffer layer is introduced to initiate the HfO_2 ALD deposition on a defined interface. As a top electrode a thin TiN electrode is used before a poly-crystalline silicon deposition. Using the Maxwell equation [35], the electric field vector and the displacement vector are perpendicular to the surface and accordingly

$$\varepsilon_B \cdot E_B = \varepsilon_{FE} \cdot E_{FE} \tag{3.1}$$

Consequently, the voltage drop U_{FE} across the ferroelectric layer is

$$U_{FE} = \frac{U}{\frac{\varepsilon_{FE} d_B}{\varepsilon_B d_{FE}} + 1} \tag{3.2}$$

with ε_B, ε_{FE} as the dielectric constant of $SiO_x(N_y)$ buffer layer and HfO_2 based ferroelectric, E_B, E_{FE} the field across the buffer or ferroelectric layer, d the thickness of the according layer, and U the voltage applied to the whole film stack.

Since in the typically FeFET configuration $\varepsilon_{FE} \sim 8 \cdot \varepsilon_B$ (with $\varepsilon_{FE} \sim 30$ and $\varepsilon_B \sim 3.9$) and $d_{FE} \sim 8 \cdot d_B$, the voltage across the ferroelectric layer is roughly reduced by a factor 2 compared to the voltage applied to the overall gate stack. To reach saturation polarization P_s a field of ~ 3 MV/cm is necessary to be applied to a MFM capacitor, which needs to be increased accordingly in a FeFET configuration. Since the field across the buffer layer is very similar to the ferroelectric layer, charge injection might occur during switching of the polarization in the device. Details will be discussed later in the reliability Sect. 3.4.2.

The memory window MW of a FeFET is defined as the difference between the threshold voltages for the two different polarization values and is roughly given by Lue et al. [36]:

$$MW = 2E_c d_{FE}\left(1 - \frac{2\delta\varepsilon_{FE}\varepsilon_0}{P_S}\right) \tag{3.3}$$

(with d_{FE} is the thickness of the ferroelectric, P_s the saturation polarization, δ factor as explained in Lue et al. [36]). If the saturation polarization of the ferroelectric layer shows a sufficient value the equation can be simplified to Lue et al. [36]:

$$MW \approx 2E_c d_{FE} \tag{3.4}$$

This means that for rather low values of E_C like in SBT, a thick ferroelectric is necessary. This can be seen in the SBT based devices from literature as summarized in Table 3.2 [19, 37]. When a memory window of >1 V is preferred, the large coercive field of hafnium oxide enables to scale down the ferroelectric layer thickness to a reasonable value [38].

Table 3.2 Parameter comparison for a SBT/PZT versus Si:HfO$_2$ FeFET device

	E_c (MV/cm)	d_{FE} (nm)	MV (V)
SBT/PZT	0.1	100	2
Si:HfO$_2$	1	10	2

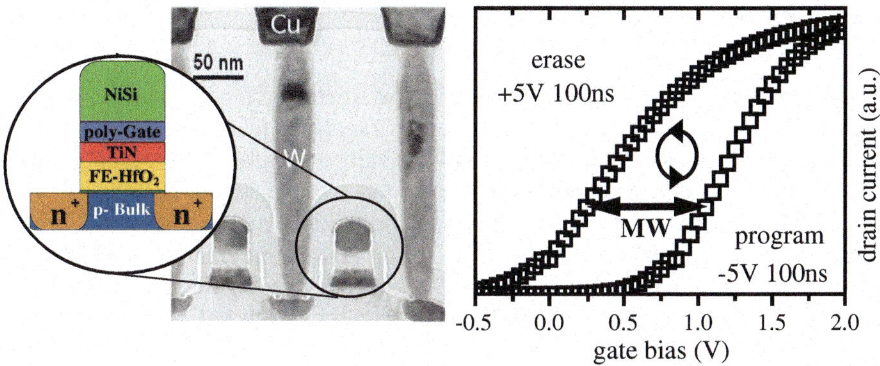

Fig. 3.7 Transmission electron microscope (TEM) cross section of a FeFET device with a 10 nm thick doped hafnium oxide fabricated using a 28 nm technology together with the transfer characteristics achieved after programming with +5 to −5 V for 100 ns

Figure 3.7 is showing the transmission electron microscope (TEM) cross section of a FeFET device with a 10 nm thick doped hafnium oxide fabricated using a 28 nm technology together with the transfer characteristics achieved after programming with +5 V/−5 V for 100 ns. A clear memory window can be seen [39]. This proves that hafnium oxide enables to scale FeFET devices much more aggressively than using conventional ferroelectrics like SBT [37, 40].

Table 3.3 compares the electrical performance characteristics of a standard NOR Flash cell and a FeFET cell. The estimated FeFET characteristics outperform the NOR Flash clearly in the write energy per bit cell and in the write/erase speed. The low energy consumption is due to the fact, that no programming current is mandatory for programming the cell. However, for a more accurate evaluation of the power consumption one would have to include all the contributions from the power supply, driver circuitry and charging/discharging of parasitic capacitances. Storage of multiple bits within one scaled FeFET cell might be difficult because of the abrupt switching of the ferroelectric polarization within one domain and the threshold voltage drift originating from the depolarization field as examined in the next section.

Table 3.3 Electrical performance of a NOR flash memory cell compared to a FeFET cell

	NOR flash	FeFET
Write/erase speed	1 µs/2 ms	**10 ns/10 ns**
Read speed	10 ns	10 ns
Retention	10 y	10 y
Endurance	10^5 cycles	10^5 cycles
Write/erase voltage	10–20 V	**5 V**
Write energy	1 nJ/bit	**1 fJ/bit**

3.4.2 Device Reliability

Figure 3.8 illustrates the mayor reliability parameters retention and endurance. Even if the device is stored at 200 °C, an extrapolated memory window >0.2 V remains after 10 years. The threshold voltage shift over time is supposedly due to charge trapping and de-trapping under the influence of the depolarization field [40]. The window can be enlarged by increasing the original memory window. Both a thicker ferroelectric and an increased E_C (see (3.4)) or a different dopant [28] with higher remanent polarization can be used for this purpose. Endurance is showing a window reduction starting at 10^4 cycles. The window is still 0.9 V after 10^5 cycles and is closing afterwards.

The trend of the memory window change with cycling is plotted in Fig. 3.9b. As expected from the simplified (3.4), the memory window should be directly correlated to the increase in the coercive field. However, differences in the trend might be

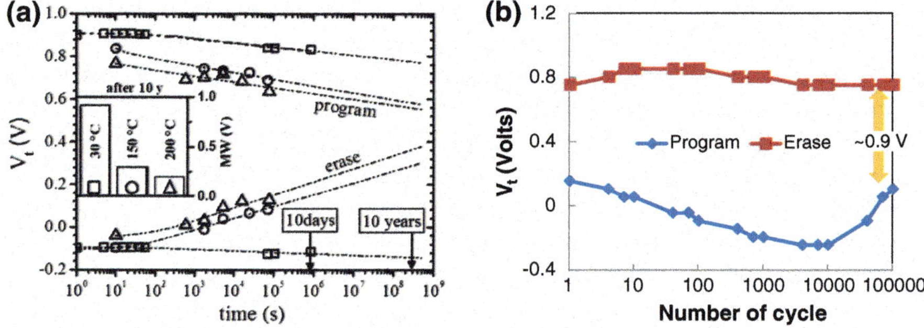

Fig. 3.8 Retention measurement at 30, 150 and 200 °C (**a**) and endurance measurement performed at room temperature (**b**) for a $Si:HfO_2$ based FeFET fabricated in 28 nm technology

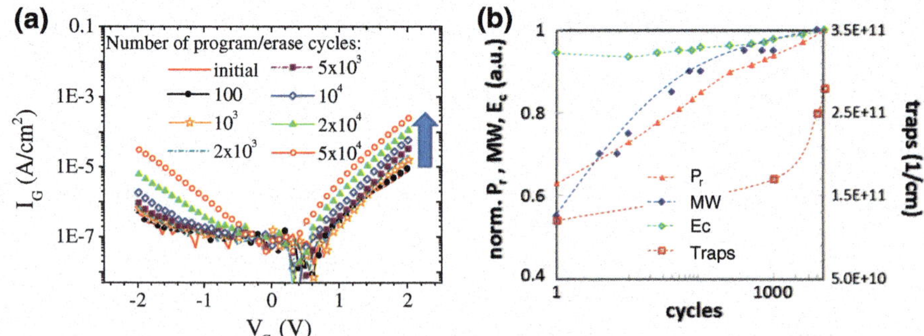

Fig. 3.9 **a** Gate leakage current determined after a certain number of program/erase cycles **b** trapped charges after a certain number of program/erase cycles in the interfacial layer between Si substrate and the ferroelectric HfO_2

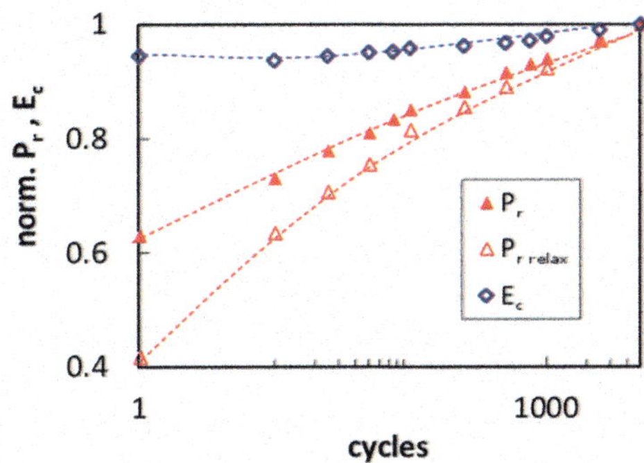

Fig. 3.10 Evolution of normalized relaxed remanent polarization P_{relax}, (one second after pulsing) the coercive field E_c of a 10 nm Si:HfO$_2$ MFM capacitor (from Fig. 3.5) as well as the memory window MW of a FeFET memory device as a function of field cycles [28]

related to the polarization relaxation as caused by the depolarization field due to trapped charges as already seen for the MFM capacitor case (relaxed polarization from Fig. 3.10 added to Fig. 3.9b. Accordingly, as described in (3.3), the maximum polarization is not available for the pristine sample and the memory window showed an impact of the depolarization field for the initial cycles. After 100 cycles, the polarization value is high enough that the memory window can follow the coercive field trend as expected from (3.4).

In addition, Yurchuk et al. performed intensive device reliability characterization to understand the closing of the memory window after 10^4–10^5 cycles [41]. Specially, since this is contradicting to results obtained for MFM capacitors. Here, an endurance of 10^9–10^{10} cycles is determined (Fig. 3.5). Gate leakage current measurements show an increase of the current after 10^4 program/erase cycles similar to the memory window decrease (see Fig. 3.9a). Charge pumping measurements to characterize interfacial traps in the SiO$_x$N$_y$ interlayer between the Si substrate and the doped HfO$_2$ are performed. Here, also an increase of the trapped charges is found, which was not detected in the ferroelectric HfO$_2$ layer (see Fig. 3.9b). Accordingly, similar to a standard high k metal gate stack, the degradation of the device is initiated in the SiO$_x$N$_y$ interfacial layer and not in the HfO$_2$ dielectric. This effect can be understood due to the high internal field on the interfacial layer as already predicted in (3.2).

Optimizing the internal field by both engineering of the stack and the operating conditions should allow reaching values comparable to NOR Flash devices using floating gate technology (see Table 3.3).

3.5 Summary and Outlook

Due to the low inherent switching energy while maintaining non-volatility, ferroelectric materials have attracted large interest for nonvolatile memories since a few decades. First attempts in the 1950s were discontinued due to disturb issues and the rapid success of competing technologies. In the late 1990s first products arrived on the market but these still suffer from limited scalability of the feature sizes and therefore are not competitive in a cost per bit. The ferroelectricity in hafnium oxide that was discovered a few years ago enables a third generation of ferroelectric solid state memories based on the ferroelectric field effect transistor that can overcome the limitations of the previous attempts. Fast progress has been made both in the material research and in the device integration segment leading to a good understanding of the control of ferroelectric properties as well as the demonstration of 28 nm based ferroelectric field effect transistors with very promising properties. Although still significant work is ahead in both basic understanding of the material effects and device and memory integration these results make hafnium oxide based ferroelectric field effect transistors a very promising alternative for next generation nonvolatile memories.

Acknowledgments The authors like to thank the FeFET team at Qimonda, Fraunhofer IPMS-CNT, GlobalFoundries, RWTH Aachen, Munich University of Applied Science, and NaMLab for their contribution to the results.

References

1. T. Mitsui, Ferroelectrics and antiferroelectrics, in *Springer Handbook of Condensed Matter and Materials Data*, ed. by W. Martienssen, H. Warlimont (Springer, Heidelberg, 2005), pp. 903–938
2. T. Mikolajick, Ferroelectric nonvolatile memories, in *Encyclopedia of Materials Science and Technology*, ed. by K.H.J. Buschow, R.W. Cahn, M.C. Flemings, B. Ilschner, E.J. Kramer, S. Mahajan (Elsevier, Oxford, 2002), pp. 1–5
3. D.A. Buck, Ferroelectrics for digital information storage and switching, master thesis, MIT Digital Computer Laboratory (1952)
4. J. Merz, J.R. Anderson, Ferroelectric storage devices. Bell Lab. Rec. **33**, 335–342 (1955)
5. J.R. Anderson, Feroelectric Materials as Storage Elements for Digital Computers and Switching Systems. Trans. Amer. Inst. Elect. Engrs. 71, Part I Commun. Electron. 395–401 (1953)
6. B. Dennard, US Patent (1968)
7. D. Bondurant, Ferroelectronic RAM memory family for critical data storage. Ferroelectrics **112**, 273–282 (1990)
8. C.A. Paz de Araujo, J.D. Cuchiaro, L.D. McMillan, M.C. Scott, J.F. Scott, Fatigue-free ferroelectric capacitors with platinum electrodes. Nature **374**, 627–629 (1994)
9. T. Mikolajick, C. Dehm, W. Hartner, I. Kasko, M.J. Kastner, N. Nagel, M. Moert, C. Mazure, FeRAM technology for high density applications. Microelectron. Reliab. **41**, 947–950 (2001)
10. C.-U. Pinnow, T. Mikolajick, Material aspects in emerging nonvolatile memories. J. Electrochem. Soc. **151**, K13–K19 (2004)
11. M. Röhner, T. Mikolajick, R. Hagenbeck, N. Nagel, Integration of FeRAM devices into a standard CMOS process—Impact of ferroelectric anneals on CMOS characteristics. Integr. Ferroelectr. **47**, 61–70 (2002)
12. A.K. Tagantsev, I. Stolichnov, E.L. Colla, N. Setter, Polarization fatigue in ferroelectric films: basic experimental findings, phenomenological scenarios, and microscopic features. J. Appl. Phys. **90**, 1387–1402 (2001)
13. K. Maruyama, M. Kondo, S.K. Singh, H. Ishiwara, New Ferroelectric Material for Embedded FRAM LSIs. Fujitsu Sci. Tech. J **43**, 502–507 (2007)

14. J.-M. Koo, B.-S. Seo, S. Kim, S. Shin, J.-H. Lee, H. Baik, J.-H. Lee, J.H. Lee, B.-J. Bae, J.-E. Lim, D.-C. Yoo, S.-O. Park, H.-S. Kim, H. Han, S. Baik, J.-Y. Choi, Y. J. Park, Y. Park, Fabrication of 3D trench PZT capacitors for 256Mbit FRAM device application. IEDM Techn. Digest. 340–343 (2005)
15. H.P. McAdams, R. Acklin, T. Blake, X.-H. Du, J. Eliason, J. Fong, W.F. Kraus, D. Liu, S. Madan, T. Moise, S. Natarajan, N. Qian, Y. Qiu, K.A. Remack, J. Rodriguez, J. Roscher, A. Seshadri, S.R. Summerfelt, A 64-Mb embedded FRAM utilizing a 130 nm 5LM Cu/FSG logic process. IEEE J. Solid-St. Circ. **39**, 667–677 (2004)
16. I.M. Ross, Semiconductive translating device, U.S. patent 2791760 A (1957)
17. J.L. Moll, Y. Tarui, I.E.E.E. Trans, Electron Devices **10**, 338 (1963)
18. T.P. Ma, J.-P. Han, Why is nonvolatile ferroelectric memory field-effect transistor still elusive? IEEE Electron Device Lett. **23**, 386–388 (2002)
19. S. Sakai, R. Ilangovan, Metal–ferroelectric–insulator–semiconductor memory FET with long retention and high endurance. IEEE Electron Device Lett. **25**, 369–371 (2004)
20. T.S. Boescke, J. Mueller, D. Braeuhaus, U. Schroeder, U. Boettger, Ferroelectricity in hafnium oxide thin films. Appl. Phys. Lett. **99**, 102903 (2011)
21. X. Sang, E.D. Grimley, T. Schenk, U. Schroeder, J.M. LeBeau, Appl. Phys. Lett. **106**, 162905 (2015)
22. International technology roadmap for semiconductors, emerging research devices (2013) http://www.itrs.net
23. M.T. Bohr, R.S. Chau, T. Ghani, K. Mistry, IEEE Spectr. **44**, 29 (2007)
24. F.M. Spiridonov, L.N. Komissarova, A.G. Kocharov, V.I. Spitsyn, Russ. J. Inorg. Chem. **14**, 1332 (1969)
25. C. Richter, T. Schenk, U. Schroeder, T. Mikolajick, Baltic ALD conference, Helsinki (2014)
26. M. Hoffmann, U. Schroeder, T. Schenk, T. Shimizu, H. Funakubo, O. Sakata, D. Pohl, M. Drescher, C. Adelmann, R. Materlik, A. Kersch, T. Mikolajick, J. Appl. Phys. (2015) accepted
27. D. Martin, J. Müller, T. Schenk, T.M. Arruda, A. Kumar, E. Strelcov, E. Yurchuk, S. Müller, D. Pohl, U. Schroeder, S.V. Kalinin, T. Mikolajick, Adv. Mat. **26**(48), 8198–8202 (2014)
28. U. Schroeder, E. Yurchuk, J. Müller, D. Martin, T. Schenk, P. Polakowski, C. Adelmann, M. I. Popovici, S. V. Kalinin, and T. Mikolajick, Jpn. J. Appl. Phys. **53**, 08LE02 (2014)
29. S. Clima, D. Wouters, C. Adelmann, T. Schenk, U. Schroeder, M. Jurczak, M. Pourtois, Appl. Phys. Lett. **104**, 092906 (2014)
30. J.F. Scott, *Ferroelectric Memories* (Springer, Berlin, 2000)
31. U. Schroeder, S. Mueller, J. Mueller, E. Yurchuk, D. Martin, C. Adelmann, T. Schloesser, R. van Bentum, T. Mikolajick, ECS J. Solid State Sci. Technol. **2**(4), N69–N72 (2013)
32. S. Müller, S.R. Summerfelt, J. Müller, U. Schroeder, T. Mikolajick, Ten-nanometer ferroelectric Si:HfO$_2$ films for next-generation FRAM capacitors. IEEE Electron Device Lett. **33**, 1300–1302 (2012)
33. T. Schenk, M. Hoffmann, J. Ocker, M. Pešić, T. Mikolajick, U. Schroeder, Adv. Funct. Mat. submitted
34. M. Pesic, S. Slesazeck, T. Schenk, U. Schroeder, T. Mikolajick, E-MRS conference, Lille (2015)
35. J. Knoch, S. Mantl, S. Feste, Chapter on HKMG/FeFET devices, in *Nanoelectronics and Information Technology*, 3rd edn, ed. by R. Waser (Wiley VCH, 2012)
36. H.-T. Lue, C.-J. Wu, T.-H. Teng, IEEE Trans. Electron Devices **49**, 10 (2002)
37. L. Van Hai, T. Mitsue, S. Shigeki, Downsizing of ferroelectric-gate field-effect-transistors for ferroelectric-NAND flash memory cells, in *Proceedings of the IMW* (2011), pp. 1–4
38. J. Müller, T.S. Böscke, S. Müller, E. Yurchuk, P. Polakowski, J. Paul, D. Martin, T. Schenk, K. Khullar, A. Kersch, W. Weinreich, S. Riedel, K. Seidel, A. Kumar, T.M. Arruda, S.V. Kalinin, T. Schlösser, R. Boschke, R. van Bentum, U. Schröder, T. Mikolajick, Ferroelectric hafnium oxide: A CMOS-compatible and highly scalable approach to future ferroelectric memories. IEDM Dig. Tech. Pap. 10.8.1–10.8.4 (2013)

39. E. Yurchuk, J. Müller, J. Paul, T. Schlösser, D. Martin, R. Hoffmann, S. Müller, S. Slesazeck, U. Schroeder, R. Boschke, R. van Bentum, T. Mikolajick, IEEE Trans. Electr. Dev, **61**(11) (2014)
40. J. Müller, J. Müller, E. Yurchuk, T. Schlösser, J. Paul, R. Hoffmann, S. Müller, D. Martin, S. Slesazeck, P. Polakowski, J. Sundqvist, M. Czernohorsky, P. Kücher, R. Boschke, M. Trentzsch, K. Gebauer, U. Schroeder, T. Mikolajick, Ferroelectricity in HfO$_2$ enables nonvolatile data storage in 28 nm HKMG, in *Proceeding of IEEE Symposia on VLSI Technology* (2012), pp. 25–26
41. E. Yurchuk, S. Mueller, D. Martin, S. Slesazeck, U. Schroeder, T. Mikolajick, J. Müller, J. Paul, R. Hoffmann, J. Sundqvist, T. Schlösser, R. Boschke, R. van Bentum, M. Trentzsch, in *Proceedings of the IRPS* (2014)

Part III
Practical Characteristics of Inorganic Ferroelectric-Gate FETs: Thin Film-Based Ferroelectric-Gate Field Effect Transistors

Chapter 4
Oxide-Channel Ferroelectric-Gate Thin Film Transistors with Nonvolatile Memory Function

Eisuke Tokumitsu

Abstract In this chapter, principle and progress of oxide-channel ferroelectric-gate transistors are reviewed. At first, it is pointed out that ferroelectric-gate insulator can induce large charge density because of the remanent polarization, in addition to non-volatile memory function. Next, using this feature of ferroelectric gate insulator, it is shown that even conductive oxide, such as indium-tin-oxide (ITO), can be used as a channel if its thickness is thin enough. Good transistor performance is demonstrated for ITO-channel ferroelectric-gate thin film transistors (TFTs). In particular, a large on-current can be obtained in such devices, because large charge density is utilized even though the channel mobility is not so high. Furthermore, transparent devices were demonstrated using ITO for both channel and electrodes. In addition, ITO channel TFTs without nonvolatile memory function are demonstrated using high-dielectric constant (high-k) material as a gate insulator.

4.1 Introduction

Since ferroelectric materials exhibit remanent polarization even in the absence of external electric field and the polarization can be switched at high speed, they have been used in nonvolatile random access memories and commercial products are now available. There are two types in ferroelectric random access memories (FeRAMs); capacitor-type and transistor-type [1, 2]. The capacitor-type FeRAM has a similar 1T1C configuration as DRAM, which consist of one switching transistor and one ferroelectric storage capacitor. Hence, to realize high-density capacitor type FeRAM, miniaturization of the storage capacitor is required which results in reduction of

E. Tokumitsu (✉)
Green Devices Research Center, Japan Advanced Institute of Science
and Technology, 1-1 Asahidai, Nomi, Ishikawa 923-1292, Japan
e-mail: e-toku@jaist.ac.jp

© Springer Science+Business Media Dordrecht 2016
B.-E. Park et al. (eds.), *Ferroelectric-Gate Field Effect Transistor Memories*,
Topics in Applied Physics 131, DOI 10.1007/978-94-024-0841-6_4

storage charge. On the other hand, transistor-type FeRAM employs ferroelectric-gate transistor, where a ferroelectric material is used as a gate insulator. One ferroelectric-gate transistor can store 1 bit and the device obeys the scaling rule, since the charge density is an important parameter. Hence, the transistor-type FeRAM is suitable for high-density memory applications. In addition, non-destructive read-out is possible. As in the Flash memory, transistor-type ferroelectric memory can utilize either NAND or NOR configuration.

Proposal of ferroelectric-gate transistor was claimed as early as 1960 [3, 4] and many efforts have been made so far [5–14], using silicon as semiconducting layer. However, we pointed out that there is charge mismatch between the ferroelectric polarization and the channel charge in the Si-MOSFET [11]. A sheet carrier density of Si-MOSFET is approximately 10^{11}–10^{12} cm^{-2}, which corresponds to 0.016–0.16 µC/cm^2. On the other hand, ferroelectric polarizations of typical ferroelectric materials such as Pb(Zr,Ti)O$_3$ (PZT) and (Bi,La)$_4$Ti$_3$O$_{12}$ (BLT) are 10–50 µC/cm^2. Because of the mismatch problem, only one of the minor loops of P-E hysteresis in ferroelectric material can be used in Si-based ferroelectric-gate transistors, which results in small memory window.

On the other hand, full polarization can be used when we used oxide semi-conductor as a channel material [15, 16]. Oxide semiconductors has attracted much interest and thin film transistors (TFTs) using In-Ga-Zn-O (IGZO) whose mobility (around 10 cm^2/Vs) is higher than that of conventional amorphous silicon, are now commercially used in flat-panel displays in some smartphones [17]. We have previously demonstrated indium-tin-oxide (ITO) channel ferroelectric-gate TFTs, here we call such devices as FGTs, and pointed out that the ferroelectric gate insulator can induce much larger charge density than conventional gate insulator such as SiO$_2$ and SiN$_x$. Using this feature, we have shown that conductive thin ITO layer can be a channel of TFTs if its thickness is sufficiently thin [15, 16].

In this paper, we first recall the features and basic characteristics of FGTs. Next, we introduce our results on ITO-channel ferroelectric-gate thin film transistors, In addition, transparent ferroelectric-gate TFTs and ITO-channel TFTs with high-k gate insulators are also described.

4.2 Features of Ferroelectric Gate Insulator

Ferroelectric materials exhibit hysteresis loops in charge (polarization)—electric filed characteristics as shown in Fig. 4.1 and are widely used to realize nonvolatile memories. This is in contrast to paraelectric materials which have Q = CV relation. If we plot ferroelectric P-E loop along with the Q-V relation of conventional paraelectric materials, one will notice that much larger charge density can be induced by ferroelectric material than conventional paraelectric materials because of the presence of spontaneous or remanent polarization. As shown in Fig. 4.1, ferroelectric PZT can induce as large as 50 µC/cm^2 even at a low electric field of 500 kV/cm. At 500 kV/cm, the induced charge density by SiO$_2$ whose relative

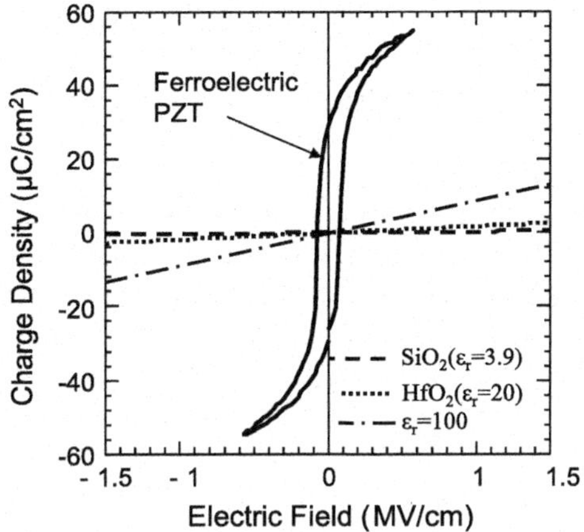

Fig. 4.1 Typical Q-E (P-E) relation of ferroelectric PZT along with SiO$_2$, HfO$_2$, and paraelectric material with $\varepsilon_r = 100$

dielectric constant is 3.9, is only 0.17 μC/cm^2 and even when a dielectric material with $\varepsilon_r = 100$ is assumed, induced charge density is 4.4 μC/cm^2. Hence, it is expected that the ferroelectric gate insulator can control much larger charge density than conventional paraelectric gate insulators.

In addition, there is a practical limitation in available charge density for paraelectric materials due to breakdown, For conventional paraelectric materials, maximum available charge density, Q_{MAX}, can be written as $Q_{MAX} = \varepsilon_0\varepsilon_r E_B$, where ε_0 is dielectric constant of vacuum, ε_r relative dielectric constant of paraelectric gate insulator, and E_B is a breakdown field, the maximum electric field we can apply. For instance, since a breakdown field of SiO$_2$ is typically 10 MV/cm and relative dielectric constant ε_r is 3.9, the maximum available charge density is calculated to be 3.5 μC/cm^2. This is much smaller than the remanent polarization of typical ferroelectric materials. Even when high-dielectric (high-k) material such as HfO$_2$ is used, the maximum available charge density cannot be increased drastically in comparison with SiO$_2$, because the breakdown field becomes smaller due to small bandgap of high-k materials.

When we apply such a ferroelectric material as a gate insulator, we have to concern about charge matching between the channel charge and the charge density which can be induced by the gate insulator. For example, in Si MOSFETs, the sheet carrier density is around 10^{12} cm^{-2}, which corresponds to 0.16 μC/cm^2. On the other hand, conventional oxide ferroelectric material such as Pb(Zr,Ti)O$_3$ (PZT), SrBi$_2$Ta$_2$O$_9$ (SBT) and (Bi,La)$_4$Ti$_3$O$_{12}$ (BLT) have remanent polarizations of more than 10 μC/cm^2. When oxide ferroelectric films are formed on Si substrate, SiO$_2$ interfacial layer is formed unintentionally because such oxide ferroelectric materials need relatively high annealing temperature to be crystallized. Since the maximum available charge density of SiO$_2$ is 3.5 μC/cm^2, it is difficult to use full polarization of ferroelectric materials in Si-based field-effect transistors. In Si-based ferroelectric-gate transistors, only minor loops in P-E hysteresis are used, which leads to small memory window. Of course, it is an interesting challenge to find out novel ferroelectric materials whose polarization matches with carrier density of Si-MOSFETs. This charge mismatch problem can be solved by using metal/ferroelectric/metal/insulator/semiconductor (MFMIS) structure in Si-based FETs by designing the small MFM capacitor compared to the MIS structure.

Another way to solve this problem is to use oxide semiconductors. When we use an oxide semiconductor as a channel, full polarization will be available. Furthermore, since large charge density can be controlled by the ferroelectric gate insulator, even conductive oxide material can be used. Hence, in terms of charge matching, oxide semiconductors or even conductors are suitable for channel material in ferroelectric-gate FETs. Large polarizations of ferroelectric materials result in another advantage of ferroelectric gate insulator, which is sharp sub-threshold slope. The subthreshold voltage swing, S, is given by $S = \ln 10 \, (kT)/q \, (1 + (C_D + C_{it})/C_{ox})$, Since the equivalent dielectric constant is much larger than that of conventional paraelectric material, the subthreshold voltage swing can be small in ferroelectric-gate FETs.

4.3 Charge Density of ITO Channel FGTs

We have fabricated ferroelectric-gate thin film transistors (FGTs) using indium-tin-oxide (ITO) as a channel to demonstrate the large charge controllability of the ferroelectric gate insulator [15, 16]. Although ITO is a conductive oxide with high carrier concentration, which is widely used as transparent electrode, it can be used as a channel of FGTs if the thickness is sufficiently thin. Roughly speaking, to deplete the channel layer completely, the charge of the carriers in the ITO channel must be smaller than the charge density induced by the gate insulator. This can be written as $q \cdot n \cdot d \cdot < P$, where n and d are carrier concentration and thickness of oxide channel material, P is a polarization of ferroelectric gate insulator, and q is the elemental charge. Figure 4.2 shows the charge density of the channel, $q \cdot n \cdot d$, as a function of carrier concentration n, when the thickness d = 1, 10 and 100 nm. The typical values of remanent polarizations of conventional oxide ferroelectric materials, SBT ($8–10 \, \mu C/cm^2$), BLT ($15–20 \, \mu C/cm^2$), and PZT ($30–50 \, \mu C/cm^2$) are also shown in the figure. If we assume a polarization of $20 \, \mu C/cm^2$ and channel thickness of 10 nm, the carrier concentration of the channel should be less than $1.25 \times 10^{20} \, cm^{-3}$. This means the ferroelectric gate insulator enables us to use conductive oxide materials with carrier concentrations of around $10^{20} \, cm^{-3}$ as TFT channel materials. Hence, when we use ITO with a carrier concentration of $10^{20} \, cm^{-3}$ as a channel, the channel thickness should be thin (thinner than 10 nm) and PZT and BLT which have large remanent polarization are suitable as gate insulators.

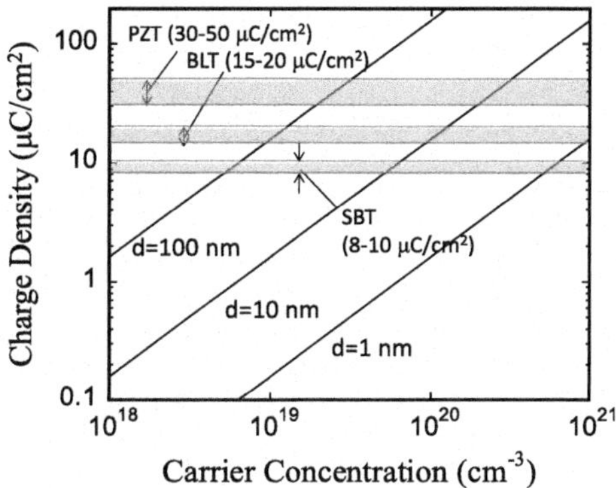

Fig. 4.2 Charge density of the ITO channel as a function of carrier concentration with thickness of 1, 10 and 100 nm

4.4   Electrical Properties of ITO Channel FGTs

The fabricated devices have conventional bottom- gate and top-contact thin film transistor structure, as shown in Fig, 4.3. Ferroelectric BLT film was prepared by the sol-gel technique with a crystallization temperature of 750 °C and ITO channel layer was deposited by sputtering at room temperature or 300 °C. La content of the BLT films used in this study is 0.75. BLT thickness is typically 200–230 nm and ITO channel thickness is as thin as 10 nm. Since the fabrication temperature of ferroelectric material is higher than that of ITO, bottom-gate structure was used. Figure 4.4 shows (a) transfer curve and (b) output characteristics of ITO/BLT FGT. The channel length (L) and width (W) are 40 and 120 µm, respectively. Because of the ferroelectric nature of the gate insulator, hysteresis loop was clearly observed and memory window, i.e. threshold voltage shift, which agrees with the P-E hysteresis loop at the same application voltage. This indicates that full polarization was used in the fabricated device. In addition, it is found in the output characteristics that a large on-current of about 1 mA was obtained. The estimated channel mobility and charge density used in the device are 4.0 cm^2/Vs and 15 µC/cm^2, respectively. Although the channel mobility is small, since induced charge density is large, a large on-current can be obtained. When the channel length L and width W are reduced to 5 and 25 µm, respectively, the on-current at V_G = 8 V was increased to 2.4 mA, which approximately corresponds to 0.1 mA/µm. It is interesting to note that this value roughly agrees with on-current of Si-MOSFET with L = 5 µm.

Fig. 4.3 Schematic illustration of fabricated ferroelectric-gate TFT with oxide channel

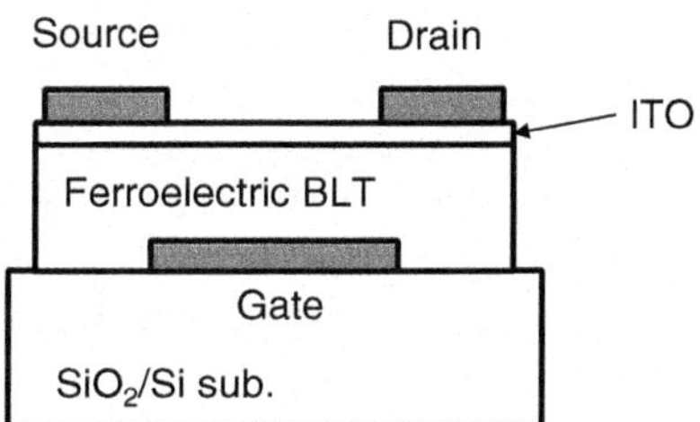

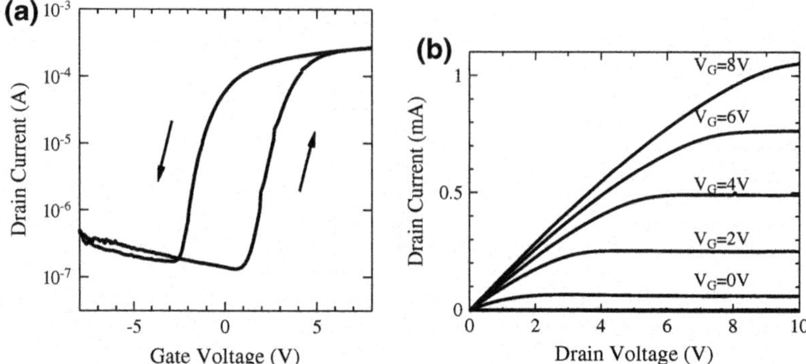

Fig. 4.4 a Transfer curve and b output characteristics of ITO channel ferroelectric-gate thin film transistor. Channel length (L) and width (W) are 40 and 120 μm, respectively

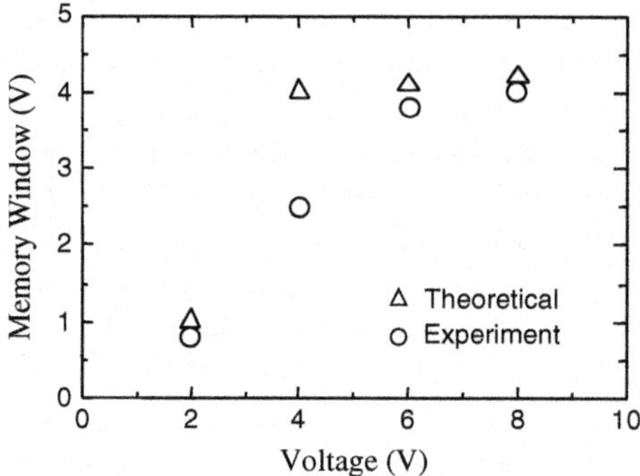

Fig. 4.5 Observed memory window of ITO/BLT ferroelectric-gate TFT as a function of applied voltage

Figure 4.5 plots the observed memory window (open circles) as a function of applied voltage. Theoretical values are calculated by $2V_C$, where V_C is a coercive voltage deduced from the P-E hysteresis loops of the BLT MFM capacitor. The experimentally observed memory window increases with applied voltage and saturate around 6 V and these agree with the theoretically calculated values except for $V = 4$ V. This indicates that the full polarization of ferroelectric BLT film was utilized for device operation when the applied voltage is larger than 6 V. In addition, we confirmed that the memory window increased with BLT thickness and that they agreed with theoretically calculated values.

To obtain the transistor operation for ITO-channel FGTs, the channel thickness is an important parameter. Since the ITO has large a carrier concentration of about 10^{20} cm^{-3}. The channel charge, $Q = q \cdot n \cdot d$, should be less than the charge density induced by the gate insulator. Figure 4.6 shows calculated carrier concentration of the ITO channel as a function of thickness, assuming a BLT polarization of 15 μC/cm^2. When we assume an induced charge density of 15 μC/cm^2 and a carrier

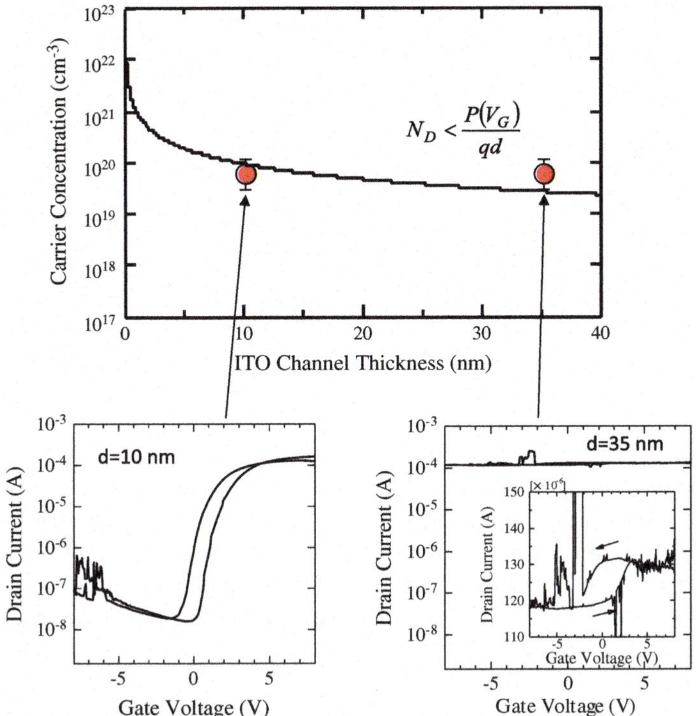

Fig. 4.6 Calculated carrier concentration of the ITO channel as a function of thickness, assuming a BLT polarization of 15 µC/cm^2. The experimentally observed data are shown by *closed circles* along with transfer characteristics

concentration of 10^{20} cm^{-3} for ITO channel layer, the channel thickness should be less than 10 nm. We fabricated ITO channel FGTs with channel thicknesses of 10 and 35 nm. When we increase the channel thickness to 35 nm, we lose on/off ratio as shown in Fig. 4.6, whereas good transistor characteristics with large on/off ratio can be obtained when the channel thickness is 10 nm. Closed circles in Fig. 4.6 show the carrier concentration and thickness of these experimentally observed data. Reasonable agreement is found between the experiments and the above simple assumption. In normal Si MOS structures, one has to consider the maximum depletion layer width which was caused by the generation of inversion layer. In the ITO/BLT structure, since the valence band maximum of the ITO is lower than that of BLT, this is not the case. However, if there is a large density of traps near the valence band of ITO, we may have to consider the similar effects as the inversion layer.

We have demonstrated large on-current of ITO channel ferroelectric-gate TFTs, however, a relatively large off-current was observed as shown in Fig. 4.3a, which is presumably due to additional current paths caused by the rough surface of crystalized BLT gate insulator prepared by the sol-gel technique. The off-current can be

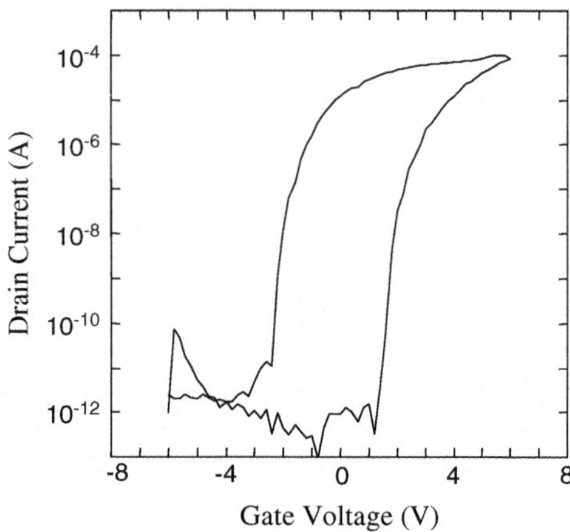

Fig. 4.7 Transfer curve of ITIO/BLT FGT, where ITO channel was deposited after publishing the BLT surface

reduced to 10^{-12} A by mechanically polishing the surface of BLT before the ITO channel deposition as shown in Fig. 4.7. We also examined electrical properties of ferroelectric BLT film with and without mechanical polishing and found that ferroelectric properties of the BLT film with polishing process is similar to that of the BLT film without polishing. Leakage current was slightly improved for the BLT film by polishing. The surface roughness is less than 0.9 nm (rms) after polishing the BLT surface.

One of the advantages of the oxide-channel ferroelectric-gate TFTs is large on-current, as explained earlier. On-current obtained in this work is about 0.1 mA/μm for L = 5 μm device. This value is much larger than the reported on-current of other oxide-channel TFTs with conventional gate insulator [18–22] and even ferroelectric-gate TFTs reported by Prins et al. [23, 24].

4.5   Transparent ITO/BLT FGT

Since both ITO and ferroelectric BLT are transparent, and ITO can be used for both electrodes and channel, we can fabricate transparent TFTs using this system. Schematic illustration of the fabricated transparent FGTs is shown in Fig. 4.8 [25]. Quarts substrate was used and ITO was used both for both channel and electrodes (gate and source/drain). Hence, only two kinds of materials, ITO and BLT were used to fabricate transparent FGTs as shown in Fig. 4.8. Since ITO is used as the bottom gate electrodes instead of conventional Pt, we first examined the electrical properties of ferroelectric BLT films fabricated on ITO electrodes. BLT film was prepared by the sol-gel technique at a crystallization temperature of 750 °C. Although it is found that the P-E hysteresis loops of the BLT film on the ITO electrode are slightly tilted, which indicates the presence of the interfacial layer, we obtained good ferroelectric properties of BLT film on ITO bottom electrode.

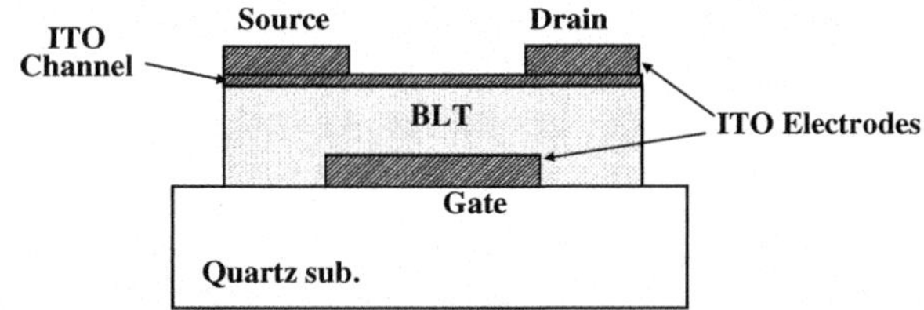

Fig. 4.8 Schematic cross section of transparent ferroelectric-gate TFT

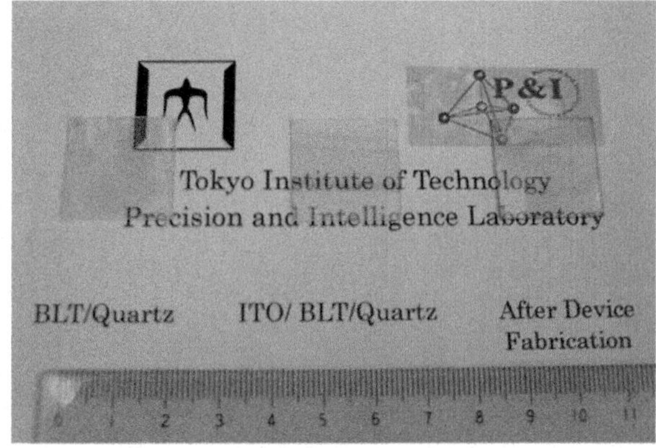

Fig. 4.9 Photograph of samples, ferroelectric BLT film, ITO/BLT layered structure on quartz substrate, and substrate after device fabrication

Remanent polarization (P_r) and coercive field (E_C) of the BLT film fabricated on the Pt electrodes are 16 μC/cm^2 and 100 kV/cm, respectively, whereas those of the BLT film on ITO electrodes are 12 μC/cm^2 and 120 kV/cm, respectively. We measured inter-diffusion of BLT and ITO layers by secondary ion mass spectrometry (SIMS), and found that the SIMS signals for the BLT elements rapidly decreases at the BLT/ITO interface and that slight diffusion is observed for Sn and In especially into the BLT film. This is probably the reason for the degradation of the electrical properties of the BLT film on ITO electrodes.

Figures 4.9 and 4.10 show photograph and optical absorption of the ferroelectric BLT film on quartz substrate, ITO/BLT layered structure, and substrate after device fabrication. For optical properties transmittance of quartz substrate is also shown for reference. The optical transmittance of the sample after device fabrication including the quartz substrate is about 70 % for the wavelength of 400-800 nm. Since main absorption of light was caused by the ferroelectric BLT film, further decrease of the BLT thickness will improve the optical transmittance.

Electrical properties, drain current—drain voltage (I_D-V_D) of transparent ferroelectric-gate TFT using BLT/ITO structure is shown in Fig. 4.11 for the device with channel length (L) of 5 μm and channel width (W) of 50 μm. It is found that the fabricated device exhibits typical n-channel transistor operation. The "on" current for $V_G = V_D = 6$ V is as large as 0.52 mA, which corresponds to 10 μA/ μm. This value is significantly larger than that of the most of the reported TFTs with oxide channel.

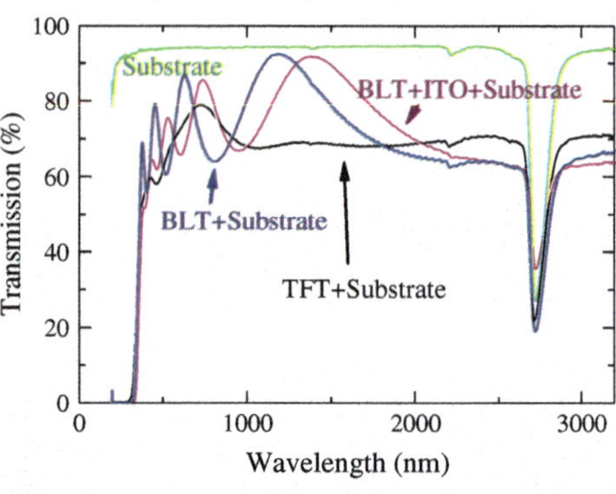

Fig. 4.10 Optical transmission of the ferroelectric BLT film on quartz substrate, ITO/BLT layered structure, and substrate after device fabrication

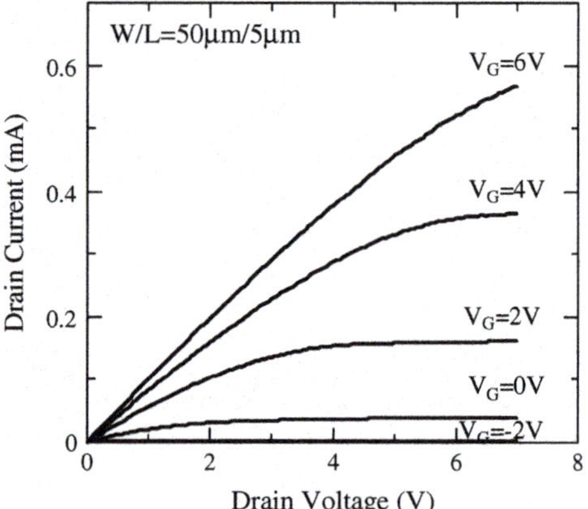

Fig. 4.11 I_D-V_D characteristics of transparent ferroelectric-gate TFT using BLT/ITO structure

Drain current—gate voltage (I_D-V_G) characteristics, of transparent ferroelectric-gate TFT using BLT/ITO structure along with retention properties are shown in Fig. 4.12. It is found that I_D-V_G curve exhibits a counterclockwise hysteresis loop due to the ferroelectric BLT film. This clearly shows the device can be used as a nonvolatile memory device. A memory window, threshold voltage shift due to the ferroelectric gate insulator, is approximately 3.8 V, which roughly agrees with the theoretically estimated memory window. In addition, we evaluated data retention characteristics by measuring drain current of on and off states with a hold voltage of 0 V, after applying the program voltage of ± 7 V. The on/off-current difference is more than two orders of magnitude after 1 day data retention.

It is worth noting that the off-current of the device is as small as 10^{-10} A at negative gate bias. This indicates that the thin ITO channel is completely depleted by the ferroelectric polarization. The on/off-current ratio is more than 10^6. Field-effect mobility of the device is estimated to be approximately 1 cm^2/Vs, which is reasonable value for poly-crystalline ITO film with high carrier concentration. In spite of the small channel mobility, we observed a large on-current in this work. This is because the ferroelectric film can induce large charge density.

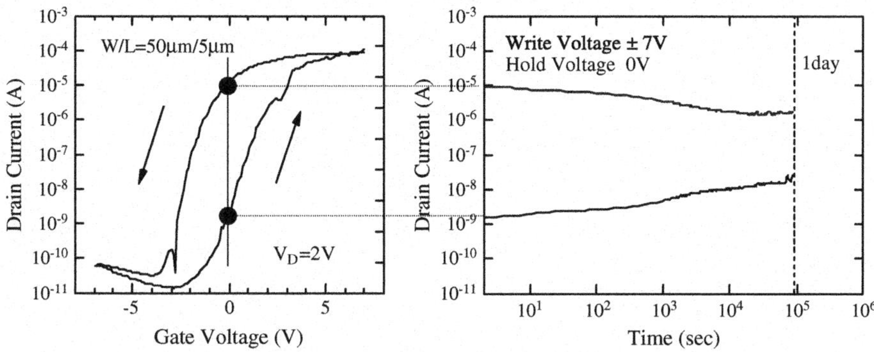

Fig. 4.12 I_D-V_G characteristics of transparent ferroelectric-gate TFT using BLT/ITO structure along with data retention properties

The charge density used in the device is estimated to be 10 $\mu C/cm^2$, which is more than 10 times larger than that of the conventional Si MOS-FET.

4.6 ITO-Channel TFTs with High-k Gate Insulator

We have shown that ferroelectric-gate insulator can induce large charge density which enables us to use conductive ITO as a channel of TFTs. The ferroelectric-gate insulator results in non-volatile memory function and transfer curve of the device exhibits hysteresis loop. It is useful for non-volatile memory applications, however, for display or logic applications, TFTs which have transfer curves without hysteresis loops are required. Even in this case, it is interesting to use ITO as a channel layer because if the device has large on-current, the device size can be small to obtain the same current. Hence, the occupied area by TFTs can be small in the display applications, which lead to high-density and bright display panels. We fabricated ITO-channel TFTs using $(Ba_{0.3}Sr_{0.7})TiO_3$ (BST) or $Bi_{1.5}Zn_{1.0}Nb_{1.5}O_7$ (BZN) as a gate insulator. It was found from Q-V characteristics measurements that the BST film can induce as large as 10 $\mu C/cm^2$ and that the relative dielectric constant is about 300. Figure 4.13 shows (a) I_D-V_G and (b) I_D-V_D characteristics of the ITO/BST TFTs. Normal transistor characteristics were obtained with an on/off-current ratio is more than 10^6 with a small charge-injection type hysteresis. However, when the gate sweep voltage was increased, charge-injection type hysteresis was observed, which indicates interface properties should be improved. We also observed a large on-current of 0.6 mA and the channel mobility was estimated 3.0 cm^2/Vs.

We also used BZN as a gate insulator. Although BZN has a relatively small relative dielectric constant of 50 compared to ferroelectric materials and BST, the breakdown field is more than 2 MV/cm, which results in about 9 $\mu C/cm^2$ of

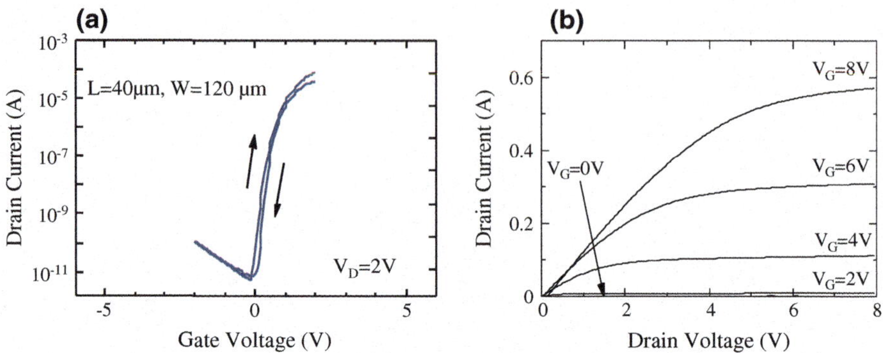

Fig. 4.13 **a** I_D-V_G and **b** I_D-V_D characteristics of the ITO/BST TFTs

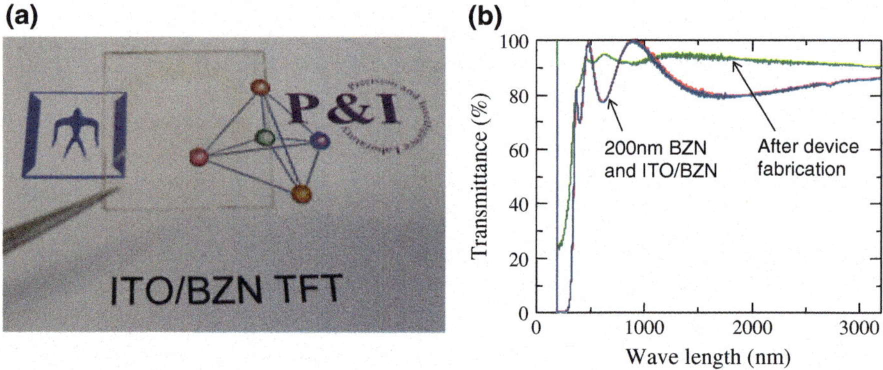

Fig. 4.14 **a** photograph and **b** optical absorption of transparent ITO/BZN TFTs

available charge density. BZN and ITO films were prepared by sputtering at room temperature. Figure 4.14 show photograph and optical absorption measurement results for the fabricated transparent ITO/BZN TFTs [26]. ITO is used for both channel and electrodes and the devices were fabricated on quartz substrate. The transmittance of the TFTs fabricated on quartz substrates is more than 90 %. Figure 4.15 show electrical properties of the transparent ITO/BZN TFTs. Although there is a charge-injection-type hysteresis, which suggests the presence of traps at interface or BZN film, n-channel transistor operation with a large on/off ratio of 10^8 was obtained. Note that the channel material is ITO. The channel mobility was estimated to be 1.6 cm^2/Vs.

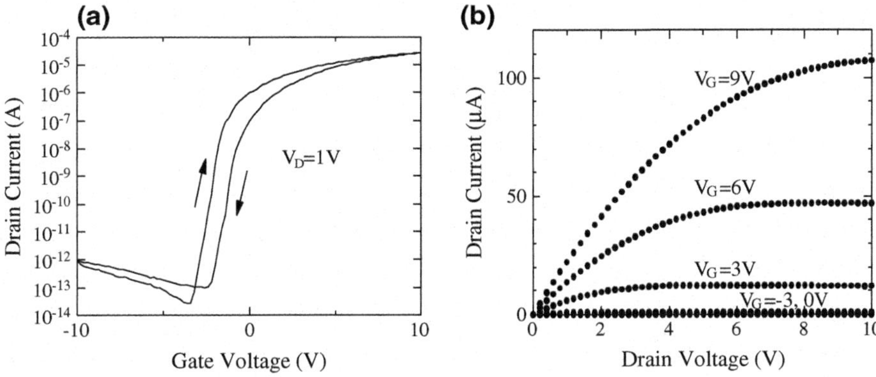

Fig. 4.15 **a** Transfer and **b** output characteristics of transparent ITO/BZN TFTs

4.7 Conclusions

Progress of oxide channel, in particular, ITO-channel ferroelectric gate thin film transistors was reviewed. It was demonstrated that ferroelectric gate insulator can induce much larger charge density than conventional gate insulator such as SiO_2 and SiNx and that full polarization can be utilized in oxide channel FGTs. Because of the large charge density which ferroelectric gate insulator induces, conductive ITO can act as a channel, if it is sufficiently thin. Good electrical properties were demonstrated for ITO-channel FGTs. In particular, large on-current was observed in such devices. This is due to large charge density used in the device in spite of small channel mobility of ITO. In addition, transparent FGTs were demonstrated using ITO for both channel and electrodes. For flat panel or logic applications, ITO channel TFTs without nonvolatile memory function can be fabricated by using high-k dielectric, such as BST and BZN.

Acknowledgments The author would like to acknowledge students in his group for their works related on the presented subject.

References

1. H. Ishiwara, M. Okuyama, Y. Arimoto (eds.), *Ferroelectric Random Access Memories, Fundamentals and Applications*. Springer (2004)
2. M. Okuyama, Y. Ishibashi (eds.), *Ferroelectric Thin Films, Basic Properties and Device Physics for Memory Applications*. Springer (2005)
3. I. M. Loss, US patent 2791760 (1957)
4. J.L. Moll, Y. Tarui, IEEE Trans. Electron Devi. **ED-10**, 333 (1963)
5. Y. Higuma, Y. Matsui, M. Okuyama, T. Nakagawa, Y. Hamakawa, Jpn. J. Appl. Phys. **17** (Suppl. 17–1), 209 (1978)
6. E. Tokumitsu, G. Fujii, H. Ishiwara, Appl. Phys. Lett. **75**, 575 (1999)

7. T. Hirai, Y. Fujisaki, K. Nagashima, H. Koike, Y. Tarui, Jpn. J. Appl. Phys. **36**, 5908 (1997)
8. Y. Nakao, T. Nakamura, A. Kamisawa, H. Takasu, Integr. Ferroelectr. **6**, 23 (1995)
9. Y. Fujimori, T. Nakamura, A. Kamisawa, Jpn. J. Appl. Phys. **38**, 2285 (1999)
10. E. Tokumitsu, R. Nakamura, H. Ishiwara, IEEE Electron Dev. Lett. **EDL-18**, 160 (1997)
11. E. Tokumitsu, G. Fujii, H. Ishiwara, Jpn. J. Appl. Phys. **39**, 2125 (2000)
12. E. Tokumitsu, K. Okamoto, H. Ishiwara, Jpn. J. Appl. Phys. **40**, 2917 (2001)
13. K. Aizawa, B.-E. Park, Y. Kawashima, K. Takahashi, H. Ishiwara, Appl. Phys. Lett. **85**, 3199 (2004)
14. M. Takahash, S. Sakai, Jpn. J. Appl. Phys. **44**, L800 (2005)
15. E. Tokumitsu, M. Senoo, T. Miyasako, J. Microelectron. Eng. **80**, 305 (2005)
16. T. Miyasako, M. Senoo, E. Tokumitsu, Appl. Phys. Lett. **86**, 162902 (2005)
17. K. Nomura, H. Ota, A. Takagi, T. Kamiya, M. Hirano, H. Hosono, Nature **432**, 488 (2004)
18. P.F. Carcia, R.S. McLean, M.H. Reilly, G. Nunes Jr., Appl. Phys. Lett. **82**, 1117 (2004)
19. J. Nishii, F.M. Hossain, S. Takagi, T. Arita, K. Saikusa, Y. Ohmaki, I. Ohkubo, S. Kishimoto, A. Ohtomo, T. Fukumura, F. Matsukura, Y. Ohno, H. Koinuma, H. Ohno, M. Kawasaki, Jpn. J. Appl. Phys. **42**, L347 (2003)
20. K. Kwon, Y. Li, Y.W. Heo, M. Jones, P.H. Holloway, D.P. Norton, Z.V. Park, S. Li, Appl. Phys. Lett. **84**, 2685 (2004)
21. E. Fortunato, P. Barquinha, A. Pimentel, A. Goncalves, A. Marques, R. Matins, L. Pereira, Appl. Phys. Lett. **85**, 2541 (2005)
22. E. Fortunato, P. Barquinha, R. Martins, Adv. Mater. **24**, 2945 (2012)
23. M.W.J. Prins, K.O. Grosse-Holz, G. Muller, J.F.M. Cillessen, J.B. Giesbers, R.P. Weening, R.M. Wolf, Appl. Phys. Lett. **68**, 3650 (1996)
24. M.W.J. Prins, S.E. Zinnemers, J.F.M. Cillessen, J.B. Giesbers, Appl. Phys. Lett. **70**, 458 (1997)
25. E. Tokumitsu, M. Senoo, E. Shin, Mater. Res. Soc. Proc. **902E**, 0902-T10-54 (2006)
26. E. Tokumitsu, Y. Kondo, J. Korean Phys, Soc. **54**, 539 (2009)

Chapter 5
ZnO/Pb(Zr,Ti)O$_3$ Gate Structure Ferroelectric FETs

Yukihiro Kaneko

Abstract We have developed a ferroelectric-gate field-effect transistor (FeFET) composed of heteroepitaxially stacked oxide materials. A semiconductor film of ZnO, a ferroelectric film of Pb(Zr,Ti)O$_3$ (PZT), and a bottom-gate electrode of SrRuO$_3$ (SRO) are grown on a SrTiO$_3$ substrate. Structural characterization shows a heteroepitaxy of the fabricated ZnO/PZT/SRO/STO structure with a good crystalline quality and absence of an interface reaction layer. When gate voltages applied to the bottom electrode are swept between -10 and $+10$ V, the ON/OFF ratio of drain currents is higher than 10^5. Such a high ratio is preserved even after 3.5 months; the extrapolation of retention behavior predicts a definite memory window over 10 years. We also switched FeFET channel conductance by applying short pulses to a gate electrode and found that the switching of the FeFET is due to domain wall motion in a ferroelectric film. Polarization reversal starts from a region located under source and drain electrodes and travels along the direction of channel length. In addition, domain wall velocity increases as the domain wall gets closer to the source and drain electrodes in the ferroelectric film. Therefore, the FeFET has the merit of high operation speeds at scale. Then, we demonstrate a 60-nm-channel-length FeFET. The drain current ON/OFF ratio was about three orders of magnitude for write pulse widths as narrow as 10 ns. Although the channel length is set at 60 nm, the conductance can be varied continuously by varying the write pulse width.

5.1 Introduction

A ferroelectric-gate field-effect transistor (FeFET) that uses a ferroelectric material as a dielectric layer for a metal-oxide-semiconductor FET is of great interest in nonvolatile memory applications because its channel conductance can be switched

Y. Kaneko (✉)
Advanced Research Division, Panasonic Corporation, 3-4 Hikaridai,
Seika-cho, Soraku-gun, Kyoto 619-0237, Japan
e-mail: kaneko.yukihiro001@jp.panasonic.com

© Springer Science+Business Media Dordrecht 2016

B.-E. Park et al. (eds.), *Ferroelectric-Gate Field Effect Transistor Memories*,
Topics in Applied Physics 131, DOI 10.1007/978-94-024-0841-6_5

and memorized with low power consumption at a high speed. FeFETs are, in principle, scalable below 10 nm, even up to the crystal unit cell size, as data are stored in the form of ferroelectric polarization [1, 2]. However, it is difficult to fabricate a FeFET that has good retention properties because of the reaction between a ferroelectric material and Si that occurs when a ferroelectric film is deposited directly on a Si substrate. This is because constituent elements of the film and substrate mix with each other during crystallization annealing. In most studies, an insulating buffer layer is inserted between the ferroelectric film and the Si substrate to avoid mixing. However, the addition of a buffer layer to the structure makes retention properties of FeFETs insufficient for practical use by introducing a potential across a ferroelectric layer [3–11]. This potential gives rise to a depolarization field and leakage currents, which are detrimental to retention properties [12].

To develop a FeFET with a good retention, we can summarize the three required properties as follows: a directly stacked structure, a flat interface, and no interface reaction layer. We have therefore undertaken to devise a perfect solution that combines these three vital requirements. For the first requirement, i.e., a directly stacked structure, we select a ZnO semiconductor film stacked on a ferroelectric Pb (Zr,Ti)O_3 (PZT) film. Compared with Si channel layers often used, it is expected that an oxide film will not react with a ferroelectric layer. For the second requirement, i.e., a flat interface, an oxide-based inverted-staggered (bottom-gate) thin-film transistor (TFT) structure is adopted. In addition, a FeFET is composed of stacked perovskite oxides with similar lattice parameters: a ferroelectric film of PZT ($a \approx 0.404$ nm), a bottom-gate electrode of SrRuO_3 (SRO; $a \approx 0.393$ nm), and a substrate of (100)-sliced SrTiO_3 (STO; $a \approx 0.391$ nm). This combination enables the heteroepitaxial growth of a PZT/SRO/STO structure with a homogeneous crystal orientation and a flat surface, which, in turn, leads to a well-oriented growth of ZnO on top of PZT. For the third requirement, i.e., no interface reaction layer, an oxide material is used as a channel layer as mentioned previously; moreover, the ZnO film is continuously grown in situ on the stacked perovskite structure under a growth temperature lower than that used during the deposition of underlying films.

Here, we would like to introduce a FeFET based on an oxide stacked structure. Section 5.2 will begin with the fabrication method of the FeFET. Next, a discussion about the basic characteristics, including retention properties, will be presented. We will then discuss a channel formation mechanism and the merits of FeFET scalability. Finally, we will discuss narrow-channel FeFETs.

5.2 Experimental Procedure

Figure 5.1 shows a schematic of a fabricated FeFET. The FeFETs produced in this study were fabricated on 15×15 mm^2 STO(001) single-crystal substrates that were polished and etched by the supplier (Shinkosha) to atomically flat surfaces. STO substrates were mounted in a vacuum chamber that was initially evacuated to a

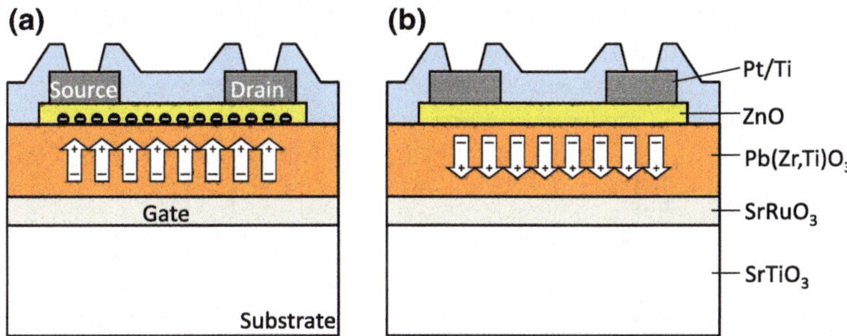

Fig. 5.1 Schematics of **a** an ON state (two-dimensional electron accumulation) and **b** an OFF state (electron depletion) of a FeFET

pressure of around 10^{-6} Torr. A 30-nm-thick SRO film (bottom-gate electrode) was deposited on the STO substrates by pulsed laser deposition (PLD) using a KrF excimer laser source (λ = 248 nm, LPX210). During SRO deposition, the oxygen pressure was 10 mTorr and the substrate temperature was 700 °C. Energy density on the surface of a ceramic ablation target was 1.0 J cm^{-2}, and the repetition rate was 5 Hz. Moreover, 450- and 675-nm-thick PZT films in Sect. (5.3.1) and Sects. (5.3.2) and (5.3.3), respectively, were deposited on SRO using the same PLD system with a Pb(Zr$_{0.52}$,Ti$_{0.48}$)O$_3$ target at 700 °C and an oxygen pressure of 100 mTorr. Energy density was 1.0 J cm^{-2}, and the repetition rate was 10 Hz. The substrate temperature was then reduced to 400 °C, and a 30-nm-thick ZnO film was deposited on the PZT/SRO/STO structure at an oxygen pressure of 10 mTorr. Energy density was 0.5 J cm^{-2}, and the repetition rate was 10 Hz. To form active regions, the ZnO film was etched with dilute nitric acid through a photoresist mask on the surface. Source and drain electrodes were formed by depositing films via electron beam evaporation followed by a lift-off process. To reveal size dependences, we fabricated FeFETs of various channel lengths (L) and widths (W).

5.3 Device Characteristics and Discussions

5.3.1 Basic Characteristics

Large-angle X-ray scans (10°–80°) show only (00l) pseudocubic reflections from a stacked perovskite structure, PZT/SRO/STO, and a (11-20) reflection from a wurtzite ZnO film (Fig. 5.2). We did not observe any reflections that would be indicative of second phases. Thus, the SRO, PZT, and ZnO films are preferentially grown along the (001), (001), and (11-20) directions, respectively. Cross-sectional transmission electron microscopy (TEM) images of a heterostructure are shown in Fig. 5.3. In particular, the high-resolution image in Fig. 5.3b reveals no undesirable layer at the ZnO/PZT interface, and ZnO appears to retain its periodic lattice

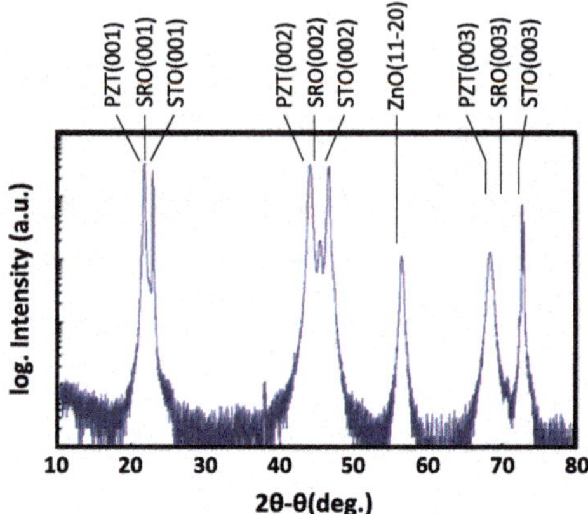

Fig. 5.2 XRD spectra of a ZnO/PZT/SRO/STO structure

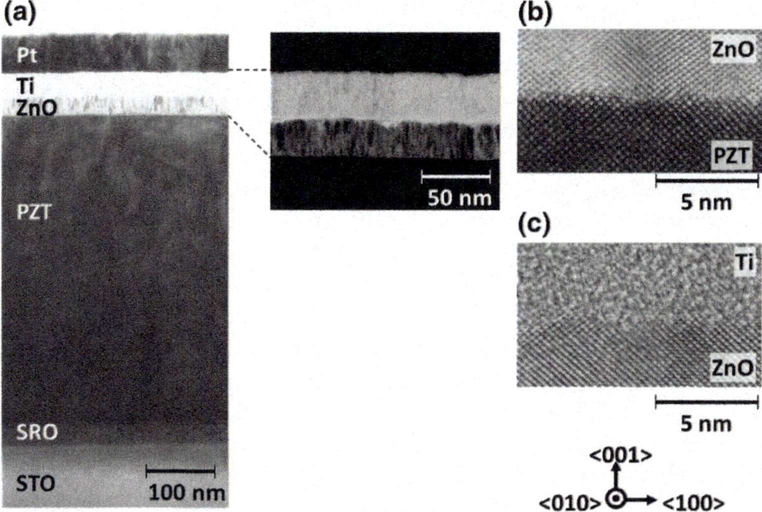

Fig. 5.3 Cross-sectional TEM images of **a** a Pt/Ti/ZnO/PZT/SRO/STO structure and magnified view of **b** the ZnO/PZT interface and **c** the Ti/ZnO interface

structure on PZT. Therefore, excellent physical and electrical properties of the interface are expected.

Selected-area electron diffraction patterns of the PZT and ZnO films were observed from a (010) cross section of the STO substrate by transmission electron diffraction (TED). Because only diffraction spots appear in the diffraction image shown in Fig. 5.4, we can confirm the single crystalline quality of PZT and ZnO. The TED patterns indicate that ZnO [1-102] is parallel to PZT [100]. From these experimental results and previously reported lattice parameters [13, 14], lattice

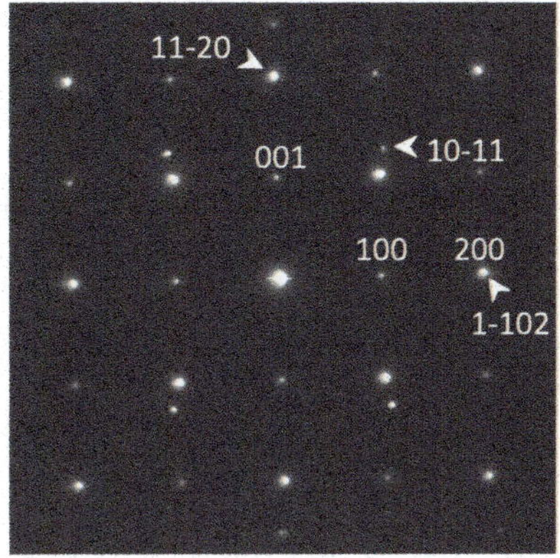

Fig. 5.4 TED *spots* related to the plane indices of ZnO (*arrows*) and PZT

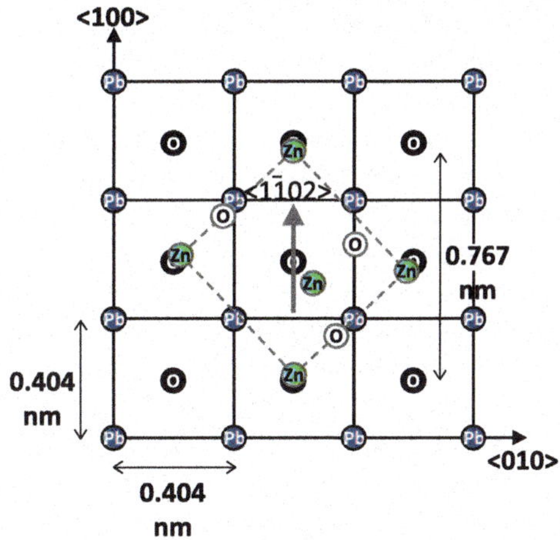

Fig. 5.5 Schematic of the lattice matching between ZnO and PZT

matching between PZT and ZnO is illustrated in Fig. 5.5. Although the mismatch is as large as 5.1 %, we did not observe any grains with different orientations by a pole figure measurement of an electron backscattering pattern (scanned area is 5 μm × 5 μm). In addition, the measured root-mean-square roughness of the surface is as small as 0.815 nm, which indicates that the heterostructure has a nearly atomically flat surface. Although the PZT film was deposited beyond the critical thickness, calculated to be ∼6 nm based on the Matthews–Blakeslee model [15], we conclude that the stacked ZnO/PZT/SRO structure is heteroepitaxially grown on the STO substrate. We consider such heteroepitaxy to be attributed to nonequilibrium growth during the deposition.

It should be emphasized that the c axis of the ZnO film, along which a spontaneous polarization of 5.7 μC cm^{-2} appears [14], is perpendicular to its polarization axis (c axis). Thus far, the c-axis orientation of ZnO has been studied for providing ultraviolet light-emitting devices and high-speed switching devices [16–19]. For instance, Tsukazaki et al. [19] reported a quantum Hall effect in polar

ZnO/Mg$_x$Zn$_{1-x}$O heterostructures, in which confined electrons in a two-dimensional layer at the interface can move within the layer with minimal scattering. In contrast, we intend to accumulate a two-dimensional electron gas at the ZnO/PZT interface using the polarization switching of PZT. To prevent the spontaneous polarization of ZnO interacting against the polarization switching of PZT, the system requires a nonpolar ZnO film. In this study, we have successfully realized the nonpolar ZnO/PZT heteroepitaxy with good crystallinity. Thus, we can expect good electrical properties and retention characteristics for FeFETs with a nonpolar ZnO/PZT/SRO structure.

Because the first deposited layer of SRO functions as a gate electrode that requires a low resistivity for fast driving, we measured the resistivity of the 30-nm-thick SRO film. The measured resistivity is 6.6×10^{-6} Ωm, which is as low as that of a bulk SRO, i.e., 2.8×10^{-6} Ωm [20].

For an electrical characterization of the channel layer in the FeFET, the carrier density and resistivity of the ZnO films with different thicknesses were measured by the van der Pauw method at room temperature (Fig. 5.6). The resistivity of ZnO increases with a reduction in film thickness, whereas the carrier density decreases. The carrier density at 30 nm was determined to be as low as about 10^{15} cm^{-3}, which is much lower than those determined in previous studies [14, 21]. We speculate that such a low carrier density is because of the epitaxial growth of ZnO with a good crystalline quality. Because the current magnitude in an OFF state depends on the resistivity of the ZnO channel, we selected a 30-nm-thick ZnO film, which exhibits the highest resistivity in our experiments, on which a FeFET is to be incorporated. Electron mobility along the c axis of the film was measured to be 26 cm^2 V^{-1} s^{-1}, which is almost equivalent to previously reported values along the a axis in thicker ZnO films than those fabricated in this study [22].

Ferroelectric hysteresis (P–V) measurements were performed using a Pt/Ti/ZnO/PZT/SRO capacitor by applying voltages to the SRO electrode with the Pt/Ti electrode grounded. The thickness of the ZnO film is 30 nm, and the area of the top electrodes is 6.2×10^{-5} cm^2. Figure 5.7 shows the obtained hysteresis loops, which indicate that the polarization of the PZT film saturates at 9 V and the remnant polarization (P$_r$) was 30 µC cm^{-2}. When the polarization direction is

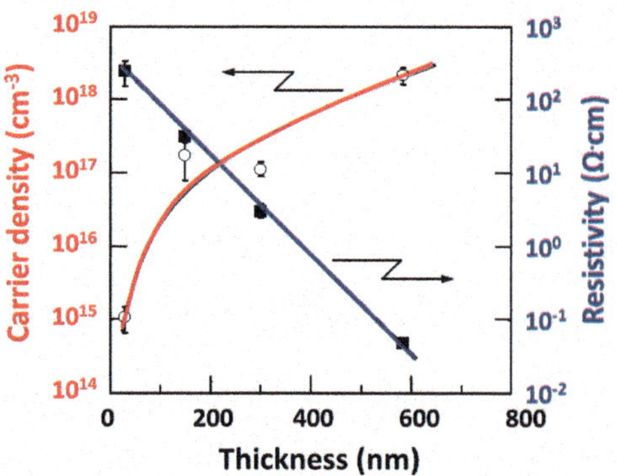

Fig. 5.6 Carrier density and resistivity of ZnO films as a function of the film thickness

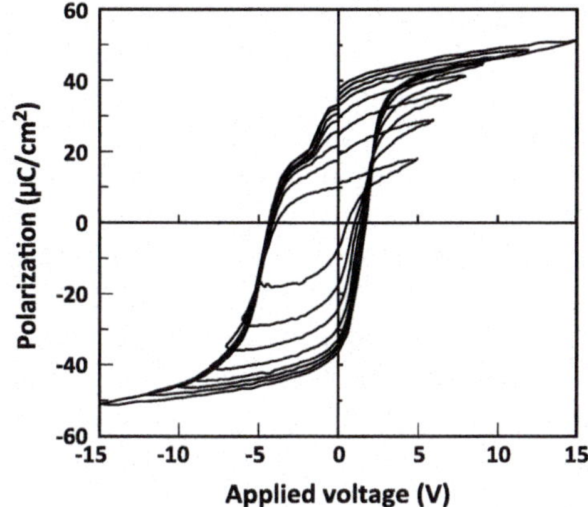

Fig. 5.7 P–V hysteresis curves of the Pt/Ti/ZnO/PZT/SRO capacitor

downward (Fig. 5.1b), the measured polarization is sufficient to deplete the entire density of electrons across the 30-nm-thick ZnO film, in which the area density of intrinsic charge is calculated to be as low as 0.0005 μC cm^{-2}. On the other hand, when the polarization direction is upward (Fig. 5.1a), polarization accumulates charges in the ZnO film with an area density of 30 μC cm^{-2}; this density is 6×10^4 times larger than the intrinsic charge density. Because the number of electrons in the depletion state is fewer than that in the intrinsic charge, the ratio of charge modulation by polarization switching is larger than 6×10^4. Thus, we can expect a high ON/OFF ratio of more than 6×10^4 for a FeFET with a 30-nm-ZnO/PZT/SRO structure.

Capacitance–voltage measurements revealed the charge response of the Pt/Ti/ZnO/PZT/SRO capacitor with respect to the applied voltage (Fig. 5.8a). We also characterized a Pt/Ti/PZT/SRO capacitor as a reference (Fig. 5.8b). When a positive voltage is applied to the SRO electrode, the capacitance of the Pt/Ti/ZnO/PZT/SRO structure is similar to that of the Pt/Ti/PZT/SRO structure. This result

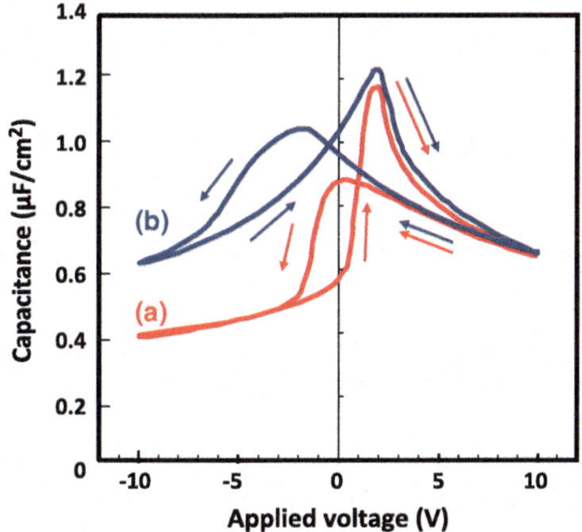

Fig. 5.8 C-V characteristics of **a** a Pt/Ti/ZnO/PZT/SRO (*red line*) and **b** a Pt/Ti/PZT/SRO (*blue line*) capacitors

indicates that quasi-two-dimensional electrons accumulate at the ZnO/PZT interface because the ZnO film has no contribution to the total capacitance in series. In contrast, when negative voltages were applied to the SRO electrode, the total capacitance decreased as a result of the full depletion of the 30-nm-thick ZnO film. These accumulation and depletion states of the ZnO film were preserved when the applied voltage was removed.

The drain current and gate voltage characteristics (I_{DS}–V_{GS}) of a FeFET with a Pt/Ti/ZnO/PZT/SRO structure were determined at V_{DS} = 0.1 V; these are shown in Fig. 5.9. The drain current shows a counterclockwise hysteresis loop corresponding to ferroelectric polarization switching and eventually splits by five orders of magnitude when a bias voltage is cycled between −10 and +10 V, according to the accumulation (V_{GS} > 0) and depletion (V_{GS} < 0) states of the ZnO film. The ratios of the drain currents higher than approximately 10^5 correspond to the calculated ratio of the charge modulation in the 30-nm-thick ZnO film between the accumulation and depletion states. The split between the ON and OFF states remains when the gate voltage is returned to zero. Furthermore, the midpoint of subthreshold swings is very close to a zero bias, suggesting that the amount of space charge in the PZT film and at the ZnO/PZT interface is remarkably low, as anticipated from the structural characterization. Therefore, good retention properties can be expected for the fabricated FeFET.

We investigated the retention characteristics of the fabricated FeFET at room temperature without applying any gate voltage during retention periods. First, as an initial write (ON-state) operation, a FeFET is biased with V_{GS} = 10 V and V_{DS} = 0 V. Then, the device state was probed by drain current measurements as a function of retention time under bias conditions of V_{DS} = 0.1 V and V_{GS} = 0 V. Similarly, an OFF-state retention was characterized by, first, biasing the FeFET with V_{GS} = −10 V and V_{DS} = 0 V and then by measuring the drain currents with previous bias conditions of V_{DS} = 0.1 V and V_{GS} = 0 V. As can be seen in Fig. 5.10, no significant change in the drain current was observed over 3.5 months.

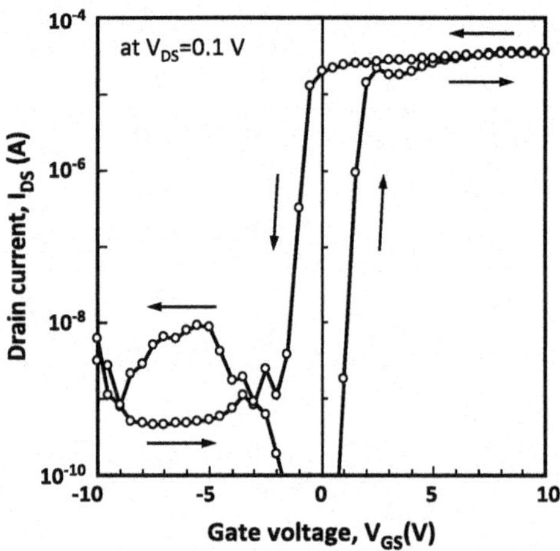

Fig. 5.9 I_{DS}–V_{GS} characteristics of a Fe FET with a 30-nm-thick ZnO film

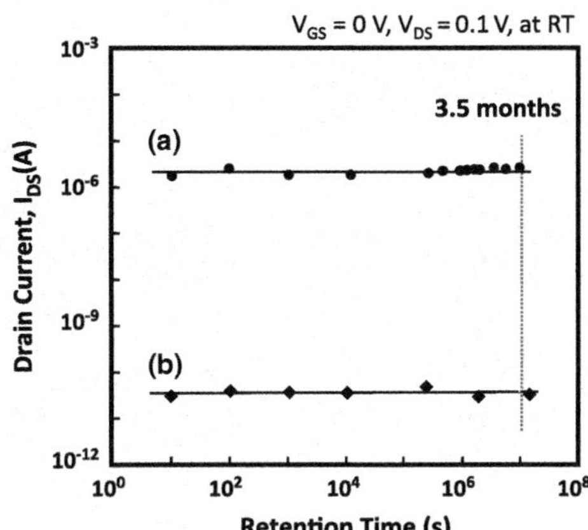

Fig. 5.10 Retention characteristics of a FeFET for **a** an ON state and **b** an OFF state

Extrapolating the retention behavior ensures a definite split in the drain current for more than 10 years (3×10^8 s) at room temperature.

The observed stable ON and OFF states may be expected from physical insights into the Pt/Ti/ZnO/PZT/SRO structure. In the ON state, electrons in the ZnO film that are strongly coupled with the polarization of the PZT film accumulate at the ZnO/PZT interface under a flat-band condition inside the PZT layer. On the other hand, because the electron density of the ZnO layer is low, the conductivity of the completed depleted ZnO layer is extremely low in the OFF state. Thus, the high ON/OFF ratio can be maintained for these long time periods.

5.3.2 Correlated Motion Dynamics of Electron Channels and Domain Walls

5.3.2.1 Transition Characteristics of Switching

We examined transition characteristics of switching by applying gate pulses to the FeFET. The circuit used for determining switching characteristics consisted of a pulse generator connected to a gate electrode and a semiconductor parameter analyzer connected to the source and drain electrodes. The transition characteristics of switching were evaluated from the drain current after applying a gate voltage pulse. Figure 5.11 shows the pulse sequences applied during the measurement. Before beginning the drain current measurements during the switching of the state from downward to upward polarization (from OFF to ON), we first performed a reset operation on the FeFET by applying a negative gate voltage as a reset pulse with the source and drain electrodes grounded. After applying the reset pulse, we applied a positive voltage pulse as a write pulse to the gate electrode, again with the source and drain electrodes grounded. We varied the pulse width from 10^{-8} to

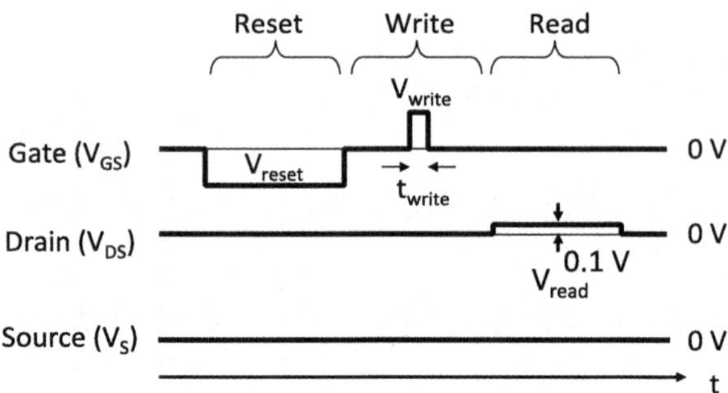

Fig. 5.11 Pulse sequences used to measure the transition characteristics of switching

10^2 s. We then probed device states by drain current (I_{DS}) measurements as a function of the write pulse width (t_{write}) under bias conditions of $V_{GS} = 0$ V and $V_{DS} = 0.1$ V. Conversely, in determining the switching of the state from upward to downward polarization (from ON to OFF), we applied a positive gate voltage as a reset pulse to the gate electrode with the source and drain electrodes grounded. We then applied a negative gate voltage as a write pulse to the gate electrode, again with the source and drain electrodes grounded. The device states were also probed by I_{DS} measurements as a function of t_{write} under the same bias conditions: $V_{GS} = 0$ V and $V_{DS} = 0.1$ V. By increasing the write pulse width, we expected the state of the FeFET to be switched from OFF to ON and from ON to OFF with polarization reversal. We measured the drain current change after applying a gate pulse by changing t_{write} and plotting it as a function of t_{write}.

Figure 5.12a, b shows the transition characteristics of switching the state from OFF to ON and from ON to OFF, respectively, for various write voltages (V_{write}) for a channel length of 10 μm and channel width of 100 μm. We used reset pulses of −15 V and +15 V for switching to upward and downward polarizations, respectively. The drain currents after applying a write pulse varied with the write pulse width. As shown in Fig. 5.12a, when the state of the FeFET was switched from OFF to ON, I_{DS} increased gradually at first and then rapidly with increasing t_{write}. Finally, I_{DS} was saturated. The transition characteristics of switching showed a clear threshold in t_{write}. As V_{write} was decreased from 15 to 3 V, the t_{write} required for switching the state significantly increased and the saturation current was reached more slowly. Conversely, when the FeFET state was switched from ON to OFF, I_{DS} decreased with increasing t_{write}, as shown in Fig. 5.12b. We again found that the transition characteristics of switching exhibited a saturation threshold with t_{write} and that the saturation current was reached more slowly with increasing V_{write}. However, results for switching from ON to OFF indicated transition characteristics, including larger threshold times and slower, more gradual transitions. Because the reversal of the polarization of ferroelectric materials depends on the magnitude and time of the applied voltage [23–28], we speculate that a decrease in an effective voltage applied to a ferroelectric layer may be a reason for the above results.

When the FeFET is switched to the ON state, the electrons accumulate at the ZnO/PZT interface. In contrast, when the FeFET is switched to the OFF state, the

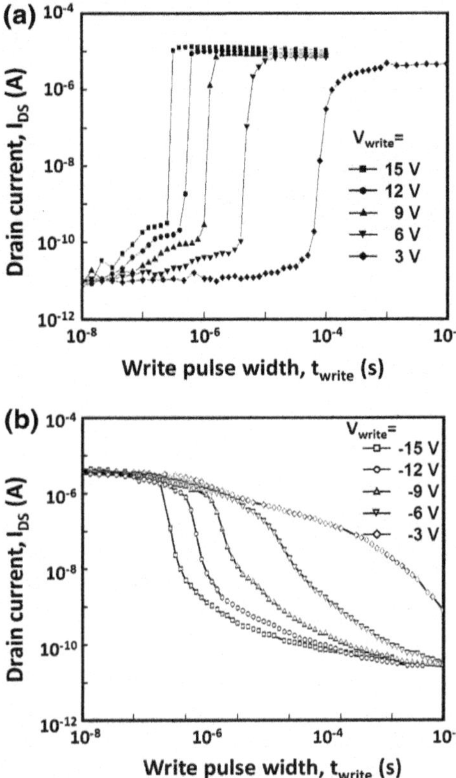

Fig. 5.12 Drain current–write pulse width (I_{DS}-t_{write}) characteristics of switching the state **a** from OFF to ON and **b** from ON to OFF for various write pulse voltages (V_{write}) for a FeFET channel length of 10 μm and channel width of 100 μm

ZnO layer is depleted. Thus, the effective voltage applied to the PZT layer is different from that applied to the states of the FeFET because of the capacitance of the ZnO layer. In the ON state, the gate voltage appears directly across the PZT layer, whereas in the OFF state, the bottom-gate voltage appears partly across the PZT layer and partly across the ZnO layer. Therefore, we conclude that the asymmetric switching characteristics observed are due to the difference in the effective voltage applied to the PZT layer.

5.3.2.2   Channel-Length Dependence

To clarify switching dynamics, the transition characteristics of switching were determined by varying the channel length. Figure 5.13 shows drain current–write pulse width characteristics (I_{DS}-t_{write}) of switching the state from OFF to ON and from ON to OFF for various FeFET channel lengths. The FeFET channel width was 100 μm. For write operations, we used pulses of +15 V for switching to ON and −15 V for switching to OFF. The measured drain currents were normalized to the channel width divided by the channel length (W/L). Solid and open symbols show

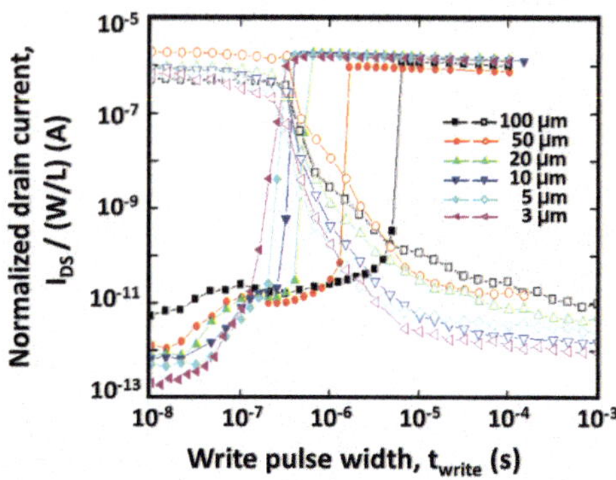

Fig. 5.13 Normalized drain current–write pulse width characteristics (I_{DS}-t_{write}) of switching the state from OFF to ON (*solid symbols*) and from ON to OFF (*open symbols*) for various channel lengths (L) of FeFETs. The channel width (W) of each FeFET is 100 μm

the transition characteristics of switching from OFF to ON and from ON to OFF, respectively. A striking transformation was observed, although the aspect ratio of the ferroelectric film's thickness to its channel length remained low, e.g., it was about 0.7/100 for a channel length of 100 μm. Because the carrier density of the ZnO film is low, the FeFET would typically show characteristics of the OFF state. Therefore, the switching observed between the ON and OFF states suggests that the polarization of the ferroelectric film at the entire channel region under the ZnO channel may be reversed by the gate voltage. This means that an electric field from fringing effects is not the only driving force of switching, other effects must also be contributors.

When the FeFET was switched from the OFF to ON state, the thresholds observed in t_{write} increased with increasing channel length. The dependence on the channel length in switching shows that polarization reversal travels along the direction of the channel length. Conversely, when the FeFET state was switched from ON to OFF, the thresholds in t_{write} were independent of the channel length, consistent with the same polarization movement. This is because the FeFET can only be switched to the OFF state if the polarization under the electrode is switched downward.

5.3.2.3 Multielectrode Analysis and Switching Model

To support our experimental findings, we fabricated a test structure with two electrodes (hereafter B and C) between a source electrode (hereafter A) and a drain electrode (hereafter D) on a ZnO film, as shown in Fig. 5.14. All electrodes have a width of 160 μm and length of 80 μm, and the distances A–B, B–C, and C–D were 40, 20, and 10 μm, respectively.

First, reset and write operations were performed using A, D, and the gate electrode, keeping electrodes B and C floating. We then measured the drain current at V_{GS} = 0 V and V_{DS} = 0.1 V. We also measured the currents between A and D to determine transition characteristics of channel regions between these two electrodes. Figure 5.15a shows these characteristics for a switching from OFF to ON.

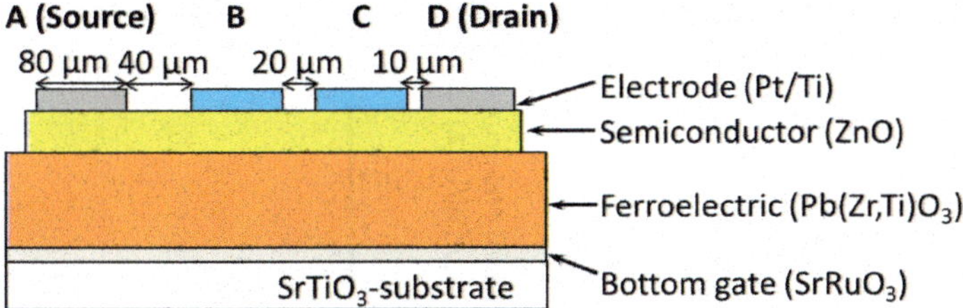

Fig. 5.14 Schematic of a test structure with two electrodes (*B* and *C*) between a source electrode (*A*) and a drain electrode (*D*) on a ZnO film

A reset pulse of −15 V and a write pulse of +15 V were used, and drain currents from A–D, A–B, B–C, and C–D, normalized to a channel width divided by a channel length (W/L), were plotted, revealing that thresholds in t$_{\text{write}}$ varied depending on electrode positions. Moreover, C–D, located nearest to the drain electrode (10 μm apart), was the first to be transformed to ON. Then, A–B, located nearest to the source electrode (40 μm apart), was transformed to ON. Finally, B–C and A–D were transformed to ON simultaneously. Thus, polarization reversal starts from the source and drain regions and then spreads to the center region.

Figure 5.15b shows the transition characteristics of switching from ON to OFF using the same four electrodes, i.e., A, B, C, and D. A reset pulse of +15 V and a write pulse of −15 V were applied. Thresholds in t$_{\text{write}}$, in an analogous manner to the previous case of switching to ON, also varied depending on electrode positions. A–D, A–B, and C–D were the first to be transformed to OFF, followed by B–C, located at the center region. Thus, in a similar manner, polarization reversal starts from the source and drain regions and spreads to the center region. Figure 5.16 shows a schematic of the motion of polarization reversal during switching. Initially, all channel regions are in the OFF state because all polarizations are downward (see upper left, Fig. 5.16). With the source and drain electrodes grounded, when a positive write pulse is applied to a gate electrode, polarization reversal starts from a region located under the electrodes. Electrons accumulate at the interface between the ZnO and PZT films, where they are located in the region of upward polarization because they are coupled to this polarization. As the positive write pulse width increases, domain walls in the PZT film, which exist between upward and downward polarizations, move toward the center of a channel. If the accumulated electrons fill the channel between each electrode, the conductance between these electrodes becomes high. Finally, upward polarization regions located on both sides of a channel connect at the center (see lower right, Fig. 5.16). Therefore, beginning from near the source and drain electrodes, the channels are sequentially switched to ON.

Similarly, when a negative write pulse is applied to a gate electrode with the source and drain electrodes grounded, polarization reversal starts from a region located under these grounded electrodes. Electrons are depleted from the interface between the ZnO and PZT films; because the electrons are decoupled to downward

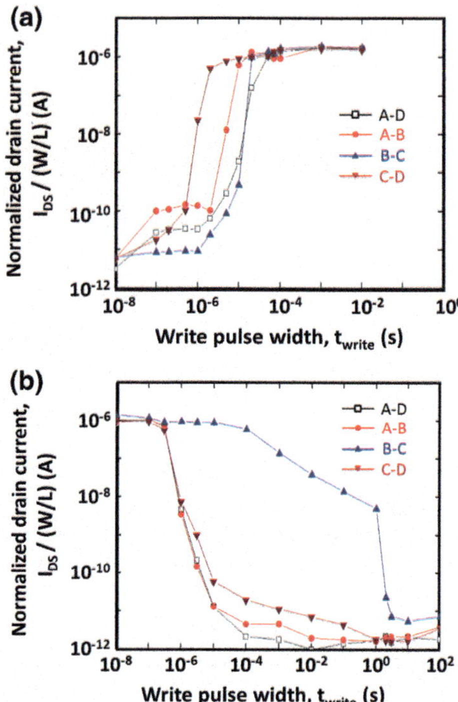

Fig. 5.15 Normalized drain current–write pulse width (I_{DS}-t_{write}) characteristics of switching the state **a** from OFF to ON and **b** from ON to OFF. The drain currents for A–D, A–B, B–C, and C–D are plotted

polarization, they are located in the region of downward polarization. As the accumulated electrons recede from the channel between each electrode, the conductance between these electrodes becomes low. Finally, downward polarization regions located on both sides of a channel connect at the center. Therefore, the channels switch sequentially to OFF from near the source and drain electrodes. Fukushima et al. have reported that switching operations of a FeFET with a ZnO/YMnO$_3$ stacked structure are dominated by polarization switching near the source electrode, consistent with our present observations of motion dynamics [29].

Domain wall motions in a ferroelectric film cause switching in a FeFET. When a channel length (L) of a FeFET is short, a domain wall in a ferroelectric film quickly arises. Thus, we expect high-speed operation in small-channel-length devices. Figure 5.17 shows the switching time from OFF to ON as a function of the channel length. We define the switching time as the period taken to reach 90 % of the saturated ON current, obtained from Fig. 5.13. The switching time decreases nonlinearly with L, suggesting that the velocity of the domain wall in a ferroelectric film is not constant across the channel length. The closer the domain wall in a ferroelectric film is to the source and drain electrodes, the higher is the velocity of the domain wall. As the channel length is decreased, the switching speed approaches the limit of a ferroelectric polarization reversal speed, i.e., of the order of sub-nanoseconds [30, 31]. Therefore, a FeFET is merited by high operation speeds at scale.

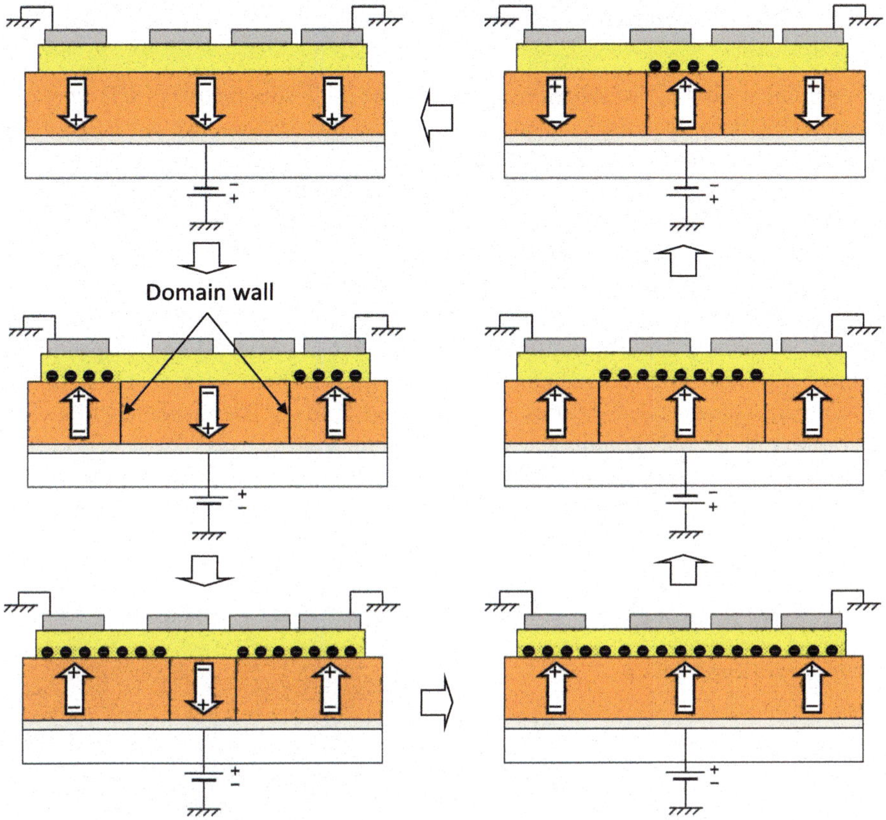

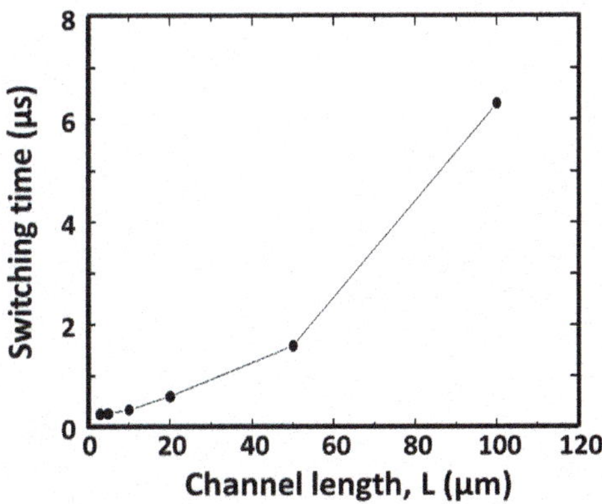

Fig. 5.16 Schematics of the motion of polarization reversal during switching

Fig. 5.17 Dependence of the OFF-to-ON switching time on channel length

5.3.3   60-nm-Channel-Length FeFET

As mentioned above, the switching speed of the TFT-structured FeFET increases dramatically with decreasing channel length. In this section, we investigate the fabrication of narrow-channel FeFETs.

To fabricate a device of sub-100-nm channel length, we first patterned a photoresist mask using an electron beam lithography technique on the ZnO film. The source and drain electrodes were then formed by depositing Pt/Ti films using electron beam evaporation, followed by a lift-off process.

Figure 5.18 shows an SEM image of the top view of the fabricated FeFET. The channel width and length of the device were 1.2 μm and 60 nm, respectively. Figure 5.19 shows the characteristics of the drain current as a function of the bottom-gate voltage (I_{DS}–V_{GS}) of the fabricated FeFET. The characteristics were determined under bias conditions where the drain voltage was $V_{DS} = 0.1$ V. The drain current of the FeFET showed a counterclockwise hysteresis loop corresponding to ferroelectric polarization switching, similar to the aforementioned FeFET with a wider channel length as discussed in Sect. 5.3.1.

Fig. 5.18 SEM image of the top view of a fabricated 60-nm-channel-length FeFET

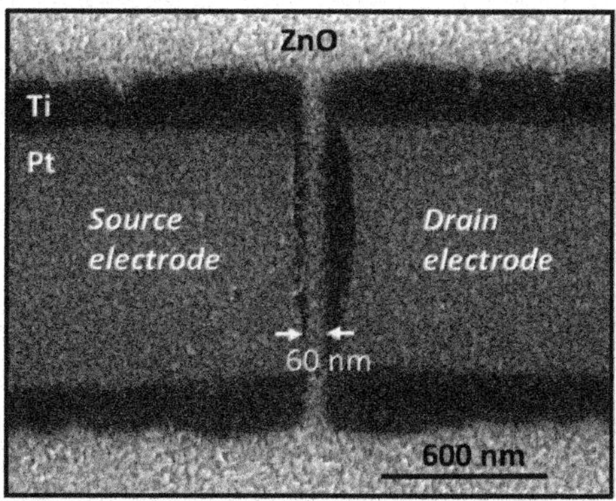

Fig. 5.19 Drain current–gate voltage (I_{DS}–V_{GS}) characteristics of the FeFET, measured under bias conditions $V_{DS} = 0.1$ V and $V_S = 0$ V

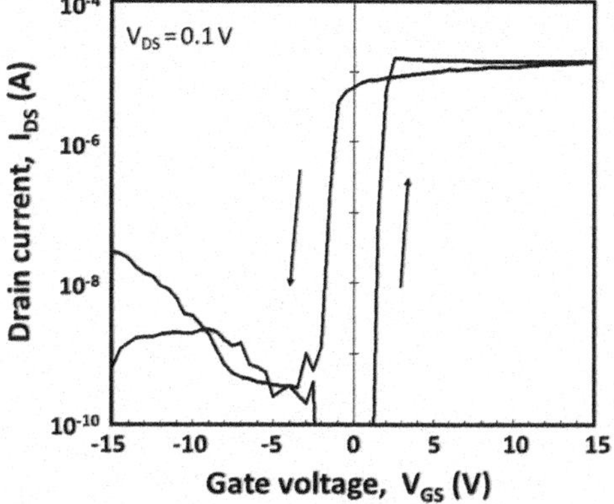

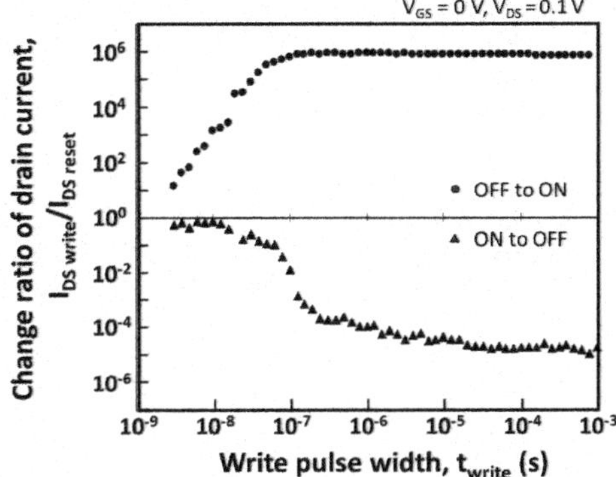

Fig. 5.20 Change ratio of drain current (I$_{DS\ write}$/I$_{DS\ reset}$)–write pulse width (t$_{write}$) characteristics for switching states from OFF to ON and from ON to OFF for 60-nm-channel-length FeFETs

To evaluate the switching speed, we applied gate pulses and examined the switching characteristics. We first reset the FeFET by applying a reset pulse voltage while the source and drain electrodes were grounded. We measured the drain current in the reset state (I$_{DS\ reset}$) under the bias conditions V$_{GS}$ = 0 V and V$_{DS}$ = 0.1 V. After applying the reset pulse, we applied a write pulse voltage to the gate electrode, again with the source and drain electrodes grounded. We then probed the device states by measuring the drain current (I$_{DS\ write}$) as a function of the write pulse width (t$_{write}$) under the same bias conditions, i.e., V$_{GS}$ = 0 V and V$_{DS}$ = 0.1 V. By increasing the write pulse width, we expect the state of the FeFET to be switched from OFF to ON and from ON to OFF with polarization reversal. Figure 5.20 shows the switching characteristics of the fabricated FeFET both from OFF to ON and from ON to OFF for a channel length of 60 nm and a channel width of 1.2 μm. We determined the ratio of the change in the drain current (I$_{DS\ write}$/I$_{DS\ reset}$) by applying write pulses of varying widths t$_{write}$ after a reset pulse was applied. We used a 1 ms reset pulse of −15 V and a write pulse of +15 V for switching from OFF to ON, and similarly, a 1 ms reset pulse of +15 V and a write pulse of −15 V for switching from ON to OFF. The drain currents determined after the application of the write pulse varied with the width of the write pulse; thus, the change ratio varied with increasing t$_{write}$, until the ratio was saturated. The switching characteristics therefore show a clear threshold for t$_{write}$. Even when t$_{write}$ was less than 5 ns, switching was observed. It was also found that the drain current ON/OFF ratio was about three orders of magnitude for OFF to ON for a write pulse width as narrow as 10 ns.

Although in this investigation the channel length was scaled down to 60 nm, the conductance could also be varied continuously to achieve rapid switching speeds, opening up the possibility of multivalued memory, and analog memory applications. The above results suggest that scaling down a FeFET does not inhibit its multivalued memory behavior, at least down to the 60 nm node. Cho et al. have reported polarization reversal at a sub-10-nm domain size using scanning nonlinear dielectric microscopy [32]. Assuming that the ferroelectric domain size is approximately 10 nm, a FeFET with 60 nm channel length has the potential for six switching domains along the channel length. The ferroelectric polarization direction

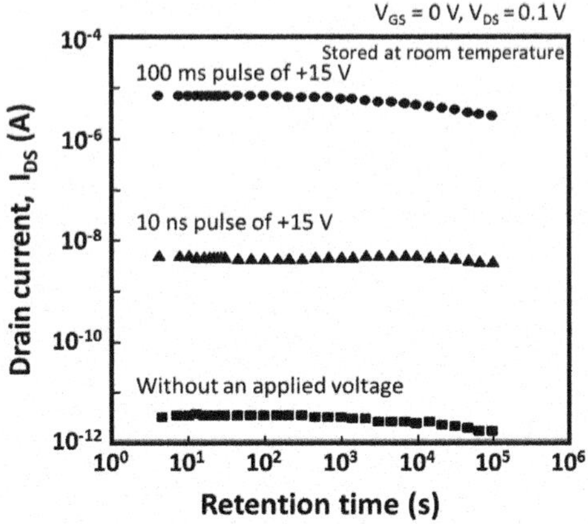

Fig. 5.21 Data retention characteristics of the FeFET at room temperature. We programmed the three-level data using a 100 ms pulse of +15 V, a 10 ns pulse of +15 V, and without an applied voltage

of the PZT film is either up or down because the film was grown epitaxially along the (001) direction. Therefore, although the channel region consists of a series of low- and/or high-resistance ZnO elements controlled by binary ferroelectric polarizations, the conductance of the FeFET may be controlled continuously. The FeFET can memorize multiple values even when the channel length was scaled down to be as narrow as the ferroelectric domain size. Further study is required to understand this discrepancy and the details of these mechanisms.

The retention characteristics of the three levels of data memorized in the FeFET were examined at room temperature. We first performed a reset operation by applying a 100 ms reset pulse of −15 V. Then, we programmed the three-level data using a 100 ms pulse of +15 V, a 10 ns pulse of +15 V, and without an applied voltage. During the retention period, we kept all electrodes grounded and measured the drain current under the bias conditions $V_{GS} = 0$ V and $V_{DS} = 0.1$ V. Figure 5.21 shows the results of the retention characterization. We observed no significant degradation in retention over 10^5 s, and we found that the channel conductance of the three-level data maintained very stable states. The extrapolation of the retention behavior over 10 years indicates a definite split in the drain current at room temperature for this time period, 3×10^8 s.

5.4 Summary

In this study, we have addressed a FeFET composed of a heteroepitaxially stacked ZnO/PZT/SRO/STO structure. A switching operation by way of electron gas accumulation and complete depletion has been demonstrated in a thin ZnO layer with an extremely low carrier density (on the order of 10^{15} cm^{-3}). As a result, a high ON/OFF ratio, greater than 10^5, of the drain current was realized. Furthermore, a retention time longer than 3.5 months with this ON/OFF ratio was achieved. We have also studied the switching dynamics of a FeFET. The transition characteristics of switching exhibited a clear threshold with the write pulse width t$_{write}$. When the state of the FeFET was switched, the thresholds in t$_{write}$ increased with increasing channel length. From a multielectrode analysis, we found that the switching of a FeFET is caused by the domain wall motion in a ferroelectric film. Polarization reversal starts from a region located under the source and drain electrodes and travels along the direction of the channel length. In addition, the domain wall velocity increases as the domain wall gets closer to the source and drain electrodes in a ferroelectric film. Therefore, a FeFET has the merit of high operation speeds at scale. We have also demonstrated the electrical properties of a 60-nm-channel-length FeFET fabricated using electron beam lithography techniques. Even if the channel length is decreased, a high ON/OFF ratio for the drain current, 10^5 at V$_{GS}$ = 0 V, was realized. It was also found that the drain current ON/OFF ratio was about three orders of magnitude for a write pulse width as narrow as 10 ns. The conductance can be changed continuously, and the device has a long retention time. We also succeeded in the fabrication of a FeFET on a Si substrate with electrical properties the same as on an STO substrate [33]. These characteristics are suitable for not only nonvolatile memory, which consists of NOR- and NAND-type FeFET memory cells, but also analog memory applications [34–45].

Acknowledgments We would like to thank Yu Nishitani, Hiroyuki Tanaka, Michihito Ueda, Atsushi Omote, Ayumu Tsujimura, Eiji Fujii, Yoshihisa Kato, Yasuhiro Shimada, Daisuke Ueda, and Eisuke Tokumitsu for valuable discussions and excellent experimental assistance.

References

1. K. Tanaka, Y. Cho, Actual information storage with a recording density of 4 Tbit/in. (2) in a ferroelectric recording medium. Appl. Phys. Lett. **97**, 092901 (2010)
2. K. Tanaka et al., Scanning nonlinear dielectric microscopy nano-science and technology for next generation high density ferroelectric data storage. Jpn. J. Appl. Phys. **47**, 3311–3325 (2008)
3. W. Shu-Yau, A new ferroelectric memory device, metal-ferroelectric-semiconductor transistor. IEEE Trans. Electron Devices **21**, 499–504 (1974)
4. M. Alexe, Measurement of interface trap states in metal-ferroelectric-silicon heterostructures. Appl. Phys. Lett. **72**, 2283–2285 (1998)
5. G. Hirooka, et al., Proposal for a new ferroelectric gate field effect transistor memory based on ferroelectric-insulator interface conduction. Jpn. J. Appl. Phys. Part 1—Regular Paper. Short Notes Rev. Papers **43**, 2190–2193 (2004)
6. S. Sakai, R. Ilangovan, Metal-ferroelectric-insulator-semiconductor memory FET with long retention and high endurance. IEEE Electron Device Lett. **25**, 369–371 (2004)
7. E. Tokumitsu et al., Use of ferroelectric gate insulator for thin film transistors with ITO channel. Microelectron. Eng. **80**, 305–308 (2005)
8. K. Takahashi, et al., Thirty-day-long data retention in ferroelectric-gate field-effect transistors with HfO2 buffer layers. Jpn. J. Appl. Phys. Part 1—Regular Paper. Short Notes Rev. Papers **44**, 6218–6220 (2005)

9. M. Takahashi, S. Sakai, Self-aligned-gate metal/ferroelectric/insulator/semiconductor field-effect transistors with long memory retention. Jpn. J. Appl. Phys. Part 1—Lett. Express Lett. **44**, L800–L802 (2005)

10. B.Y. Lee, et al., Fabrication and characterization of ferroelectric gate field-effect transistor memory based on ferroelectric-insulator interface conduction. Jpn. J. Appl. Phys. Part 1—Regular Paper. Short Notes Rev. Papers **45**, 8608–8610 (2006)

11. Q.H. Li, S. Sakai, Characterization of Pt/SrBi2Ta2O9/Hf-Al-O/Si field-effect transistors at elevated temperatures. Appl. Phys. Lett. **89** (2006)

12. H. Ishiwara, Current status of ferroelectric-gate Si transistors and challenge to ferroelectric-gate CNT transistors. Curr. Appl. Phys. **9**, S2–S6 (2009)

13. S. Yokoyama et al., Dependence of electrical properties of epitaxial Pb(Zr, Ti)O3 thick films on crystal orientation and Zr/(Zr+Ti) ratio. J. Appl. Phys. **98**, 094106 (2005)

14. Ü. Özgür et al., A comprehensive review of ZnO materials and devices. J. Appl. Phys. **98**, 041301 (2005)

15. J.W. Matthews, A.E. Blakeslee, Defects in epitaxial multilayers. I. Misfit dislocations. J. Cryst. Growth **27**, 118 (1974)

16. Z.K. Tang et al., Self-assembled ZnO nano-crystals and exciton lasing at room temperature. J. Cryst. Growth **287**, 169–179 (2006)

17. E.M.C. Fortunato et al., Wide-bandgap high-mobility ZnO thin-film transistors produced at room temperature. Appl. Phys. Lett. **85**, 2541–2543 (2004)

18. P.F. Carcia et al., Transparent ZnO thin-film transistor fabricated by rf magnetron sputtering. Appl. Phys. Lett. **82**, 1117–1119 (2003)

19. A. Tsukazaki, et al., Quantum hall effect in polar oxide heterostructures. Science **315**, 1388–1391 (2007)

20. N. Tsuda, et al., Electronic conduction in oxides, 2nd ed. (Shokabo, 1993)

21. B.L. Zhu et al., Effect of thickness on the structure and properties of ZnO thin films prepared by pulsed laser deposition. Jpn. J. Appl. Phys. **45**, 7860 (2006)

22. E. Bellingeri et al., High mobility in ZnO thin films deposited on perovskite substrates with a low temperature nucleation layer. Appl. Phys. Lett. **86**, 012109 (2005)

23. Y. Ishibashi, Y. Takagi, Note on ferroelectric domain switching. J. Phys. Soc. Jpn. **31**, 506 (1971)

24. J.F. Scott et al., Switching kinetics of lead zirconate titanate submicron thin-film memories. J. Appl. Phys. **64**, 787–792 (1988)

25. T. Tybell et al., Domain wall creep in epitaxial ferroelectric $Pb(Zr_{0.2}Ti_{0.8})O_3$ thin films. Phys. Rev. Lett. **89**, 097601 (2002)

26. Y.-H. Shin et al., Nucleation and growth mechanism of ferroelectric domain-wall motion. Nature **449**, 881–884 (2007)

27. E. Tokumitsu et al., Partial switching kinetics of ferroelectric $PbZr_xTi_{1-x}O_3$ thin films prepared by sol-gel technique. Jpn. J. Appl. Phys. **33**, 5201 (1994)

28. J.Y. Jo et al., Composition-dependent polarization switching behaviors of (111)-preferred polycrystalline $Pb(Zr_xT_{1-x})O_3$ thin films. Appl. Phys. Lett. **92**, 012917-3 (2008)

29. T. Fukushima, et al., Impedance analysis of controlled-polarization-type ferroelectric-gate thin film transistor using resistor–capacitor lumped constant circuit. Jpn. J. Appl. Phys. **50**, 04DD16 (2011)

30. J. Li et al., Ultrafast polarization switching in thin-film ferroelectrics. Appl. Phys. Lett. **84**, 1174–1176 (2004)

31. H. Ishii et al., Ultrafast polarization switching in ferroelectric polymer thin films at extremely high electric fields. Appl. Phys. Express **4**, 031501 (2011)

32. Y. Cho, Ultrahigh-density ferroelectric data storage based on scanning nonlinear dielectric microscopy. Jpn. J. Appl. Phys. **44**, 5339 (2005)

33. H. Tanaka et al., A ferroelectric gate field effect transistor with a ZnO/Pb(Zr, Ti)O3 heterostructure formed on a silicon substrate. Jpn. J. Appl. Phys. **47**, 7527–7532 (2008)

34. Y. Kaneko, et al., NOR-type nonvolatile ferroelectric-gate memory cell using composite oxide technology. Jpn. J. Appl. Phys. **48**, 09ka19 (2009)

35. Y. Kaneko et al., A dual-channel ferroelectric-gate field-effect transistor enabling nand-type memory characteristics. IEEE Trans. Electron Devices **58**, 1311–1318 (2011)
36. M. Ueda et al., A neural network circuit using persistent interfacial conducting heterostructures. J. Appl. Phys. **110**, 086104-3 (2011)
37. Y. Nishitani et al., Three-terminal ferroelectric synapse device with concurrent learning function for artificial neural networks. J. Appl. Phys. **111**, 124108-6 (2012)
38. Y. Kaneko, et al., *Neural network based on a three-terminal ferroelectric memristor to enable on-chip pattern recognition*. In 2013 Symposium on VLSI Technology (VLSIT) (2013), pp. T238–T239
39. Y. Nishitani, et al., Dynamic observation of brain-like learning in a ferroelectric synapse device. Jpn. J. Appl. Phys. **52**, 04CE06 (2013)
40. Y. Kaneko et al., Ferroelectric artificial synapses for recognition of a multishaded image. IEEE Trans. Electron Devices **61**, 2827–2833 (2014)
41. Y. Nishitani, et al., Supervised learning using spike-timing-dependent plasticity of memristive synapses. IEEE Trans. Neural Netw. Learn. Syst. (accepted)
42. M. Ueda et al., Battery-less shock-recording device consisting of a piezoelectric sensor and a ferroelectric-gate field-effect transistor. Sens. Actuators A: Phys. **232**, 75–83 (2015)
43. Y. Kato et al., Nonvolatile memory using epitaxially grown composite-oxide-film technology. Jpn. J. Appl. Phys. **47**, 2719–2724 (2008)
44. Y. Kaneko et al., Correlated motion dynamics of electron channels and domain walls in a ferroelectric-gate thin-film transistor consisting of a ZnO/Pb(Zr, Ti)O$_3$ stacked structure. J. Appl. Phys. **110**, 084106–084107 (2011)
45. Y. Kaneko et al., A 60 nm channel length ferroelectric-gate field-effect transistor capable of fast switching and multilevel programming. Appl. Phys. Lett. **99**, 182902–182903 (2011)

Chapter 6
Novel Ferroelectric-Gate Field-Effect Thin Film Transistors (FeTFTs): Controlled Polarization-Type FeTFTs

Norifumi Fujimura and Takeshi Yoshimura

Abstract Controlled-polarization-type ferroelectric-gate thin film transistors (FeTFTs), which utilize the interaction between the polarizations of a polar semiconductor and a ferroelectric layer, have been proposed. When the polarizations align head to head, electrons that correspond to the sum of the polarizations are induced at the interface between the polar semiconductor and the ferroelectric layer. When the semiconductor is depleted, however, the polarization in the polar semiconductor aligns in the same direction as the polarization in the ferroelectric layer, whereas the polarization of the ferroelectric layer remains stable even under a depolarization field. This chapter describes the non-volatile operation of the controlled-polarization-type FeTFTs resulting from the ferroelectric polarization reversal.

6.1 Introduction

6.1.1 Field-Effect Control of Carrier Concentration

Recently, surface and interface phenomena have attracted much attention because of the difference between their physical properties and those of the corresponding bulk materials. In actual devices, the ability to control the carrier concentration is an important factor to be considered when using these distinguishing properties. The field effect can be effectively employed for carrier doping at heterointerfaces because the carrier concentration can be varied continuously and independently. In particular, there have been significant developments in Si metal–oxide-semiconductor field-effect-transistors (MOSFETs), whose main advantage of Si MOSFETs is their clean and comfortable interface formation with SiO_2. Moreover, the effects of electrostatic fields on carrier transport in Si MOSFETs have been

N. Fujimura (✉) · T. Yoshimura
Graduate School of Engineering, Department of Physics and Electronics,
Osaka Prefecture University, Sakai, Osaka 599-8531, Japan
e-mail: fujim@pe.osakafu-u.ac.jp

© Springer Science+Business Media Dordrecht 2016
B.-E. Park et al. (eds.), *Ferroelectric-Gate Field Effect Transistor Memories*,
Topics in Applied Physics 131, DOI 10.1007/978-94-024-0841-6_6

extensively examined. Takagi et al. proposed three scattering mechanisms, namely Coulomb (impurity) scattering, phonon scattering, and surface-roughness scattering. Coulomb scattering was shown to decrease in the presence of strong electric fields, whereas surface scattering degraded the carrier mobility in stronger electric fields [1–3]. These principles have been applied to other material systems, including the high-k materials of the hafnium family and composite semiconductors such as GaAs [4–7]. Field-effect control at heterojunctions with compound semiconductors enables modulation of two-dimensional electron gase (2DEG) and has contributed to the development of high-electron-mobility transistors (HEMTs) [8]. For instance, in the HEMT $Al_xGa_{1-x}As/GaAs$, a modulation doping system triggers the flow of electrons from $Al_xGa_{1-x}As$ to GaAs using the potential difference between the two materials. Using this modulation doping technique, high electron-mobility values exceeding 10^6 cm^2/Vs were reported to be compatible with an interface carrier concentration of approximately 10^{12} cm^{-2}.

The use of a spontaneous (piezoelectric) polarization charge has also been proposed as a new carrier-doping method in $Al_xGa_{1-x}N/GaN$ HEMT devices. The polarization discontinuity in $Al_xGa_{1-x}N/GaN$ results in carrier generation of 10^{13} cm^{-2}. Polarization-mediated 2DEGs obtained using oxide heterostructures with ZnO-related systems also achieved high electron mobility values of approximately 10^6 cm^2/Vs, which are close to the values obtained for other HEMT systems [9, 10]. For these systems, realization of precise control of the interface electrical properties required an understanding of the relationship between the carrier concentrations in the 2DEGs and the differences in the polarization charges at the heterointerfaces.

Recently, heterointerfaces with oxide materials have attracted significant interest due to their novel capabilities. Developments in the area of thin-film technologies, such as molecular-beam epitaxy, enable the fabrication of epitaxial films with atomic-level control and provide insight into emerging properties such as metallic conduction at insulator interfaces [11], tunable spin-orbit interactions through external electric fields [12], and the formation of low-dimensional electronic structures [13–15]. These properties can be controlled via carrier doping based on chemical composition, carrier modulation based on the field effect, and strains based on lattice mismatches with substrates. Gate-stacking systems with ionic liquids, organic materials, correlated oxides, and ferroelectrics have also attracted some interest. Charge injection and transfer at inorganic-organic interfaces [16, 17], as well as control of the Mott transition [17, 18] and superconducting transition temperatures using electric double-layer transistors (EDLTs) [19], have been reported. EDLTs, which are field-effect devices with ionic-liquid gates, have been commonly used in studies involving interfacial physical properties and have the advantage of realizing high carrier doping of up to 10^{15} cm^{-2}. Ferroelectrics are also attractive materials in this field because they have large and switchable spontaneous polarizations. Typical ferroelectrics such as Pb(Zr, Ti)O$_3$, BaTiO$_3$, and

$BiFeO_3$ exhibit polarization charges that range from 10 to 100 $\mu C\ cm^{-2}$ and correspond to the surface charges of the EDLTs. These polarizations can be controlled at relatively low electric fields below 100 $kV\ cm^{-1}$ [20–26]. Thus, these ferroelectric properties enable the development of electronic devices using ferroelectric polarization-mediated interfaces. Field-effect transistors (FETs) with ferroelectric gate stacks are believed to be among the next generation of nonvolatile memories.

6.1.2 Ferroelectric Field-Effect

6.1.2.1 Ferroelectric-Gate Field-Effect Transistors

Ferroelectric-gate FETs (FeFETs) have been studied in practical nonvolatile memory applications because of their nondestructive readout, high-speed operation, and low power consumption [27–34]. FeFETs operate by performing carrier modulation in channel semiconductors using polarization switching. Many challenges were faced in order to enable FeFETs to obtain high-quality ferroelectric gate layers on Si and other conventional semiconductors. Oxidation and interdiffusion at the heterointerfaces degraded the ferroelectricity, and initial devices exhibited poor retention and endurance. To overcome these problems, Sakai et al. proposed the use of a high-k dielectric buffer layer [27]. They succeeded in obtaining a retention time of more than 10^5 s. Kato et al. proposed metal-ferroelectric-semiconductor (MFS) structures with highly crystalline ferroelectrics by using epitaxial oxide heterostructures [28]. FeTFTs with ZnO/Pb(Zr,Ti)O_3/SrRuO$_3$ heterostructures achieved large on/off ratios on the order of 10^5 and retention times greater than 10^5 s.

To realize the practical application of FeTFTs as nonvolatile memory devices, a low-temperature fabrication process using organic ferroelectrics [29], a complete solution process [30, 31], the use of a sub-100-nm channel-length FeTFT [32], and the demonstration of NOR and NAND memory cells using FeTFTs [33] and FeFETs [34] have all been achieved. The devices prepared using these methods were successfully operated by employing polarization switching of their ferroelectric gates. However, the FeTFTs were found to have low field-effect mobility values. Kaneko et al. reported a memory cell composed of a FeTFT and an insulator (SiN$_x$)-gate FET [33]. In that study, the field-effect mobilities were 25 and 0.1 in the FET and FeTFT, respectively. The authors reported that the former value corresponded to the Hall effect values for ZnO films (26 cm^2/Vs), although the latter was degraded by interface-roughness scattering.

These results revealed two problems that are associated with FeTFTs: mobility degradation due to the higher field effect, which is caused by large polarizations, and the potential impact of incorrect assumptions on the calculated values. The field-effect mobility is typically calculated using the polarization–voltage curve, because ferroelectrics have nonlinear dielectric (capacitance) characteristics when a voltage is applied. This method assumes that ferroelectric polarization completely induces the carriers at the heterointerfaces, although the carrier concentration of the channel has a large charge mismatch with the ferroelectric polarization charge. Therefore, to understand electron transport in the channels of FeTFTs, it is necessary to conduct both a detailed investigation of the relationship between the ferroelectric polarization and the interface carriers and to develop an experimental method for evaluating FeTFT performance based on existing techniques, such as Hall-effect measurements.

6.1.2.2 Controlled-Polarization-Type FeTFTs

To reduce the operation voltage in FeTFTs, it is important to reduce the charge mismatch at the interface between the ferroelectric film and the oxide semiconductor. In this study, to reduce the charge mismatch, the use of a semiconductor with a spontaneous polarization (polar semiconductor) was proposed. FeTFTs with polar oxide semiconductors are hereinafter referred to as controlled-polarization-type (CP-type) FeTFTs. The effects of the spontaneous polarization of a polar semiconductor (P_{SPS}) on the characteristics of TFTs are schematically presented in Fig. 6.1. Figure 6.1a shows the band diagram for a heterostructure consisting of a ferroelectric film and an n-type polar oxide semiconductor with a top-gate structure. In the off-state, the carriers are depleted by spontaneous polarization of ferroelectrics (P_{SFE}). However, P_{SFE} can be compensated for by P_{SPS}, because the direction of P_{SFE} is parallel to that of the polar semiconductor. Therefore, the charge mismatch at the interface between the oxide semiconductor and the ferroelectric layer can be reduced.

Because the charge mismatch may cause difficulty in polarization reversal and consequently suppress the formation of a single polarization domain (see Sects. 6.1 and 6.2), the operation voltage can be reduced and a uniform direction can be achieved for the P_{SFE}. On the other hand, in the on-state, carriers are accumulated by the P_{SFE}, and this carrier accumulation can be enhanced because the P_{SPS} direction is toward the ferroelectric layer. Figure 1.1b shows a combination of the eight possible P_{SFE}/P_{SPS} interaction patterns when the CP-type FeTFT is in the on-state.

In this chapter, fabrication of CP-type FeTFTs using P_{SPS} is described. The results of investigations of the carrier modulation, transport, and scattering at the ferroelectric/polar semiconductor interface, and the effects of the interaction between different polarizations (P_{SPS}, P_{SFE}) and the band alignment are also discussed [35, 36]. In addition, the novelty of CP-type FeTFTs is highlighted.

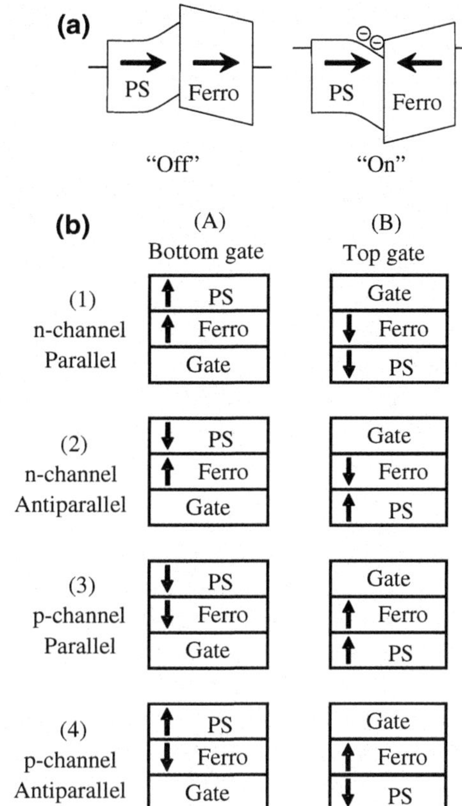

Fig. 6.1 **a** Band diagrams of heterostructure of ferroelectric film and polar semiconductor and **b** possible configurations of FeTFTs using polar semiconductor channel. Conditions are as follows: structure of TFT, (*A*) bottom-gate or (*B*) top-gate; conduction type of polar semiconductor, (1, 2) *n*, or (3, 4) *p*; direction of P_{SFE} and P_{SPS}, (1, 3) parallel or (2, 4) antiparallel. *Arrow* indicates spontaneous polarization. *PS* polar semiconductor; *Ferro* ferroelectric

6.2 Fabrication and Properties of CP-Type FeTFTs

6.2.1 Bottom-Gate-Type TFTs (1-a in Fig. 6.1)

For bottom-gate-type TFTs, $YMnO_3$ was employed as the ferroelectric film because it has been reported to have a high Curie temperature of 640 °C and a spontaneous polarization of 5.5 μC cm^{-2}, which is similar to the spontaneous polarization of ZnO [37–40]. We also previously investigated the ferroelectric properties of $YMnO_3$ thin films [41–43]. It is important to investigate the polarization interactions when the spontaneous polarizations of the ferroelectric films and polar semiconductor films are similar. As the substrate, sapphire (0001) single crystals with epitaxially grown (111) Pt electrodes were used. The $YMnO_3$ and ZnO films were deposited via pulsed laser deposition (PLD) at thicknesses of 100 and 30 nm, respectively. Details of the deposition methods are described elsewhere [41–45]. The epitaxially grown ZnO (0001)/$YMnO_3$(0001) heterostructures were used to fabricate TFTs via photolithography. The capacitance–voltage (*C–V*) characteristics of the ferroelectric layers were determined using an inductance–capacitance–resistance (LCR) meter

(HP4284A), while the polarization–electric-field (P–E) characteristics were measured using a Sawyer-Tower circuit. The drain-current–drain-voltage (I_D–V_D) and drain-current–gate-voltage (I_D–V_G) characteristics of the TFTs were determined using a picoammeter (HP4140B) and a digital oscilloscope (HP54603B), repectively. In addition, the dynamic I_D–V_G characteristics of the TFTs were determined using a function generator (NF WF1943) in order to evaluate their behaviors under the applied pulse and AC voltages.

The electrical characteristics of the bottom-gate-type TFTs with ZnO/YMnO$_3$/Pt/sapphire structures are presented in Fig. 6.2 for the bottom-gate-type FET for which the YMnO$_3$ thin film was deposited under optimized conditions. In addition to the change in the butterfly-shaped C–V curve, a difference between the capacitances for negative and positive bias voltages is observed, which indicates carrier modulation in the ZnO channel. The effects of the carrier density in the ZnO layer on the carrier modulation in the channel and the ferroelectric polarization switching were investigated (not shown). In this polarization configuration (1-A), the ferroelectric polarization switching occurred more readily if the channel included a slightly higher carrier concentration. Therefore, the ZnO was doped with 2 wt% Al (used for the results shown in Fig. 6.2) in order to increase the carrier density to 1.0×10^{13} cm^{-2} layer, which is less than the controllable sheet carrier concentration due to saturated polarization in YMnO$_3$ (8 µC cm^{-2}) of 4.8×10^{13} cm^{-2}. To verify carrier accumulation and depletion, quantitative analysis was then performed. The relative permittivity of YMnO$_3$ at an electric field of 800 kV cm^{-1} is 27. From the thickness of the YMnO$_3$ thin films (100 nm) and the capacitance in the saturated region under positive bias voltage, the effective electrode area was calculated to be 8.0×10^{-4} cm^2. In contrast, the actual areas of the drain electrode, source electrode, and channel region were 2.95×10^{-4} cm^2, 2.95×10^{-4} cm^2, and 1.0×10^{-5} cm^2, respectively. The effective electrode area was thus larger than the total of these three areas (6.0×10^{-4} cm^2) and greater than the total area of the channel region (1.0×10^{-5} cm^2) and the gate-drain overlapped region (2.95×10^{-4} cm^2), the latter of which was less than half of the effective electrode area. These results indicate that carriers accumulated in the ZnO channel at a positive bias voltage and that the drain and source electrodes were electrically connected. Moreover, carriers accumulated in the ZnO layer outside the drain and source electrodes. However, the decrease in capacitance upon the application of a negative bias voltage suggests that the drain and source electrodes were disconnected due to the formation of a depletion layer.

The I_D–V_D characteristics for the same FET are shown in Fig. 6.2b. I_D increases with increasing V_G and V_D and is saturated when V_D is greater than the applied V_G, indicating that the bottom-gate-type FET operated as an n-channel transistor. The I_D–V_G characteristics and gate leakage current are shown in Fig. 6.2c. The gate leakage current is smaller than I_D, and counterclockwise hysteresis loops and an abrupt change in I_D when the gate voltage is 0 V are observed. The memory window width of the I_D–V_G characteristics is 0.5 V, which is nearly equal to the memory window width of the C–V characteristics. In addition, the memory window width as a function of the voltage sweep is nearly saturated at a voltage sweep of 6 V. Therefore, this memory operation originated from ferroelectric polarization

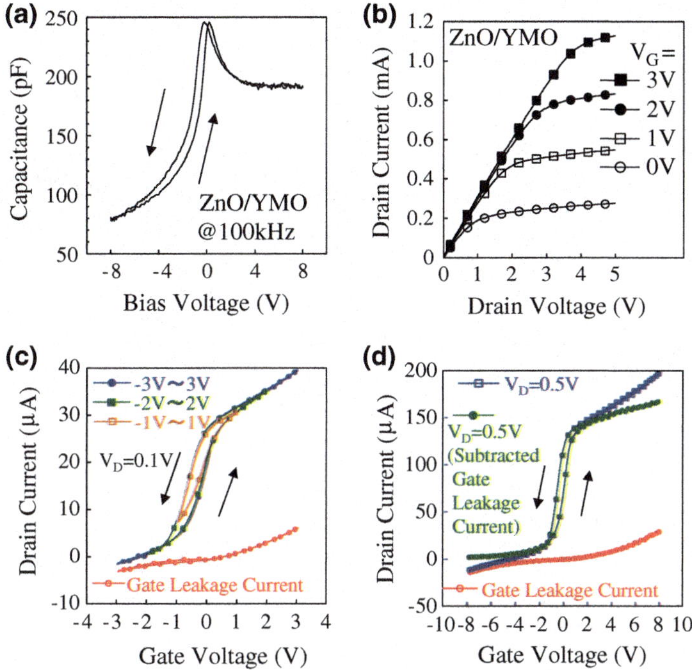

Fig. 6.2 **a** C–V, **b** I_D–V_D, and **c** I_D–V_G characteristics of FET with ZnO:2%Al/YMnO$_3$ structure grown under optimized conditions. **d** I_D–V_G characteristics after subtracting gate leakage current. Channel length and width were 10 and 100 μm, respectively

switching of the YMnO$_3$ film, and other charge effects, such as space charge rearrangement and charge injection, were negligible. Although the YMnO$_3$ thin film has a relatively large leakage current (1.0×10^{-2} Acm^{-2} at 3 V), carrier modulation via ferroelectric polarization is verified by this figure.

Next, the field-effect mobility and accumulated charge density in the channel region of the FET were calculated in order to investigate the P_{SPS} effect in detail. Given the electric field dependence of the capacitances of ferroelectric films, the I_D–V_G characteristics were determined at small V_D and large V_G, because the electric field between the gate and channel regions became nonuniform when a large V_D was applied. Under this condition, the relationship between I_D and V_G can expressed as:

$$\frac{\partial I_D}{\partial V_G} = \frac{Z\mu_n \varepsilon_0 \varepsilon_F}{Ld} V_D, \tag{6.1}$$

where μ_n represents the field-effect mobility, Z is the channel width, L is the channel length, d is the thickness of the ferroelectric thin film, ε_0 is the vacuum permittivity, and ε_F is the relative dielectric permittivity of the ferroelectric thin film [46]. The Z and L values for this FET were 100 μm and 10 μm, respectively. The I_D–V_G

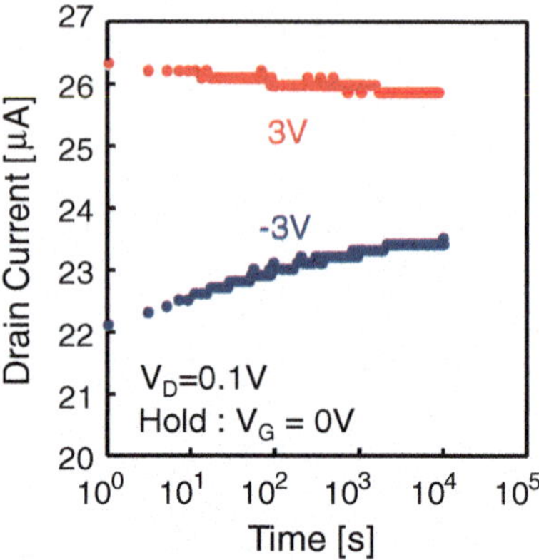

Fig. 6.3 Retention property of FET used in Fig. 6.2. Measurements were conducted at 0 V after applying bias voltage of ±3 V

characteristics were determined at $V_D = 0.5$ V and V_G values ranging from −8 and 8 V in order to calculate the field effect mobility from the saturated region ($V_G = 6$–8 V), which was found to be 3.3 cm^2/Vs and is comparable to that of other oxide semiconductor-channel TFTs [47, 48]. The accumulated charge density (Q) was then calculated using:

$$I_D = ZQ\mu_n E, \tag{6.2}$$

where E is the electric field applied between the source and drain electrodes. To eliminate the effect of the leakage current, the gate leakage characteristics were subtracted from the actual I_D–V_G characteristics. The accumulated charge density of the bottom-gate-type FET was 10.2 µC cm^{-2}, which is larger than the saturated polarization of YMnO$_3$ films (8 µC cm^{-2}). Because the effects of space-charge rearrangement and charge injection were not observed, these results suggest that the increase in accumulated charge density was caused by the effect of the P_{SPS} of the ZnO layer. The difference between the accumulated charge density and the saturated polarization of the YMnO$_3$ film indicates that the direction of P_{SPS} for the ZnO layer was aligned partially downward and suggests that this bottom gate FET did not have the ideal 1-A type polarization structure shown in Fig. 6.1. The retention properties of the FET are depicted in Fig. 6.3. While the retention property in the off-state appears to be poorer than that in the on-state, it remains within 10 % of its initial value, even after 10^4 s, likely due to the partial domains with downward directions. Although further investigation of the direction of the P_{SPS} of the ZnO film is required, it can be concluded that the observed memory operation was caused by the ferroelectricity of the YMnO$_3$ thin film and that the effect of the P_{SPS} of the ZnO layer was observed.

6.2.2 Impedance Analysis of Channel Conduction Underneath the Bottom-Ferroelectric-Gate Using an RC Lumped Constant Circuit

The nonvolatile memory operation of a CP-type FeTFT with a ZnO/YMnO$_3$ structure was discussed in Sect. 6.2.1 through evaluation of its I_D–V_G and C–V characteristics. Notably, the on/off ratios at a gate voltage of 0 V and the electron field-effect mobility were determined from the I_D–V_G characteristics. However, the obtained on/off ratios were smaller than those expected considering $P_{SFE,}$ and the obtained field effect mobility was one order of magnitude smaller than the Hall effect mobility. The nonuniformity of the channel conductance caused by P_{SFE} may be a possible origin of these differences, because ferroelectrics, unlike paraelectrics, have geometrical polarization domain structures that originate from their insufficient poled domain structures (NOT fully polarized state).

When a semiconductor exhibits n-type conduction, the carriers in the channel accumulate only in the domains with upward P_{SFE}. However, uniform channel conductance was assumed in the previous investigations. Therefore, a study of the direction of P_{SFE} was important to analyze the characteristics of the FeTFT, because its channel conductance changed dramatically at the ferroelectric domain wall. In addition, because the CP-type FeTFT had a low charge mismatch, it was expected that P_{SFE} was readily switched by applying a gate voltage. Unfortunately, it was difficult to determine the direction of P_{SFE} below the channel region of the FeTFT from the I_D–V_G and C–V characteristics because they are based on static quantities. The electrical structures of inhomogeneous materials or heterostructures are often investigated using impedance spectroscopy [49, 50]. Analysis of the impedance spectrum using an equivalent circuit provides information on the electrical components of a sample. Therefore, to investigate the change in the channel conductance of the CP-type FeTFT, impedance analysis was performed. Impedance spectra were obtained using an RC lumped constant circuit, which was used as an equivalent for a conventional FET circuit in the high frequency region. However, RC lumped constant circuits are not typically used in FeTFTs because the conductance of the channel is significantly affected by P_{SFE}. Therefore, it was first necessary to investigate the suitability of using the RC lumped constant circuit for the CP-type FeTFT. From the above discussion, various conditions for the channel conductance were assumed, and the impedance between the source electrode and gate electrode for each condition was calculated using a Simulation Program with Integrated Circuit Emphasis (SPICE). The calculated impedance spectra were then compared to the spectra obtained experimentally under the different P_{SFE} directions with an applied gate voltage.

The thicknesses of the YMnO$_3$ and ZnO:Al 0.001 % (4.8×10^{10} cm^{-2}) films used in this study were 100 and 20 nm, respectively. A FET with a bottom-gate structure was fabricated via photolithography. The length and width of the channel of the FET were 10 and 50 µm, respectively. The carrier concentration of the ZnO channel was determined by performing Hall effect measurements at room temperature (TOYO Corporation ResiTest8300). The operational characteristics of the FET were investigated by determining the I_D–V_G characteristics using a picoammeter (Hewlett Packard 4140B). The impedance spectra of the CP-type FeTFT were obtained using an LCR meter (Hewlett Packard 4284A).

In the high-frequency region, a paraelectric-gate FET is described by many resistances and capacitances, as shown in Fig. 6.4a [51]. This circuit is called an RC lumped constant circuit. When the impedance is measured between the source and gate electrodes for a paraelectric-gate FET, the circuit equivalent to that in Fig. 6.4a is shown in Fig. 6.4b. R, C, and L are the resistance, capacitance, and inductance, respectively. The subscripts S, D, GS, GD, and CH represent the channel regions near the source electrode and the drain electrode, between the source and gate electrodes, between the drain and gate electrodes, and the entire channel region, respectively. Thus, the total of C_S and C_D is the capacitance of the entire channel region. In the present study, the RC lumped constant circuit was applied to the FeTFT, for which the channel conductance depended on the direction of P_{SFE}, and the C_S/C_D ratio depended on the channel conductance. Therefore, it was expected that the direction of P_{SFE} would be related to the C_S/C_D ratio. However, unlike in a paraelectric-gate FET, the channel conductance in a FeTFT undergoes a steep change due to P_{SFE}. Therefore, it was necessary to confirm whether the circuit in Fig. 6.4b could be adopted for analysis of the channel of the FeTFT.

As mentioned above, various channel conditions (Fig. 6.5a–e) were assumed for the RC lumped constant circuit shown in Fig. 6.4a, and the impedance spectra were calculated using Simulation Program with Integrated Circuit Emphasis (SPICE). The calculated impedance spectra were then fitted to the experimentally obtained spectra for the circuit shown in Fig. 6.4b. In Fig. 6.5a, a channel with low resistivity is assumed, which corresponds to all upward P_{SFE} directions, and it can be seen in Fig. 6.5a′ that this model agrees well with the experimental data for the circuit shown in Fig. 6.4b. The impedance spectra calculated under other channel conditions are shown in Fig. 6.5b′–e′. In Fig. 6.5b, the channel has a high resistivity, which corresponds to the depletion condition formed by all downward P_{SFE} directions. In Fig. 6.5c, the region near the source electrode has low resistivity, while the other regions have high resistivities. Thus, it was assumed that the gate voltage increased from negative to positive. Figure 6.5d shows the opposite condition to that in Fig. 6.5c. In Fig. 6.5e, the channel resistivity gradually changes, which corresponds to the depletion condition of a paraelectric-gate FET. All of the calculated results are in good agreement with the experimental data for the circuit shown in Fig. 6.4b, as can be seen in Fig. 6.5b′–e′. These results indicate that the circuit shown in Fig. 6.4b can be adopted for analysis of the channel in a FeTFT, including detection of steep changes in the channel resistivity such as those shown in Fig. 6.5c, d due to different P_{SFE} directions.

Thus, to investigate the P_{SFE} direction in the CP-type FeTFT with a ZnO/YMnO$_3$ bottom-gate structure, the relationship between the C_S/C_D ratio and the channel condition was investigated. The C_S and C_D values obtained from the fitting analysis are summarized in Fig. 6.6. The C_S/C_D ratio for Fig. 6.6a is 3.9. Thus, C_S is larger than C_D when all of the P_{SFE} directions are upward. The C_S/C_D

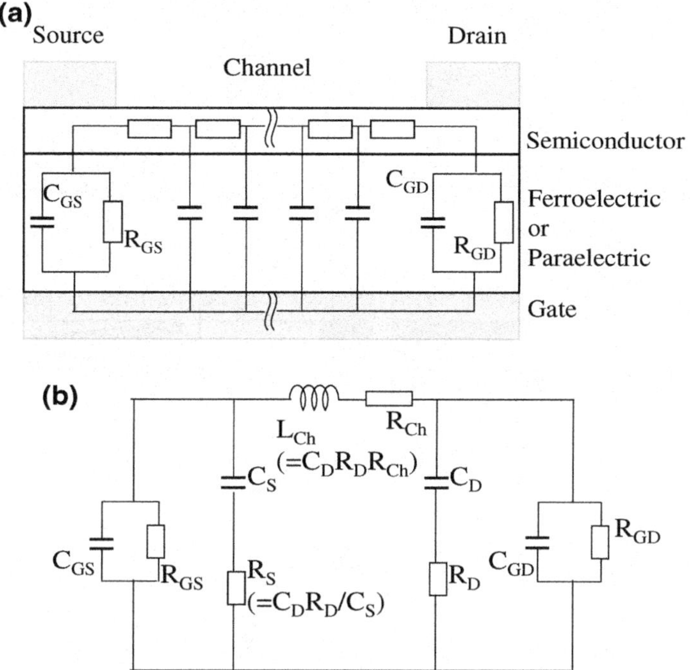

Fig. 6.4 **a** RC lumped constant circuit and **b** equivalent circuit

ratios for Fig. 6.6b, c, e are all nearly 1. Thus, each of these C_S values is nearly equal to the corresponding C_D value when the direction of the P_{SFE} is downward throughout the entire channel region and when it is upward only near the source electrode. On the other hand, the C_S/C_D ratio for Fig. 6.6d is 0.09; thus, C_S is smaller than C_D when the P_{SFE} direction is downward near the source electrode.

The results shown in Fig. 6.6 indicate that the direction of the P_{SFE} below the channel region in a FET can be investigated using C_S/C_D ratios obtained via analysis of the calculated impedance spectra. Figure 6.7 show the impedance spectra for the CP-type FeTFT that were obtained between the source and gate electrodes at gate voltages ranging from -4 to -1 V, (a), and from 1 to 4 V, (b). These experimental spectra were fitted using the circuit shown in Fig. 6.4b, and the results are indicated in Fig. 6.7a, b by the solid lines. The obtained C_S and C_D values are shown in Fig. 6.7c. C_S is two orders of magnitude larger than C_D at positive gate voltages. Given the calculated results shown in Fig. 6.6a, this data suggests that all of the P_{SFE} directions were upward at positive gate voltages. When the gate voltage decreases, however, C_S decreases and C_D increases to approximately -1 V. This result indicates that the P_{SFE} direction was downward only near the source electrode when the gate voltage is a small negative value. Moreover, when the gate voltage decreases to below -1 V, C_S and C_D are nearly the same, indicating that the channel conditions shown in Fig. 6.5b, c, e were likely to occur

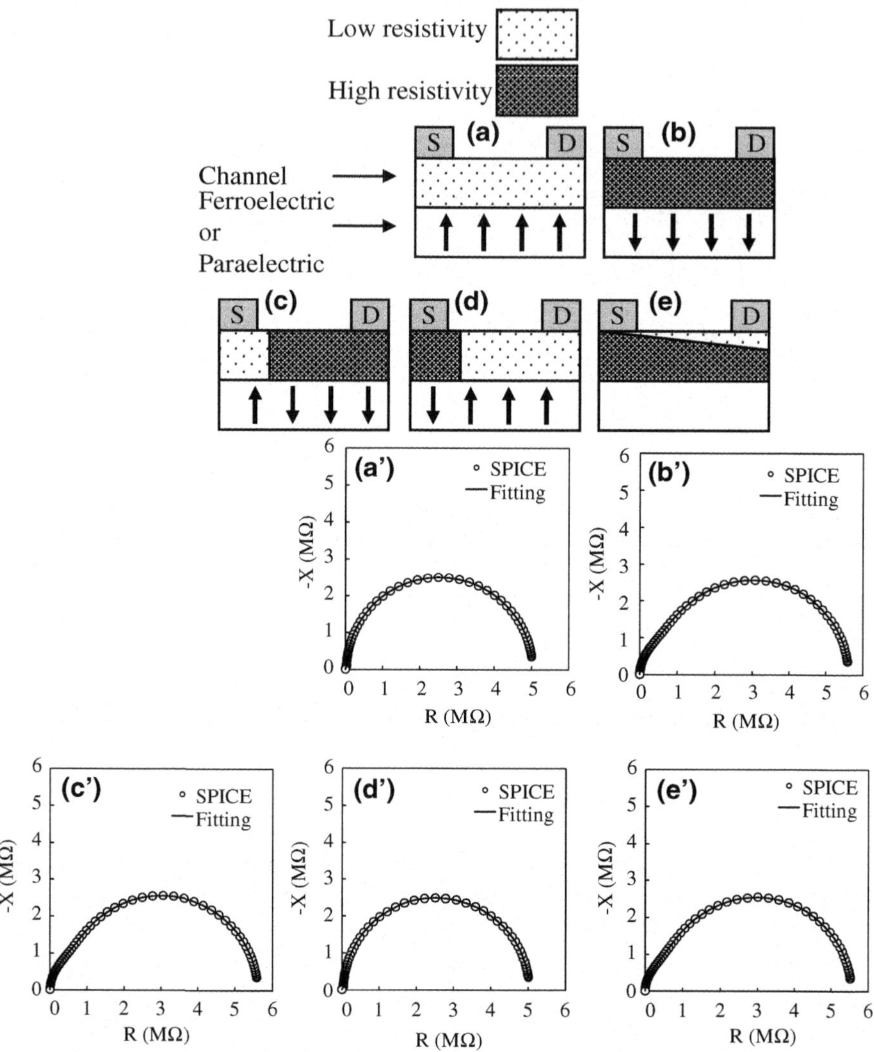

Fig. 6.5 Comparison of calculated impedance spectra for RC lumped constant circuit and experimental data for equivalent circuit for different assumed channel conditions: **a** all upward P_{SFE} directions, **b** all downward P_{SFE} directions, **c** downward P_{SFE} only near source electrode, **d** upward P_{SFE} only near source electrode, and **e** gradual change in P_{SFE} direction. Calculated and fitted results are shown in **a'–e'**

at larger negative gate voltages. However, it is not possible for an upward P_{SFE} to switch to a downward direction near the source electrode due to an increase in the gate voltage; therefore, the channel condition shown in Fig. 6.5c can be discounted. Moreover, the channel condition shown in Fig. 6.5e is also unlikely because this

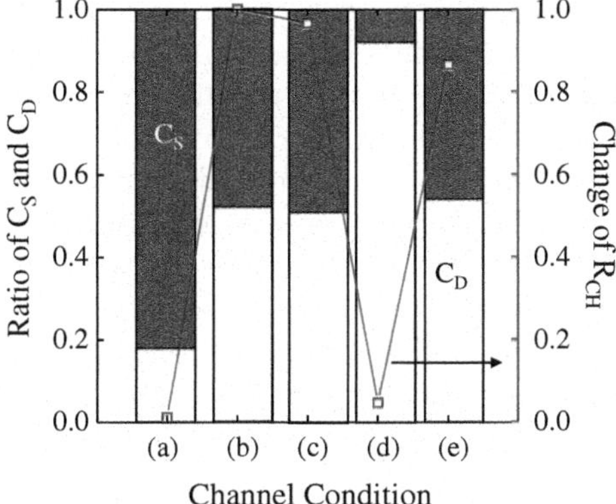

Fig. 6.6 Relationship between C_S/C_D ratio and channel condition. **a–c** Correspond to the states of channel shown in Fig. 6.5a–c

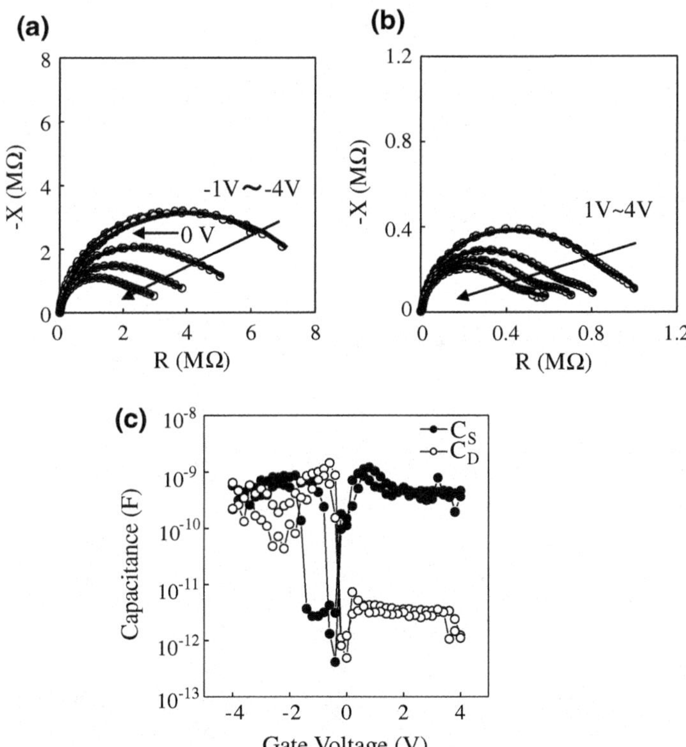

Fig. 6.7 Impedance spectra and fitting results under the bias voltage **a** from −4 to −1 and **b** from 1 to 4 and **c** the gate voltage dependences of C_S and C_D

condition corresponds to paraelectric-gate TFTs. Therefore, the results suggest that the directions of all of the P_{SFE} directions were downward below the channel region (Fig. 6.5b) of the CP-type FeTFT at large negative gate voltages.

Based on the above analyses, it can be concluded that the direction of P_{SFE} for the CP-type FeTFT changed through the process shown in Fig. 6.5a–e, depending on the applied gate voltage and starting with all upward P_{SFE} directions in the channel region when no gate voltage was applied. Moreover, the I_D–V_G characteristics shown in Fig. 6.2c, d indicate that the FET normally exhibited on-type operation. Because the accumulation state was formed by upward P_{SFE} directions, the normally on-type operation also suggests that the P_{SFE} directions were upward at 0 V and indicates that a driving force to switch P_{SFE} to the upward direction existed. It appears that the origin of the driving force was the effect of P_{SPS}, the charge mismatch of the ZnO/YMnO$_3$ interface, or the imprint of the YMnO$_3$ thin film. Moreover, the results presented above indicate that the nonvolatile memory operation was dominated by P_{SFE} near the source electrode. The decrease in the on/off ratio and field-effect mobility are thus likely caused by the nonuniformity of the channel conductance.

To investigate the effect of the nonuniform ferroelectric domain structure on the carrier transport in the channel region, the temperature dependence of the conduction properties of YbMnO$_3$/ZnO heterostructures were carefully examined using Hall-effect measurements and analyzed using various scattering effects, including polar optical phonon scattering, acoustic phonon scattering through deformation and piezoelectric potentials, ionized impurity scattering, and grain boundary scattering. The analyses revealed that the predominant scattering mechanism changes from grain boundary scattering to ionized impurity scattering upon insertion of a partially polarized YbMnO$_3$ film [52].

6.2.3 Top-Gate-Type TFTs (2-B in Fig. 6.1)

In Sects. 6.2.1 and 6.2.2, the carrier transport and FET operation were analyzed using bottom-gate-type TFTs. As shown in Fig. 1.1a, because the available interfacial charge in the channel during the on-state equals P_{SFE}–P_{SPS} if it has an ideal polarization structure, it should be quite small. In the off-state, on the other hand, the polarization in the ferroelectric layer should become unstable, because it stays in a tail-to-tail state. However, in a (2)-B-type structure, the situations are completely opposite. The available interfacial charge in the channel during the on-state equals $P_{SFE} + P_{SPS}$, which increases the on-current and field effect mobility.

To determine the effects of the ferroelectric domain structure and P_{SPS} on the carrier transport properties in the channel of a ferroelectric gate thin-film transistor, the electrical properties of the ferroelectric gate stack structure consisting of an organic ferroelectric copolymer of vinylidenefluoride and tetrafluoroethylene [P(VDF-TeFE)] and ZnO were investigated. In this study, neither YMnO$_3$ nor YbMnO$_3$ ferroelectrics were used due to their high deposition temperature (approximately 700 °C), which leads to interdiffusion of Y and Mn ions into the ZnO channel. Therefore, organic ferroelectrics, which are typically deposited at low temperatures (<150 °C), were used without any special preparation processes [53–55]. In particular, P(VDF-TeFE) has a remnant polarization (4.0 μC cm^{-2}) comparable to the spontaneous polarization of ZnO [56].

CP-type FeTFTs were constructed by first depositing 20- to 500-nm-thick films of the oxide polar semiconductor ZnO onto (111) yttria-stabilized zirconia (YSZ) substrates via pulsed laser deposition at a substrate temperature of 650 °C and an oxygen pressure of 3 × 10^{-5} Torr. The ZnO films were then annealed in air at 1050 °C to relax the lattice misfit strain originating from the large lattice mismatch ($\sim$10 %) between ZnO and YSZ. All of the ZnO surfaces had step-and-terrace structures. A Au/Ti source and drain electrodes were then formed using the liftoff process and the electron beam (EB) evaporation method. The samples were processed via photolithography into a Hall bar shape with a channel width (W) and length (L) of 100 and 320 μm, respectively. P(VDF-TeFE) with an 80:20 VDF:TeFE molar ratio was dissolved in methyl ethyl ketone at a solution concentration of 7 wt%. This solution was spin-coated on the ZnO Hall bar structures, and the resulting films were annealed in air at 145 °C for 30 min in order to achieve crystallization. The thickness of the obtained P(VDF-TeFE) layers was approximately 400 nm. Finally, Au films were deposited as gate electrodes on the P(VDF-TeFE)/ZnO heterostructures via the liftoff process using the EB evaporation method. A schematic of the sample and a micrograph of the TFT structure with the Hall measurement system are shown in Fig. 6.8. The details of the fabrication and electrical characteristics of the P(VDF-TeFE)/ZnO heterostructures are described elsewhere [57]. The poling voltage dependence of the carrier transport properties was evaluated using a Hall effect measurement system (ResiTest 8300, Toyo Technica). Analysis of the 100-nm-thick channel FeTFT after positive and negative poling was performed in the temperature range from 80 K to 300 K.

Figure 6.9 shows the P–V characteristics of a Au/P(VDF-TeFE)/AuTi capacitor fabricated with the same P(VDF-TeFE)/ZnO/YSZ heterostructure (400-nm-thick P(VDF-TeFE) film) obtained using a Sawyer-Tower circuit. A rectangular hysteresis loop due to the ferroelectricity of the P(VDF-TeFE) is observed. Although quite a high voltage was required to saturate the polarization due to the thickness of the P(VDF-TeFE) layer, the values of the remnant and saturated polarizations and the coercive voltage of the P(VDF-TeFE) film were 2.7 μC cm^{-2}, 3.0 μC cm^{-2}, and 30 V, respectively.

Figure 6.10a shows the C–V characteristics of the P(VDF-TeFE)/ZnO Hall bar structures with various channel thicknesses. The C–V characteristics at channel thicknesses of 500 and 370 nm exhibit butterfly-shaped counterclockwise hysteresis curves due to polarization switching of the P(VDF-TeFE) and slightly smaller capacitance values at negative bias voltages than at positive bias voltages, which are attributed to the depletion of the ZnO channel. The calculated depletion widths were 50 and 80 nm for the samples with channel thicknesses of 500 and 370 nm, respectively. On the other hand, samples with channel thicknesses of 100 and 20 nm exhibit C–V curves typical for metal-ferroelectric-semiconductor

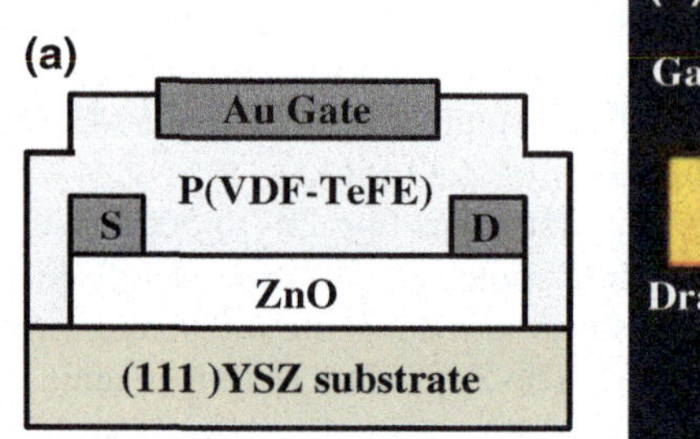
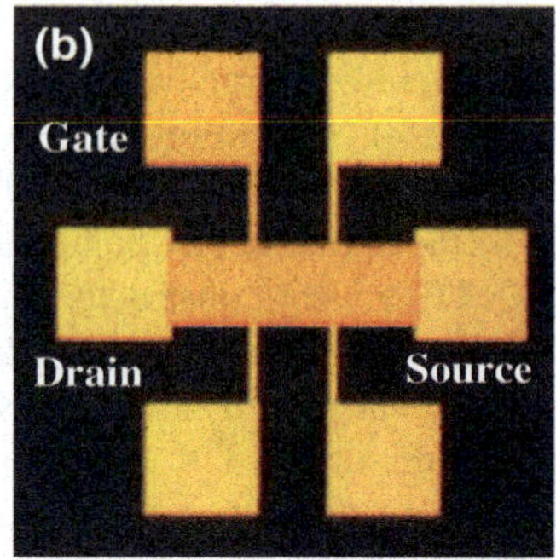

Fig. 6.8 **a** Schematic of sample and **b** micrograph of TFTstructure with Hall measurement system

Fig. 6.9 *P–V* characteristics of Au/P(VDF-TeFE)/AuTi capacitor with various bias voltages

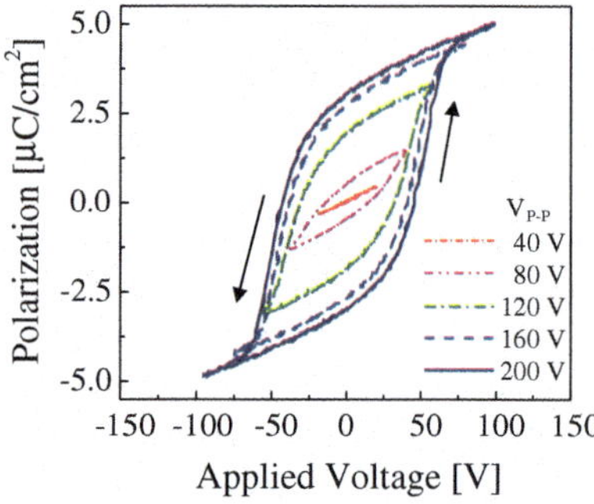

structures. The counterclockwise hysteresis loops indicate the ferroelectricity of the P(VDF-TeFE) and the complete depletion of the ZnO channel.

The I_D–V_G characteristics of the P(VDF-TeFE)/ZnO Hall bar structures with various channel thicknesses are shown in Fig. 6.10b. Interestingly, the I_D values for the samples with channel thicknesses of 500 and 370 nm remain nearly constant, while for the samples with channel thicknesses of 100 and 20 nm, each I_D value changes significantly with V_G, which is consistent with complete depletion of the channel, as described above. The memory window at a drain current of approximately 2 μA for the 20-nm-thick channel FeTFT was 57 V, which was twice the value of the coercive voltage for the P(VDF-TeFE) capacitor. This result indicates that the counterclockwise hysteresis loop originated from the ferroelectric nature of the P(VDF-TeFE) layer. The smaller memory window in the 100-nm-channel FeTFT also indicates the existence of a trapping state. The on/off ratio for the I_D at a gate voltage of 0 V reached the order of 10^5, which is much larger than that obtained in the bottom-gate FET described in Sect. 6.2.1. The change in the field-effect mobility as a function of the ZnO thickness calculated from the I_D–V_G curve is shown in Fig. 6.11. The values of $\partial V_G / \partial Q_i$ and $\partial I_D / \partial V_G$ were estimated from the saturated regions of the *P–V* and I_D–V_G characteristics shown in Figs. 6.9 and 6.10b, respectively. Although the field-effect mobility decreases with

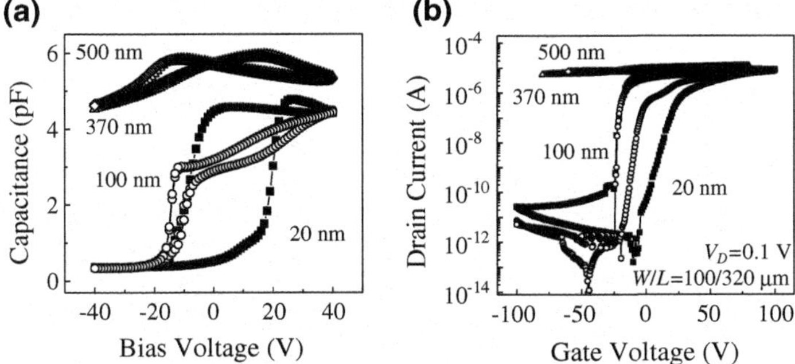

Fig. 6.10 a C–V and b I_D–V_G characteristics of P(VDF-TeFE)/ZnO/YSZ heterostructures with channel thicknesses ranging from 20 to 500 nm

decreasing channel thickness, these values are also much higher than that obtained for the bottom-gate FET.

To determine the channel conductions of the FeTFTs in detail, Hall-effect measurements for the ZnO films with and without the P(VDF-TeFE) layer were performed. Figure 6.12a shows the carrier concentrations as a function of the channel thickness. For P(VDF-TeFE)/ZnO, the analysis was performed at $V_G = 0$ V after positive poling treatment so that the channels were set to the ON state. Although the carriers in the P(VDF-TeFE)/ZnO should accumulate in the positively polarized ferroelectric layer, the carrier concentration of the P(VDF-TeFE)/ZnO structure is smaller than that of ZnO. This result may be due to the existence of an upward polarization domain in the P(VDF-TeFE) layer. When the P(VDF-TeFE) layer is not a single domain, formation of a depleted region in the ZnO channel under the upward domain can be expected, and this reduction of the conductive region leads to the calculated carrier concentration value being lower than the actual value. It is also possible that surface adsorption affected the carrier concentration, because the ZnO films without P(VDF-TeFE) layers did not receive any passivation treatment. Several reports have indicated that adsorbed hydrogen acts as a donor and enlarges the conductivity of ZnO [58–60].

Fig. 6.11 Channel thickness dependence of field-effect mobility of P(VDF-TeFE)/ZnO/YSZ heterostructures

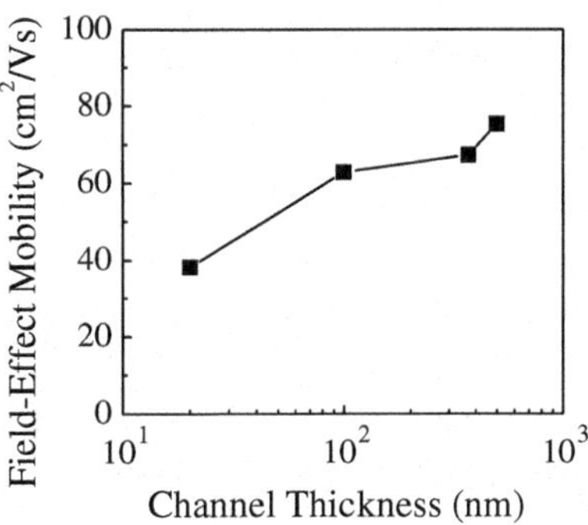

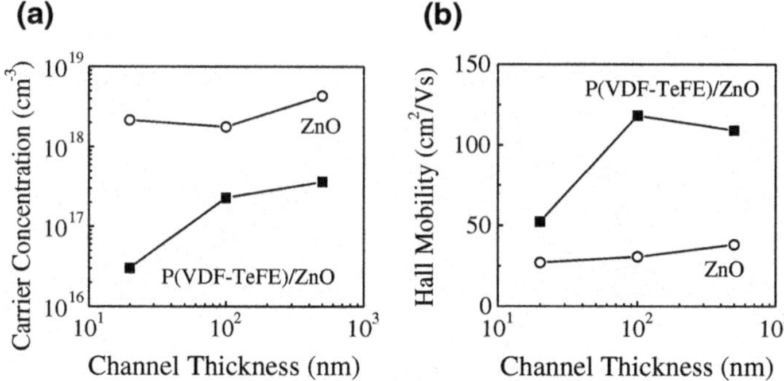

Fig. 6.12 Channel thickness dependences of **a** carrier concentrations and **b** Hall mobilities of ZnO films (*open circles*) and P(VDF-TeFE)/ZnO/YSZ heterostructures (*filled squares*)

The Hall mobility of the ZnO films with and without P(VDF-TeFE) layers are shown in Fig. 6.12b. While the Hall mobilitites of the ZnO films without P(VDF-TeFE) layers increase slightly with increasing film thickness, those of the ZnO films with P(VDF-TeFE) layers increase drastically. These results suggest that the field effect due to ferroelectric polarization enhanced the mobility. When the carriers were accumulated due to ferroelectric polarization, they were transported near the P(VDF-TeFE)/ZnO interface. Reduction of the influence of misfit dislocations at the ZnO/YSZ interface on carrier transport is a possible reason for the enlargement of the Hall mobility. It is known that a degenerate layer exists at the ZnO/YSZ interface, even after annealing [61]. In this degenerate layer, the Hall mobility is lower than that in the bulk region due to the high defect density [62, 63]. Because the conductivity near the interface between the P(VDF-TeFE) and ZnO was increased via carrier induction due to ferroelectric polarization, the influence of the degenerate layer was reduced. The higher Hall mobility also ensured an interface with reduced thermal interdiffusion between the P(VDF-TeFE) and ZnO.

Notably, the Hall mobility of P(VDF-TeFE)/ZnO is higher than the field-effect mobility shown in Fig. 6.11. Several factors may contribute to this difference. First is the effect of V_G. It is known that the field-effect mobility is smaller than the effective mobility due to the influence of the gate voltage [64]. While the Hall-effect measurement was performed at $V_G = 0$ V, the field-effect mobility was calculated from the saturation regions of the I_D–V_G curve for $V_G = 40$ V. Second is the Hall factor, which is the proportional constant for the Hall and effective mobilities and can be written as [65]

$$r = \frac{\langle \tau^2 \rangle}{\langle \tau \rangle^2} = \frac{\int_0^\infty \tau^2(E) \cdot E^{3/2}(\partial f_0/\partial E)dE / \int_0^\infty E^{3/2}(\partial f_0/\partial E)dE}{\left[\int_0^\infty \tau(E) \cdot E^{3/2}(\partial f_0/\partial E)dE / \int_0^\infty E^{3/2}(\partial f_0/\partial E)dE\right]^2},$$

where τ is the relaxation time of the conduction electrons and f_0 is the Fermi-Dirac distribution function. Although the Hall factor changes due to the energy dependence of the relaxation time, it is generally greater than 1. Therefore, the Hall mobility tends to be overestimated compared to the field-effect mobility. However, it has also been suggested that the field-effect mobility and Hall mobility both tend to increase with increasing channel thickness. Although the influence of the misfit dislocations at the ZnO/YSZ interface are stronger, the present results indicate that the field effect of the ferroelectric polarization promoted conduction close to the P(VDF-TeFE)/ZnO interface and enhanced the mobility.

The poling voltage dependence of the carrier transport properties of the P(VDF-TeFE)/ZnO heterostructures was also investigated to provide greater insight into the effects of ferroelectric polarization on carrier transport in the FeTFTs. Analyses were performed on the 100- and 500-nm-thick FeTFTs at a gate voltage of 0 V after poling treatment. Figure 6.13 shows the results for the Hall-effect measurements of the sample with a channel thickness of 500 nm. The poling voltage was varied from −40 to 40 V in 20 V steps. The carrier concentration at a channel thickness of 500 nm, which is shown in Fig. 6.13a, increases with increasing poling voltage from −40 to 40 V and exhibits counterclockwise hysteresis. This behavior indicates carrier modulation in the channel due to the ferroelectric polarization. Figure 6.13b shows the Hall mobility at a channel thickness of 500 nm. Unlike the carrier concentration, the Hall mobility remains nearly constant at approximately 108 cm^2/Vs, which is much higher than that of the ZnO film without the P(VDF-TeFE) layer (38 cm^2/Vs). This constant, high Hall mobility of the 500-nm-thick FeTFT can be explained by a change in the conduction path and reduced depletion behavior, as described above.

Figure 6.14a shows the carrier concentration at a channel thickness of 100 nm. Here again, the carrier concentration increases with increasing poling voltage, but without clear hysteretic behavior, which may be attributed to charge injection due to the existence of trapping states and the poling treatment at voltage above the coercive voltage. Unlike for the 500-nm-thick sample, the Hall mobility of the 100-nm-thick channel is modulated by the ferroelectric polarization, as shown in Fig. 6.14b. When the P(VDF-TeFE) layer is positively polarized, the Hall mobility

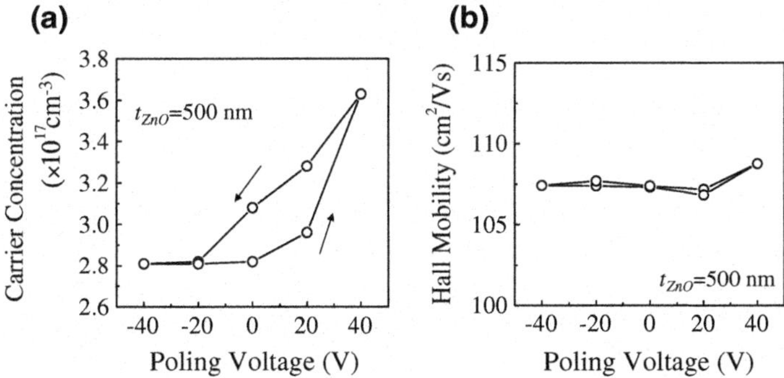

Fig. 6.13 Poling voltage dependences of **a** carrier concentration and **b** Hall mobility at channel thickness of 500 nm

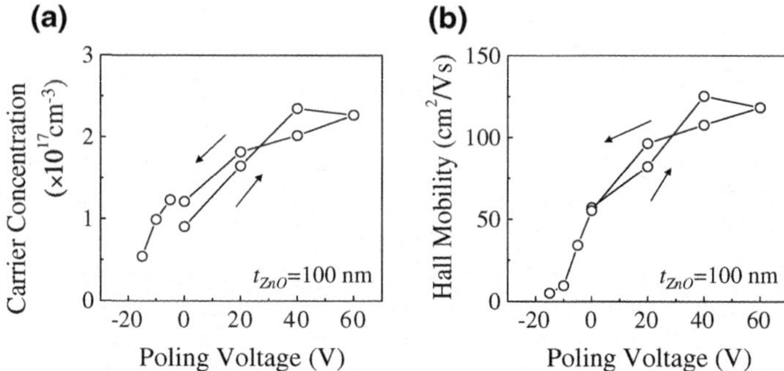

Fig. 6.14 Poling voltage dependences of **a** carrier concentration and **b** Hall mobility of P(VDF-TeFE)/ZnO heterostructure at channel thickness of 100 nm

increases to 125 cm^2/Vs, which is higher than that for the 500 nm-thick channel sample. On the other hand, after negative poling, the Hall mobility decreases to below 5 cm^2/Vs.

The sheet carrier concentration at a channel thickness of 500 nm is 2.8×10^{13} cm^{-2}, which is larger than that at a channel thickness of 100 nm (2.4×10^{12} cm^{-2}). The available sheet carrier concentration modulated by the remnant polarization of the P(VDF-TeFE) (2.7 μC cm^{-2}) was calculated to be approximately 1.7×10^{13} cm^{-2}, which is larger and smaller than the sheet carrier concentrations at channel thicknesses of 100 and 500 nm, respectively. The slight change in the Hall mobility in the 500 nm-thick channel appears to be due to a small ferroelectric polarization. However, the Hall mobility of the FeTFT with a thin channel (20 nm) was modulated by ferroelectric polarization, even when it had a high conductivity and carrier concentration. Thus, the effect of ferroelectric polarization on the mobility was large in the region close to the ferroelectric/polar semiconductor interface. In the thick-channel FeTFTs, the ferroelectric polarization also accumulated or depleted the carriers near the ferroelectric/polar semiconductor interface. However, the ferroelectric polarization had little influence over the carrier transport across the entire channel layer because the depletion width was less than 50 nm and was much smaller than the channel thickness. In contrast, the carriers in thin-channel FeTFTs transport near the ferroelectric/semiconductor interface; therefore, the influence of ferroelectric polarization on carrier transport is strong, and the Hall mobility can be modulated by poling treatment.

The poling voltage dependences of the carrier transport properties of the FeTFTs reveal that the carrier accumulation due to ferroelectric polarization can increase the mobility. Next, to investigate the changes in the carrier scattering mechanism, the temperature dependence of the carrier transport properties was characterized. In the 100-nm-thick channel FeTFT, two poling states were formed by applying positive (+50 V) and negative (−50 V) voltages, and Hall-effect measurements were performed at temperatures ranging from 80 to 300 K. When the P(VDF-TeFE) layer was positively polarized, carriers accumulated due to the ferroelectric polarization. Thus, the carrier concentration with positive poling was larger than that with negative poling over the entire temperature range. This result suggests that the two distinct poling states were maintained during the measurements. In addition, the Hall mobility increased when the P(VDF-TeFE) layer was positively polarized, particularly below 200 K. The difference between the temperature dependences of the Hall mobility in the two poling states at low temperatures may be due to the presence of ionized impurities, which can scatter carriers through their screened Coulomb potentials with associated mobilities that are proportional to two-thirds of the temperature [65–67]. Additionally, the field effect is known to reduce the scattering effect of the Coulomb potential [68, 69]. These results suggest that carriers accumulated due to screening of the influence of the Coulomb potential from the ionized impurities by the ferroelectric polarization. Thus, it can be concluded that the FeTFT mobilities increased due to reduction of ionized impurity scattering by the field effect of the ferroelectric polarization without an applied gate voltage.

6.3 Effect of Spontaneous Polarization of the Polar Semiconductor on the Electronic Structure of the Poly(Vinylidene Fluoride-Trifluoroethylene)/ZnO Heterostructures

In Sect. 6.2, to study the advantages of CP-type FeTFTs, the effects of P_{SFE} and P_{SPS} were carefully examined. It was demonstrated that the interaction between the two polarizations plays an extremely important role in determining the electrical properties and can improve FET operation. However, it is important to know the electronic states of these interfaces, because the band offset also plays an important role. Therefore, the electrical properties of heterostructures composed of poly (vinylidene fluoride-trifluoroethylene) (P(VDF-TrFE)) and ZnO with different crystallographic polarities, i.e., O- and Zn-polar ZnO, were investigated. The band configurations were also determined via X-ray photoelectron spectroscopy (XPS) using a synchrotron radiation beam in order to analyze the differences in the electrical properties of the P(VDF-TrFE)/O- and Zn-polar ZnO structures.

Thick P(VDF-TrFE) films (85 nm) and thin P(VDF-TrFE) (8 nm)/O- and Zn-polar ZnO heterostructures were fabricated for XPS analysis, and thick P(VDF-TrFE) (85 nm)/O- and Zn-polar ZnO heterostructures were prepared to investigate the electrical properties. Double-polished ZnO single-crystalline substrates were used to obtain the distinct crystallographic polarities of the O-polar and Zn-polar (0001) faces. These substrates were processed using HF chemical etching and organic cleaning, followed by annealing at 950 °C in air to obtain flat surfaces with step and tetras structure. P(VDF-TrFE) with a VDF/TrFE molar ratio of 75/25 was dissolved in diethyl carbonate. The thickness of the P(VDF-TrFE) films was controlled by adjusting the solution concentrations to 1 and 3 wt% for the thin

(8 nm) and thick (85 nm) films, respectively. The P(VDF-TrFE) films were spin-coated onto the O- and Zn-polar ZnO and Pt/Ti/SiO$_2$/Si substrates. Annealing was performed at 120 °C in order to crystallize the P(VDF-TrFE) layers. Top Au electrodes and backside Au/Ti ohmic electrodes were formed on the samples intended for electrical measurements using electron beam evaporation. Au was deposited on the samples intended for XPS analysis via pulsed laser deposition at room temperature in order to calibrate the energy shift caused by the charge effect. The XPS analyses were performed at Beam Line 27A of the Photon Factory at the High Energy Accelerator Research Organization, Tsukuba, Japan, using a synchrotron radiation beam with an energy of 3.1 keV.

The C–V characteristics of the P(VDF-TrFE)/O- and Zn-polar ZnO heterostructures are shown in Fig. 6.15. The measurements were performed by applying a bias voltage from −15 to +15 V. The AC signal was 10 mV and 10 kHz. Butterfly-type C–V curves attributed to ferroelectric polarization reversal are observed for both the P(VDF-TrFE)/O- and Zn-polar ZnO heterostructures. A distinct difference in the C–V curves is observed on the negative voltage side, while the C–V curves on the positive voltage side, which corresponded to the dielectric properties of the P(VDF-TrFE) layers, are similar. The distinct decrease in the capacitance of the P(VDF-TrFE)/O-polar ZnO indicates the formation of a depletion layer due to ferroelectric polarization reversal. However, the P(VDF-TrFE)/Zn-polar ZnO sample exhibits only a slight decrease in capacitance, even at −15 V. The depletion widths of the P(VDF-TrFE)/O- and Zn-polar ZnO heterostructures calculated from the capacitances at ± 15 V were found to be 280 nm and 33 nm, respectively.

The remnant and saturated polarizations and coercive voltage were respectively found from the P–V characteristics (results not shown) to be 3.9 μC cm^{-2}, 4.5 μC cm^{-2}, and 7.0 V for the P(VDF-TrFE)/O-polar ZnO sample and 4.1 μC cm^{-2}, 5.3 μC cm^{-2}, and 8.0 V for the P(VDF-TrFE)/Zn-polar ZnO heterostructure. The formation of depletion layers in the ZnO was indicated by changes in the slopes of the saturation regions of the P–V characteristics, which corresponded to the capacitance values of the heterostructures. The calculated relative permittivities for the P(VDF-TrFE)/O- and Zn-polar ZnO samples were 7.2 and 8.7 on the positive side and 1.9 and 5.9 on the negative side, respectively. Notably, the apparent permittivity of the O-polar ZnO in the negative voltage region

Fig. 6.15 C–V characteristics of P(VDF-TrFE)/O—(*open circles*) and Zn-polar ZnO (*filled squares*)

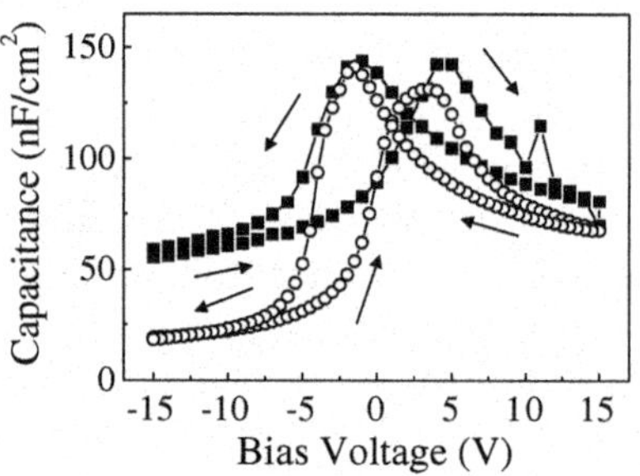

was lower than those of the other samples, in agreement with the $C–V$ analysis results.

Next, to further explore the influence of spontaneous polarization on the electrical properties in detail, the band diagram, which reflects the charge variation at the heterointerface, was analyzed using XPS. A ZnO substrate, a thick P(VDF-TrFE) film on a Pt/Ti/SiO2/Si substrate, and the two P(VDF-TrFE)/ZnO heterostructures with different ZnO polarities were investigated in order to reveal the energy band relationships between P(VDF-TrFE) and ZnO. Figure 6.16 shows the results of the XPS analyses, including the core level (CL) spectra for the Zn $2p3/2$ and F 1 s orbitals and the valence band edge spectra. The spectra were recorded with energy steps of 0.08 eV, which defines the energy resolution of the spectra. All of the spectra were calibrated using the Au $4f7/2$ CL peak, which was assumed to be 83.8 eV. The CL spectra were also fitted using the Voigt (mixed Lorentzian-Gaussian) functional form and the Shirley background. The binding energies for each CL were determined from the positions of the maximum intensities of the fitted lines. The valence band maximum (VBM) was determined via linear extrapolation of the valence band edges to the baselines in order to account for the instrument resolution-induced tail. The analysis was performed with the assumption that the highest occupied molecular orbital (HOMO) level corresponded to the VBM. The band alignment at the P(VDF-TrFE)/ZnO interface was analyzed by calculating the valence band offset (E_{VBO}) using the following equation [70]:

$$E_{VBO} = \left(E_{F1s}^{PVDF} - E_V^{PVDF} \right) - \left(E_{Zn2p}^{ZnO} - E_V^{ZnO} \right) - \Delta E_{CL},$$

where ΔE_{CL} is the separation of the binding energy of the CL spectra for the F 1 s and Zn 2p3/2 orbitals at the P(VDF-TrFE)/ZnO heterointerface. The former two components were determined from the CL and valence band edge spectra for the P(VDF-TrFE) thick film and ZnO (Fig. 6.16) and were found to be 687.4 and 1019.9 eV, respectively. The ΔE_{CL} values for the P(VDF-TrFE)/O- and Zn-polar ZnO were determined to be 335.2 and 335.4 eV, respectively. Therefore, the E_{VBO} values for the P(VDF-TrFE)/O- and Zn-polar ZnO heterostructures were calculated to be 2.7 and 2.9 eV, respectively, as schematically shown in Fig. 6.17a, b. In this scheme, the band gap energies of the P(VDF-TrFE) and two ZnO substrates were assumed to be 5.628, 2.9, and 3.4 eV, respectively. The results indicate that the heterostructures with P(VDF-TrFE) and ZnO have a staggered band alignment with a large E_{VBO}.

The origin of the observed differences in the electrical properties (Fig. 6.15) of the heterostructures with different ZnO crystallographic polarities was then considered. The P(VDF-TrFE)/Zn-polar ZnO heterostructure exhibited only slight depletion compared to that of the P(VDF-TrFE)/O-polar ZnO heterostructure. The carrier concentrations of these two samples estimated from the corresponding maximum depletion widths were 3.5×10^{16} cm^{-3} and 2.7×10^{18} cm^{-3}, respectively. Because the Hall-effect measurements indicate that the O- and Zn-polar ZnO substrates had nearly the same numbers of carriers (1.1×10^{17} cm^{-3} and

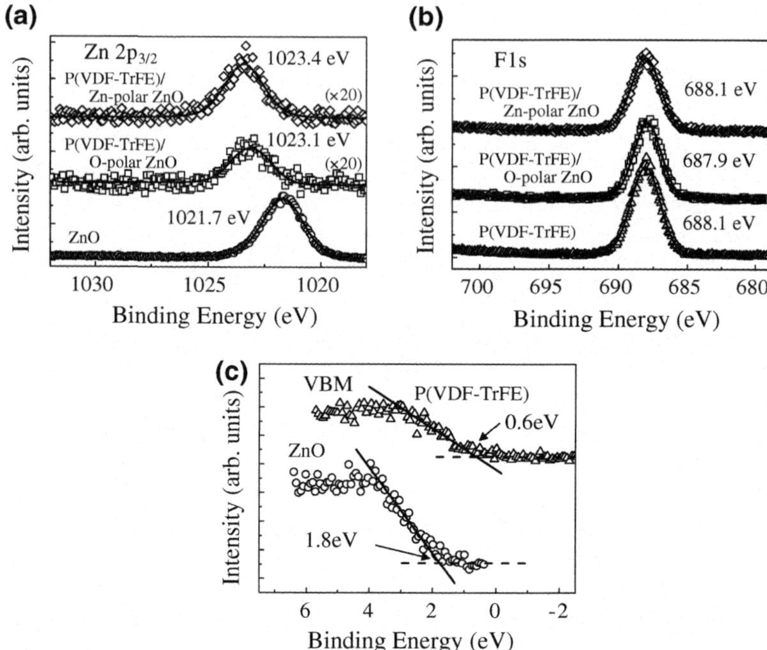

Fig. 6.16 CL spectra of **a** Zn 2p$_{3/2}$ and **b** F 1s and **c** valence band edge spectra. Measured results for ZnO substrate (*circles*), P(VDF-TrFE) film (*triangles*), P(VDF-TrFE)/O—(*squares*), and Zn-polar ZnO (*rhomboids*) heterostructures are shown. *Solid lines* correspond to the lines fitted using Voigt function in (**a**) and (**b**) and linear extrapolation forms in (**c**), and *dashed lines* are base lines

9.8×10^{16} cm^{-3}, respectively), the difference in the electrical properties likely originated from the polarization charges and band lineup. Figure 6.17c–f show schematics of the band diagrams for the heterointerfaces in the P(VDF-TrFE)/ZnO heterostructures, taking into account the directions of the polarizations. When the polarization of P(VDF-TrFE) is directed downward (Fig. 6.17c, e), both ZnO polarities at the heterointerfaces form accumulation states and bend the energy band downward. Conversely, when the polarization of the P(VDF-TrFE) is aligned upward (Fig. 6.17d, f), both ZnO layers are depleted and bend the energy band upward.

This influence of the spontaneous polarization of ZnO was also observed in the differences in the ZnO surface potentials. In the accumulation state, the O-polar surface required a large number of electrons to compensate the head-to-head polarizations, while the Zn-polar surface required fewer electrons due to the charge compensation between the polarizations. ZnO is a typical n-type semiconductor; thus, the interface charge was neutralized by the electrons. When the ZnO was depleted, conversely, the O-polar surface held the charge compensation due to the polarizations, and the Zn-polar surface required a large number of holes to

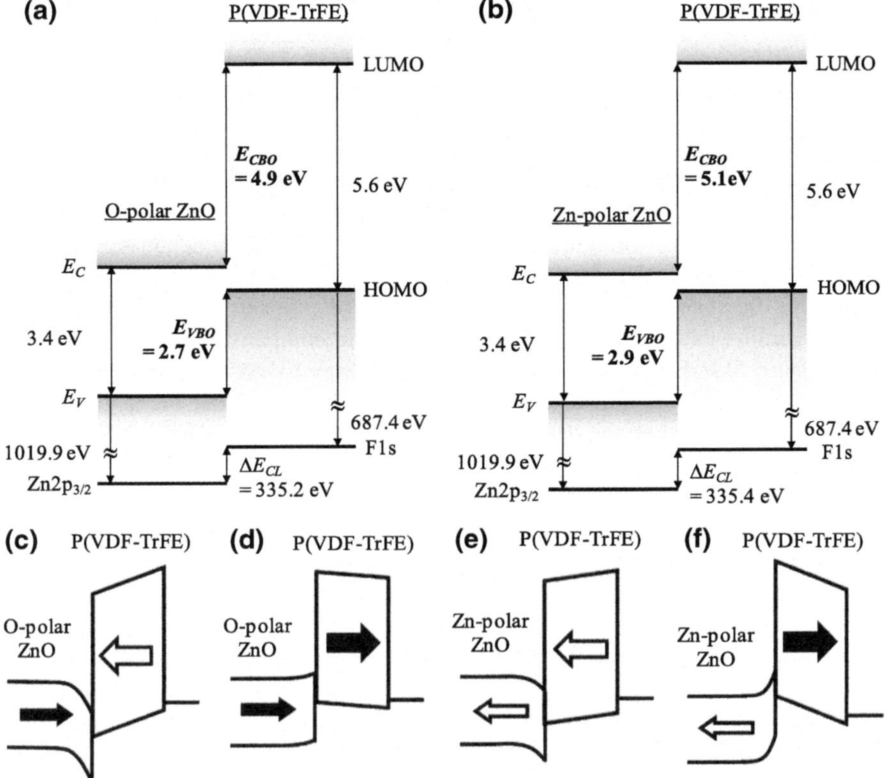

Fig. 6.17 Energy band diagrams of **a** P(VDF-TrFE)/O- and **b** Zn-polar ZnO as determined by XPS measurements. Schematics of band diagrams of (**c**), **d** P(VDF-TrFE)/O- and (**e**), **f** Zn-polar ZnO considering polarization directions of P(VDF-TrFE) layers to be downward in (**c**) and (**e**) and upward in (**d**) and (**f**)

compensate for the mismatched charges arising from the tail-to-tail geometry of the polarizations. As a result of the lack of p-type carriers in the ZnO layer, the Zn-polar ZnO generated a high surface potential to induce holes. Consequently, a high electric field was applied to the P(VDF-TrFE) layer. Although P(VDF-TrFE) has an insulating nature, the XPS spectrum of the P(VDF-TrFE) film shown in Fig. 6.16c indicates that the Fermi level in the P(VDF-TrFE) existed near the HOMO level (up to 0.6 eV). This result suggests that holes were generated in the P(VDF-TrFE) and screened the electrical contribution of the ZnO layer. This explanation is consistent with the difference in the depletion behaviors of the P(VDF-TrFE)/O- and Zn-polar ZnO heterostructures. The results presented here therefore indicate that the relationship between the interface charge induction by spontaneous polarization and the energy band alignment with a staggered and large gap had a significant impact on the electrical properties of the heterostructures consisting of P(VDF-TrFE) and ZnO.

6.4 Conclusions

In this chapter, the electronic transport properties of CP-type FeTFTs with top-gate and bottom-gate structures were investigated.

The bottom-gate type CP ferroelectric-gate TFTs were fabricated using a ZnO channel as a polar oxide semiconductor and an $YMnO_3$ film as a ferroelectric gate and they show good TFT operation and memory characteristics even in the bottom-gate structure, which the P_{SPS} of the ZnO should act as negative. Based on the analysis using C–V and the Cole-Cole plots between the source and the gate electrodes, the state of channel conductance was also discussed. The direction of P_{SFE} at various gate voltages is determined by the comparisons of the calculated results using SPICE and experimental results. It was found that the directions of P_{SFE} are aligned downward below underneath the channel under the negative bias voltage suggesting that the P_{SFE} is easily switched if the effect of the charge mismatch between which P_{SFE} and P_{SPS} small.

To clarify the effect of ferroelectric polarization on the carrier transport of FeTFTs with top-gate structure, Hall-effect measurements after poling treatments were performed using FeTFTs with various channel thicknesses. While a carrier modulation effect was observed in all of the samples, modulation of the Hall mobility was only detected in FeTFTs with channel thicknesses less than 100 nm. These results indicate that modulation of the mobility was confined to the region close to the ferroelectric/semiconductor interface; therefore, the ferroelectric polarization strongly influenced the interface. It was also found that electronic transport in the ZnO thin films was disturbed by the formation of a multidomain structure in the ferroelectrics. However, this effect is not a significant issue, because the enhancement of the mobility due to ferroelectric polarization is rather prominent, and the disturbing effect can be reduced by fully polarizing the ferroelectric. Furthermore, the temperature dependences of Hall measurements with positive and negative poling revealed that modulation of the Hall mobility due to ferroelectric polarization resulted from reduced ionized impurity scattering. Thus, it can be concluded that ferroelectric polarization can enhance mobility by screening the potentials of impurities through the field effect.

References

1. S. Takagi, A. Toriumi, M. Iwase, H. Tango, IEEE Trans. Electron Devices **41**, 2357 (1994)
2. S. Takagi, A. Toriumi, M. Iwase, H. Tango, IEEE Trans. Electron Devices **41**, 2363 (1994)
3. Y. Zhao, H. Matsumoto, T. Sato, S. Koyama, M. Takenaka, S. Takagi, IEEE Trans. Electron Devices **57**, 2057 (2010)
4. R. Chau, J. Brask, S. Datta, G. Dewey, M. Doczy, B. Doyle, J. Kavalieros, B. Jin, M. Metz, A. Majumdar, M. Radosavljevic, Microelectron. Eng. **80**, 1 (2005)
5. M. Yokoyama, T. Yasuda, H. Takagi, N. Miyata, Y. Urabe, H. Ishii, H. Yamada, N. Fukuhara, M. Hata, M. Sugiyama, Y. Nakano, M. Takenaka, S. Takagi, Appl. Phys. Lett. **96**, 142106 (2010)

6. Y. Xuan, Y.Q. Wu, H.C. Lin, T. Shen, P.D. Ye, IEEE trans. **28**, 935 (2007)
7. V. Tilak, K. Matocha, G. Dunne, IEEE Trans. Electron Devices **54**, 2823 (2007)
8. T. Mimura, S. Hiyamizu, T. Fujii, K. Nanbu, Jpn. J. Appl. Phys. **19**, L225 (1980)
9. T. Edahiro, N. Fujimura, T. Ito, J. Appl. Phys. **93**, 7673 (2003)
10. J. Falson, D. Maryenko, Y. Kozuka, A. Tsukazaki, M. Kawasaki, Appl. Phys. Express **4**, 091101 (2011)
11. A. Ohtomo, H.Y. Hwang, Nature **427**, 423 (2004)
12. H. Yuan, M.S. Bahramy, K. Morimoto, S. Wu, K. Nomura, B.-J. Yang, H. Shimotani, R. Suzuki, M. Toh, C. Kloc, X. Xu, R. Arita, N. Nagaosa, Y. Iwasa, Nat. Phys. **9**, 563 (2013)
13. A. Tsukazaki, S. Akasaka, K. Nakahara, Y. Ohno, D. Maryenko, A. Ohtomo, M. Kawasaki, Nat. Mater. **9**, 889 (2010)
14. G. Cheng, P.F. Siles, F. Bi, C. Cen, D.F. Bogorin, C.W. Bark, C.M. Folkman, J.-W. Park, C.-B. Eom, G.M. Ribeiro, J. Levy, Nat. Nanotechnol. **6**, 343 (2011)
15. H. Ohta, S. Kim, Y. Mune, T. Mizoguchi, K. Nomura, S. Ohta, T. Nomura, Y. Nakanishi, Y. Ikuhara, M. Hirano, H. Hosono, K. Koumoto, Nat. Mater. **6**, 129 (2007)
16. Y. Yuan, T.J. Reece, P. Sharma, S. Poddar, S. Ducharme, A. Gruverman, Y. Yang, J. Huang, Nat. Mater. **10**, 296 (2011)
17. C.H. Ahn, A. Bhattacharya, M.D. Ventra, J.N. Eckstein, C.D. Frisbie, M.E. Gershenson, A.M. Goldman, I.H. Inoue, J. Mannhart, A.J. Millis, A.F. Morpurgo, D. Natelson, J.-M. Triscone, Rev. Mod. Phys. **78**, 1185 (2006)
18. S. Okamoto, A.J. Millis, Nature **428**, 630 (2004)
19. K. Ueno, S. Nakamura, H. Shimotani, A. Ohtomo, N. Kimura, T. Nojima, H. Aoki, Y. Iwasa, M. Kawasaki, Nat. Mater. **7**, 885 (2008)
20. W.L. Warren, D. Dimos, B.A. Tuttle, G.E. Pike, R.W. Schwartz, P.J. Clews, D.C. Mclntyre, J. Appl. Phys. **77**, 6695 (1995)
21. H. Funakubo, T. Watanabe, H. Morioka, A. Nagai, T. Oikawa, M. Suzuki, H. Uchida, S. Kouda, K. Saito, Mater. Sci. Eng. B **118**, 23 (2005)
22. X.X. Wang, X.G. Tang, H.L.W. Chan, Appl. Phys. Lett. **85**, 91 (2004)
23. D.J. Kim, J.Y. Jo, Y.S. Kim, Y.J. Chang, J.S. Lee, J.-G. Yoon, T.K. Song, T.W. Noh, Phys. Rev. Lett. **95**, 237602 (2005)
24. K.Y. Yun, D. Ricinschi, T. Kanashima, M. Noda, M. Okuyama, Jpn. J. Appl. Phys. **43**, L647 (2004)
25. J. Yan, M. Gomi, T. Yokota, H. Song, Appl. Phys. Lett. **102**, 222906 (2013)
26. K.J. Choi, M. Biegalski, Y.L. Li, A. Sharan, J. Schubert, R. Uecker, P. Reiche, Y.B. Chen, X.Q. Pan, V. Gopalan, L.-Q. Chen, D.G. Schlom, C.B. Eom, Science **306**, 1005 (2004)
27. S. Sakai, R. Ilangovan, M. Takahashi, Jpn. J. Appl. Phys. **43**, 7876 (2004)
28. Y. Kato, Y. Kaneko, H. Tanaka, Y. Shimada, Jpn. J. Appl. Phys. **47**, 2719 (2008)
29. S. Fujisaki, H. Ishiwara, Y. Fujisaki, Appl. Phys. Express **1**, 081801 (2008)
30. T. Miyasako, M. Senoo, E. Tokumitsu, Appl. Phys. Lett. **86**, 162902 (2005)
31. T. Miyasako, B.N.Q. Trinh, M. Onoue, T. Kaneda, P.T. Tue, E. Tokumitsu, T. Shimoda, Jpn. J. Appl. Phys. **50**, 04DD09 (2011)
32. Y. Kaneko, Y. Nishitani, M. Ueda, E. Tokumitsu, E. Fujii, Appl. Phys. Lett. **99**, 182902 (2011)
33. Y. Kaneko, H. Tanaka, Y. Kato, Jpn. J. Appl. Phys. **48**, 09KA19 (2009)
34. T. Hatanaka, M. Takahashi, S. Sakai, K. Takeuchi, IEICE Trans. electron. **94**, 539 (2011)
35. T. Fukushima, T. Yoshimura, K. Masuko, K. Maeda, A. Ashida, N. Fujimura, Jpn. J. Appl. Phys. **47**, 8874 (2008)
36. T. Fukushima, K. Maeda, T. Yoshimura, A. Ashida, N. Fujimura, Jpn. J. Appl. Phys. **50**, 04DD16 (2011)
37. E.F. Bertant, F. Forrast, P.H. Fang, C. R. Acad. Sci. **256**, 1958 (1963)
38. V.A. Bokov, G.A. Smolenski, S.A. Kizhaev, I.E. Myl'nilova, Sov. Phys. Solid State **5**, 2646 (1964)
39. G.A. Smolenski, V.A. Bokov, J. Appl. Phys. **35**, 915 (1964)
40. I.G. Ismailzade, S.A. Kizhaev, Sov. Phys. Solid State **7**, 236 (1965)

41. D. Ito, N. Fujimura, T. Yoshimura, T. Ito, J. Appl. Phys. **93**, 5563 (2003)
42. N. Shigemitsu, H. Sakata, D. Ito, T. Yoshimura, A. Ashida, N. Fujimura, Jpn. J. Appl. Phys. **43**, 6613 (2004)
43. K. Maeda, T. Yoshimura, N. Fujimura, Jpn. J. Appl. Phys. **48**, 09KB05 (2009)
44. T. Edahiro, N. Fujimura, T. Ito, J. Appl. Phys. **43**, 6613 (2004)
45. K. Masuko, A. Ashida, T. Yoshimura, N. Fujimura, J. Appl. Phys. **103**, 043714 (2008)
46. S.M. Sze, *Physics of Semiconductor Devices*, 2nd edn. (Wiley, New York, 1981), pp. 486–495
47. T. Miyasako, M. Senoo, E. Tokumitsu, Appl. Phys. Lett. **86**, 162902 (2005)
48. J. Siddiqui, E. Cagin, D. Chen, J.D. Phillips, Appl. Phys. Lett. **88**, 212903 (2006)
49. N. Fujimura, N. Shigemitsu, T. Takahashi, A. Ashida, T. Yoshimura, H. Fukumura, H. Harima, Philos. Mag. Lett. **87**, 193 (2007)
50. H.C.F. Martens, H.B. Brom, P.W.M. Blom, Phys. Rev. B **60**, R8489 (1999)
51. J.R. Hauser, IEEE Trans. Electron Device **12**, 605 (1965)
52. H. Yamada, T. Fukushima, T. Yoshimura, N. Fujimura, J. Korean Phys. Soc. **58**, 792 (2011)
53. H. Kodama, Y. Takahashi, T. Furukawa, Jpn. J. Appl. Phys. **38**, 3589 (1999)
54. T. Fukuma, K. Kobayashi, T. Horiuchi, H. Yamada, K. Matsushige, Jpn. J. Appl. Phys. **39**, 3830 (2000)
55. K. Tashiro, H. Kaito, M. Kobayashi, Polymer **33**, 2915 (1992)
56. P. Ruterana, M. Abouzaid, A. Bere, J. Chen, J. Appl. Phys. **103**, 033501 (2008)
57. K. Masuko, A. Ashida, T. Yoshimura, N. Fujimura, J. Appl. Phys. **103**, 043714 (2008)
58. C.G. Van de Walle, Phys. Rev. Lett. **85**, 1102 (2000)
59. Y. Wang, B. Meyer, X. Yin, M. Kunat, D. Langenberg, F. Traeger, A. Birkner, C. Woll, Phys. Rev. Lett. **95**, 266104 (2005)
60. D.C. Look, H.L. Mosbacker, Y.M. Strzhemechny, L.J. Brillson, Superlattices Microstruct. **38**, 406 (2005)
61. S. Sakamoto, T. Oshio, A. Ashida, T. Yoshimura, N. Fujimura, Appl. Surf. Sci. **254**, 6248 (2008)
62. D.C. Look, R.J. Molnar, Appl. Phys. Lett. **70**, 3377 (1997)
63. H. Tampo, A. Yamada, P. Fons, H. Shibata, K. Matsubara, K. Iwata, S. Niki, K. Nakahara, H. Takasu, Appl. Phys. Lett. **84**, 4412 (2004)
64. S.M. Sze, K.K. Ng, *Physics of Semiconductor Devices*, 3rd edn. (Wiley, New York, 2007), p. 308
65. D.C. Look, *Electrical Characterization of GaAs Materials and Devices* (Wiley, New York, 1989) pp. 54, 76
66. K. Hess, Appl. Phys. Lett. **35**, 484 (1979)
67. D.C. Look, D.C. Reynolds, J.R. Sizelove, R.L. Jones, C.W. Litton, G. Cantwell, and W.C. Harsch, Solid State Commun. **105**, 399–401 (1998)
68. S. Takagi, A. Toriumi, M. Iwase, H. Tango, I.E.E.E. Trans, Electron Devices **41**, 2357 (1994)
69. S.V. Vandebroek, E.F. Crabbe, B.S. Meyerson, D.L. Harame, P.J. Restle, J.M.C. Stork, A.C. Megdanis, C.L. Stanis, A.A. Bright, G.M.W. Kroesen, A.C. Warren, IEEE Electron Device Lett. **12**, 447 (1991)
70. E.A. Kraut, R.W. Grant, J.R. Waldrop, S.P. Kowalczyk, Phys. Rev. Lett. **44**, 1620 (1980)

Part IV
Practical Characteristics of Organic Ferroelectric-Gate FETs: Si-Based Ferroelectric-Gate Field Effect Transistors

Chapter 7
Non-volatile Ferroelectric Memory Transistors Using PVDF and P(VDF-TrFE) Thin Films

Byung-Eun Park

Abstract In this work, metal–ferroelectric–semiconductor field effect transistors (MFSFETs) have been fabricated for the first time using poly(vinylidene fluoride) (PVDF) and polyvinylidene fluoride trifluoroethylene [P(VDF-TrFE)] thin films as ferroelectric layers. PVDF and P(VDF-TrFE) thin films were fabricated by a sol-gel method on Si(100) wafers. The drain current–gate voltage (I_D–V_G) characteristics of both MFSFETs fabricated with PVDF and P(VDF-TrFE) thin films exhibited very good ferroelectric hysteretic curves with a counterclockwise loop that is the same as those of other ferroelectric materials. It also demonstrates the realization of a one-transistor type (1T-type) ferroelectric memory without a buffer layer using thin organic material. The absence of a buffer layer presents many advantages such as the elimination of the depolarization field, leakage current influence of the thin buffer layer, reduction of the process steps, low operational voltage, and low power consumption. MFSFETs using PVDF and P(VDF-TrFE) thin films as ferroelectric layers have promising potential for use in low-voltage and flexible 1T-type ferroelectric random access memory (FeRAM) using organic material.

7.1 Introduction

Ferroelectric random access memory (FeRAM) [1–3], consisting of one transistor and one ferroelectric capacitor for storing data (1T1C-type), can be nonvolatile, can perform high-speed read/write operations, and requires low power consumption. In particular, ferroelectric gate field effect transistors (FeFETs), in which the gate insulator film is composed of a ferroelectric material (1T-type), has drawn much attention because of its nondestructive data read-out and the possibility of being proportionally scaled down. A 1T-type FeRAM is one of the most promising candidates for the creation of high-capacity non-volatile memory because the

B.-E. Park (✉)
School of Electrical and Computer Engineering, University of Seoul,
163 Seoulsiripdae-ro, Dongdaemun-ku, Seoul 130-743, Korea
e-mail: pbe@uos.ac.kr

© Springer Science+Business Media Dordrecht 2016
B.-E. Park et al. (eds.), *Ferroelectric-Gate Field Effect Transistor Memories*,
Topics in Applied Physics 131, DOI 10.1007/978-94-024-0841-6_7

memory cell can be scaled smaller than dynamic RAM (DRAM). The original idea for 1T-type ferroelectric memory was proposed in 1957 [4]. Despite the potential advantages, 1T-type ferroelectric memory has not been developed because of poor interface properties between the ferroelectric film and silicon (Si) substrate. If a ferroelectric film is directly deposited on a Si substrate, constituent atoms easily diffuse into the Si substrate, and thus the interfacial electrical properties between them become very poor. In order to solve this problem, a buffer layer with a high dielectric constant is inserted between the ferroelectric film and Si substrate forming a metal-ferroelectric-insulator-semiconductor (MFIS) structure, as shown in Fig. 7.1. Therefore, there has been great interest in making MFIS structures [5, 6] as a gate stack of the FeFET because it is expected that the insulating buffer layer can prevent chemical reaction and/or thermal diffusion of the constituent atoms between the ferroelectric film and semiconductor substrate and improve the interface properties. However, MFIS structures present problems such as the generation of the depolarization field in the ferroelectric film and an increase in operation voltage requirements because of the application of additional voltage to the buffer layer. An equivalent circuit of this structure is a series connection of ferroelectric and dielectric capacitors; when the power supply is off and the gate terminal of the FET is grounded, the top and bottom electrodes of the two capacitors are short-circuited. At the same time, electronic charges Q appear on the electrodes of both capacitors because of the remanent polarization of the ferroelectric film and the charge neutrality condition at a node between the two capacitors. The charge-versus-voltage (Q–V) relation for the buffer layer capacitor is $Q = CV$, and thus the relation in the ferroelectric capacitor becomes $Q = -CV$ under the short-circuited condition, where C is the capacitance of the buffer layer, as shown in Fig. 7.1. Specifically, the direction of the electric field in the ferroelectric film is opposite the polarization. This field is known as a depolarization field and it reduces the data retention time significantly, particularly when C is small. In order to make the depolarization field low in the ferroelectric film, it is necessary to make the buffer layer capacitance C as large as possible. This condition means that a thin buffer layer with a high dielectric constant is desirable. Another important point is to reduce the leakage current of both the ferroelectric film and buffer layer. If the charge neutrality at a node between the two capacitors is destroyed by the leakage current, electric charges on the electrodes of the buffer layer capacitor disappear, which means that carriers on the semiconductor surface disappear and the stored data cannot be read by the drain current of the FET, even if the polarization of the ferroelectric film is retained. Specifically, the thinnest limit of the buffer layer thickness is determined by the leakage current running through it [3]. Some papers have reported good data retention characteristics in MFIS structures, in which high dielectric constant materials, such as Si_3N_4 [7] HfO_2 [8] and $HfAlO_3$ [9] are used as buffer layers. Although the MFIS structure seems to have solved the problems associated with the MFS structure, the retention characteristics of the MFIS structure are not reliable for commercialization despite recent reports indicating considerable advancement in data retention time [10, 11]. Nevertheless, the depolarization field, which causes the degradation of data retention and an increase in the operational voltage requirement, is always generated in MFIS structures.

To solve the problems of the MFIS structures, this work attempts to make an MFS structure using a poly(vinylidene fluoride) (PVDF) thin film as a ferroelectric layer for the first time. This combines the advantages of the MFS structure, such as high density, simple circuit, and non-destructive readout, and PVDF thin film, such as low cost, low thermal budget, and advantages of organic materials [12–14]. PVDF and its copolymer PVDF/trifluoroethylene (TrFE) are well-known ferroelectric polymers. The general features of PVDF and P(VDF-TrFE) are described thoroughly in many studies [13–16]. PVDF and P(VDF-TrFE) have attracted much attention and their applications have been thoroughly studied in non-volatile polymer ferroelectric random access memory and ferroelectric devices. Naber et al. [17] used P(VDF-TrFE) solutions to fabricate a non-volatile organic ferroelectric field effect transistor. Fujisaki et al. [18] succeeded in fabricating low-voltage operational P(VDF-TrFE) capacitors and diodes. With respect to PVDF films, the capacitance–voltage (C–V) and current density–voltage (J–V) properties of MFS diodes formed by spin-coating on Si substrates have also been reported [13, 14]. Furthermore, the fabrication and electrical properties of an organic ferroelectric field effect transistor with spin-coated PVDF film on a gold bottom gate were reported by Kang et al. [19].

To the best of our knowledge, the fabrication of an MFS field effect transistor (MFSFET) with PVDF films deposited on a silicon substrate has never been reported. For the first time, this study demonstrates the realization of 1T-type ferroelectric memory without a buffer layer using thin organic PVDF films [20, 21]. The absence of a buffer layer presents many advantages such as elimination of the depolarization field, elimination of the leakage current influence on the thin buffer layer, reduction of the process steps, low operational voltage, and low power consumption. Furthermore, thin polymer films offer the potential for very low fabrication costs, can serve as building blocks of larger devices, can be used with flexible substrates, and have a versatile chemical structure [22–24].

7.2 Experimental Procedure

To make the PVDF thin film, commercially available PVDF pellets and powders were mainly obtained from Sigma-Aldrich and dissolved in dimethyl acetamide (DMA), methyl ethyl ketone (MEK), and dimethylformamide (DMF) at 60 °C on a hot-plate for 24 h. Film thickness was controlled by changing the weight concentration of PVDF pellets and powders in the solvent with an otherwise constant speed and coating time using a spin coater. Various conditions of thickness were introduced by controlling the concentration. The PVDF films were deposited by a sol-gel spin-coating method. The deposited films were dried in the temperature

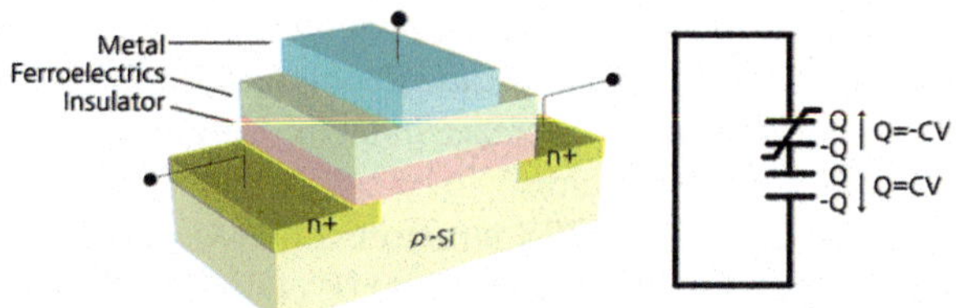

Fig. 7.1 a Schematic drawings for the metal-ferroelectric-insulator-semiconductor (MFIS) structure and **b** equivalent circuit of the MFIS structure

range of 160–180 °C for 30 min on a hot plate to remove solvent materials. To obtain the β-phase PVDF, the fabricated PVDF films were cooled rapidly to 60–70 °C and incubated for longer than 60 min [25–27]. The typical thickness was 10 to 600 nm. For measurement of the electrical properties, dot-shaped upper Au electrodes were formed on the PVDF/Si structures at room temperature by vacuum evaporation using a metal mask. Capacitance-voltage and current-voltage characteristics were measured by an LCR meter and a precision semiconductor parameter analyzer, respectively.

Alternatively, in order to fabricate MFSFET, p-type Si (100) wafers were dipped in a buffered oxide etchant (BOE) to remove the native silicon oxide (SiO_2) layer film from the surface. 100-nm thick SiO_2 layers were grown by thermal oxidation as a passivation layer. Source and drain regions were formed by diffusing phosphorus atoms from a deposited glass film into the silicon substrate. After removing the residual silicon oxide layer, 2 and 6 wt% PVDF solutions were spin-coated on the substrates at 3000 rpm. The spin-coated PVDF layer was dried at 160 °C for 60 min to remove the solvent. To make contact holes, the PVDF films were etched with a reactive ion etcher (RIE) in an O_2 atmosphere. Then, Au electrodes were thermally evaporated on the entire surface of the PVDF films. After patterning the contact electrodes, unnecessary Au residuals were removed with a Au etching solution. The typical channel width and length of the MFSFETs are 26 and 5 μm, respectively. The surface morphology of the PVDF films was observed by atomic force microscopy (AFM). The C–V characteristics were investigated with an HP 4280A capacitance meter. Drain current–gate voltage (I_D–V_G) and drain current–drain voltage (I_D–V_D) characteristics were measured with an HP 4155C semiconductor parameter analyzer. All measurements were performed at room temperature in ambient air.

MFSFETs were also attempted with P(VDF-TrFE) films in order to compare the PVDF and P(VDF-TrFE) films. In this work, P(VDF-TrFE) ferroelectric polymer containing VDF (75 mol%) and TrFE (25 mol%) was used as a ferroelectric material. The P(VDF-TrFE) copolymer powders were dissolved in the DMF at room temperature for 24 h. A p-type Si (100) substrate was dipped into a diluted HF solution to remove the surface oxide. The P(VDF-TrFE) solution was spin-coated on the cleaned Si (100) substrate. The as-deposited film was heat-treated at 165 °C for half an hour to remove the solvents and improve crystallization. In order to fabricate MFSFETs with P(VDF-TrFE) films. Field oxide was formed with a thermal oxidation process as a diffusion barrier. After definition of the source and drain regions, phosphorus glass (PSG) thin film was spin-coated as a doping source of phosphorus. Diffusion was performed at 1050 °C for 45 s with rapid thermal annealing (RTA).

Before deposition of the P(VDF-TrFE) thin films, PSG and a diffusion barrier oxide layer were removed from the Si substrate. After deposition of the P(VDF-TrFE) thin film, contact holes in the drain and source were etched by reactive ion etching at 150 mTorr and the RF power was fixed at 100 W for 30 s. A gold electrode was deposited on the surface of the P(VDF-TrFE) film using thermal evaporation and then patterned with gold etchant to define the gate and make contact with the drain and source regions. The channel length and width were 30 and 60 µm, respectively.

7.3 Results and Discussion

7.3.1 *Electrical Properties of the Poly(Vinylidene) (PVDF) Thin Films*

A typical capacitance-voltage characteristic for a Au/PVDF/Si structure, which was measured at 1 MHz, is shown in Fig. 7.2a. The thickness of the PVDF film was approximately 60 nm. In this figure, the PVDF film showed clockwise hysteresis, as indicated by the arrows, and the value of the memory window width was approximately 1.5 V for the sweep range of ±2 V. To check the effect of mobile ionic charges on the hysteretic loop, the scanning speed of the bias voltage was changed from 0.01 to 0.75 V/s in the $C–V$ measurement. However, the memory window width did not change within the scanning speed range, which suggests that the hysteresis characteristics are due to the ferroelectric property of the PVDF film. Figure 7.2b shows measured retention characteristics of the Au/PVDF/Si structures. In this measurement, after the write voltage of +2 V or −2 V was applied to the sample, the time dependence of the capacitance was measured while keeping the bias voltage at 0.2 V. The high and low capacitance values in this sample were nearly the same as those observed before and after with no change, even after 3 days had elapsed. This result illustrates that this sample has very excellent data retention characteristics.

Figure 7.3a shows the $C–V$ characteristic for another sample of the Au/PVDF/Si structure with a 10-nm thick PVDF film. As shown in the figure, the $C–V$ curve of this sample also has a clockwise hysteresis loop as indicated by the arrows. The memory window width is approximately 0.2 V for the bias voltage sweep from −1 V to +1 V. Since the capacitance changes steeply with applied voltage, the high and low capacitance values are well distinguished at a certain applied voltage in the hysteresis loop, in spite of the relatively narrow memory window width.

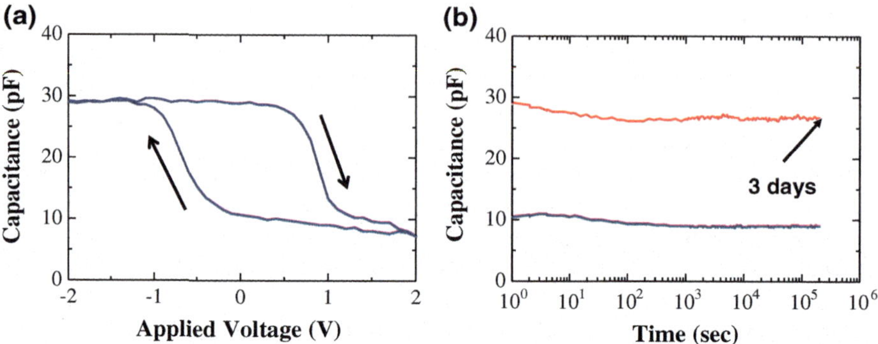

Fig. 7.2 **a** A typical capacitance-voltage (*C–V*) characteristic for a Au/PVDF/Si structure, which was measured at 1 MHz. **b** Retention characteristic of this Au/PVDF/Si structure. In this measurement, after the write voltage of +2 V or −2 V was applied to the sample, the time dependence of the capacitance was measured while maintaining the bias voltage at 0.2 V

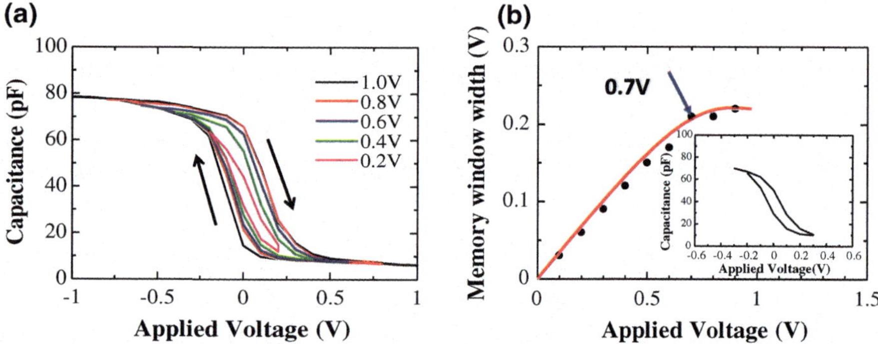

Fig. 7.3 **a** A typical *C–V* characteristic for another sample of the Au/PVDF/Si structure with a 10-nm thick PVDF film. **b** Memory window width as a function of the applied voltage for the Au/PVDF/Si structure. The measurement was performed with 0.1 V units at 1 MHz

The memory window width as a function of the applied voltage is plotted in Fig. 7.3b. The memory window width gradually increases with the applied voltage and saturates at approximately 0.2 V when the applied voltage is greater than 0.7 V. The memory window widths of the applied voltages of 0.7 and 1.0 V are 0.21 and 0.23 V, respectively. It is worth noting that typical ferroelectric hysteresis loops are obtained with an applied voltage greater than 0.3 V. Since the memory window width is approximately 0.1 V at an applied voltage of 0.3 V, a new possibility of low-voltage memory devices is expected that operate at 0.3 V. The retention characteristic of this sample is shown in Fig. 7.4a. In this measurement, after write voltage pulses of 1 V in height and 10 ms in width were applied to the sample, the time dependencies of the high and low capacitance values were measured. As shown in Fig. 7.4a, this sample also has nearly the same tendency. The capacitance

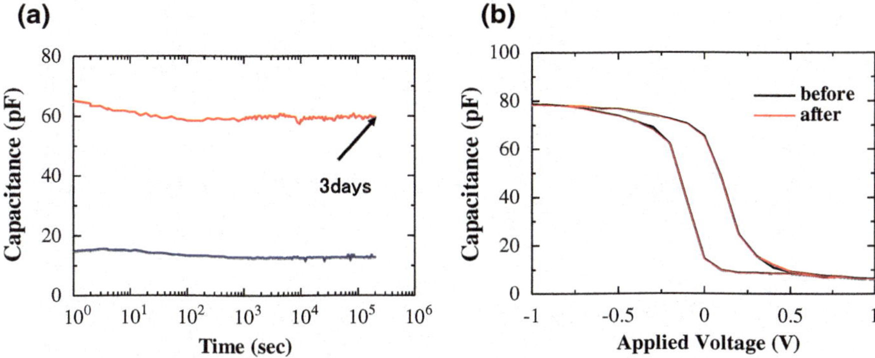

Fig. 7.4 **a** The retention characteristic of the sample shown in Fig. 7.3. In this measurement, after the write voltage pulses were applied to the sample, the time dependencies of the high and low capacitance values were measured. **b** *C–V* curve of this sample after the retention measurement

values hardly changed over 3 days, before and after retention measurement. Figure 7.4b shows the *C–V* curve of this sample after the retention measurement. Black and red lines in the figure show the *C–V* characteristic before and after retention measurements for the 10-nm thick PVDF film, respectively. The *C–V* curve did not change, even for the 3-day retention measurement. A simple extrapolation of the capacitance values indicates that the retention time is sufficiently long for multipurpose non-volatile memories. It appears that the very excellent data retention characteristic is due to the negligible effect of the depolarization field because of the absence of the buffer layer.

Figure 7.5 shows the surface morphological images of the PVDF thin films measured by atomic force microscopy. The measured area was $1 \times 1 \ \mu m^2$. The thicknesses of the measured samples were approximately 10 and 60 nm, respectively. As shown in the images, surface lamellae were observed, indicating that each chain of molecules was ordered in crystalline lamella. In the case of the 10-nm thick PVDF film, complex interlaced domains with rod-shaped were observed. Alternatively, in the PVDF film with 60 nm, slender stem domains aligning in one direction were observed. Additionally, the lamella size gradually increased as the thickness increased. It appears that many domain nuclei were generated and oriented in crystallization for a relatively thick film.

7.3.2 Electrical Properties of MFSFETs with PVDF Thin Films

Based on the electrical properties of MFS capacitors, there was an attempt to fabricate the n-channel MFISFETs. A *p*-type Si (100) wafer was dipped in a buffered oxide etchant to remove the native silicon oxide layer film from the surface.

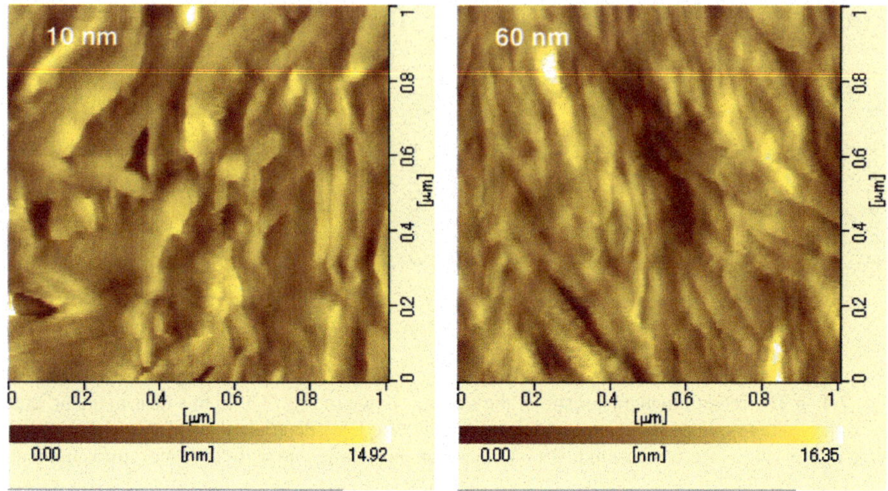

Fig. 7.5 The surface morphological images of the PVDF thin films measured by atomic force microscopy. The measured area was $1 \times 1\ \mu m^2$

A passivation layer with a 100-nm thick SiO_2 film was grown by a thermal oxidation method. Definition of the source and drain region was performed through PSG diffusion into the silicon substrate in a rapid thermal annealing apparatus. After removing the residual silicon oxide layer, 2 and 6 wt% PVDF solutions were spin-coated on the substrates, and the spin-coated PVDF layer was dried in order to remove the solvents. The thicknesses of the PVDF films spin-coated from 2 and 6 wt% solutions were approximately 25 and 180 nm, respectively. For contact holes in the drain and source, the PVDF film was etched with a reactive ion etcher in an O_2 atmosphere. Au electrodes were thermally evaporated on the surface of the PVDF film. After patterning the contact electrodes, the unnecessary Au residuals were removed with a Au etching solution. The schematic of the MFSFETs fabrication process is shown in Fig. 7.6.

Figure 7.7 shows an optical microscope image of the fabricated MFSFETs and the surface AFM image of the 6 wt% PVDF film of the MFSFETs. In the PVDF films with 2 wt% films, crystalline lamellae with thin platelet-shapes were observed, which were ordered and stacked. Some crevices between the lamellae were also observed, and they appeared dark due to their varying thickness. The average roughness range of the fabricated PVDF films was approximately 10–15 nm.

The capacitance–gate voltage (C–V_G) characteristics of the fabricated MFSFETs with the 2 and 6 wt% PVDF films were measured at 1 MHz. In order to verify if the fabricated PVDF thin films were formed in the β-phase, measurement of the PVDF films between the gate and bottom electrodes was performed. As shown in Fig. 7.8, the typical C–V_G curve of the 6 wt% PVDF MFSFETs shows a hysteretic loop with a clockwise trace, as indicated by the arrows. Moreover, the scanning speed of the gate voltage was changed from 0.1 to 0.5 V/s to verify whether the memory

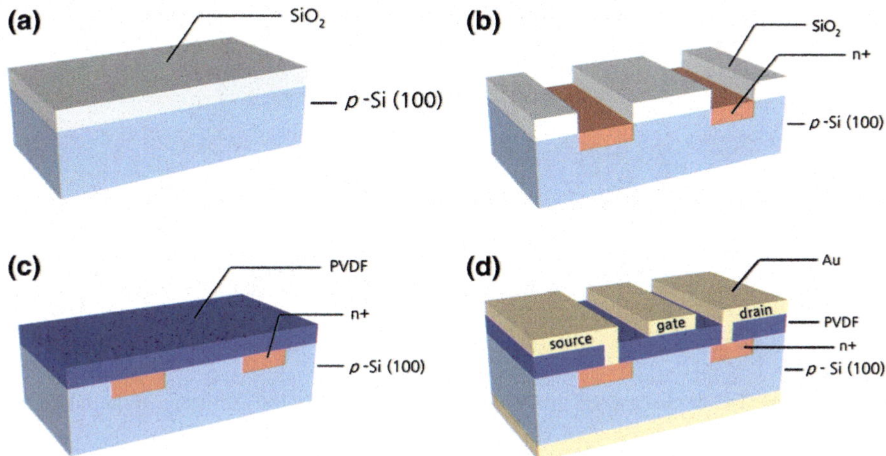

Fig. 7.6 The schematic of the MFSFETs fabrication process: **a** thermal oxidation (dry), **b** phosphorus diffusion and SiO₂ etching out, **c** PVDF deposition, and **d** Au metallization

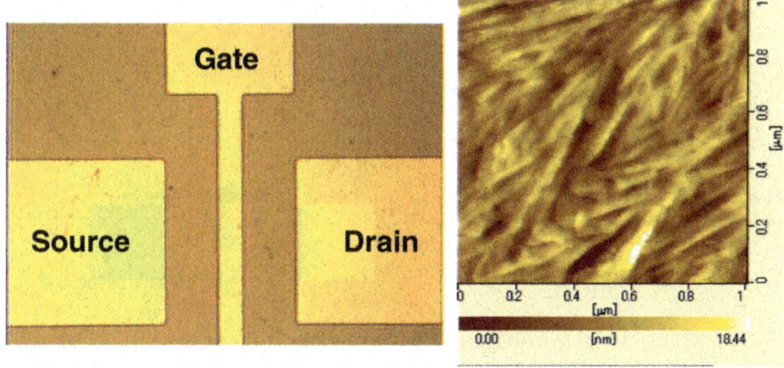

Fig. 7.7 **a** An optical microscope image of the fabricated MFSFETs and **b** surface AFM image of the 6 wt% PVDF film of the MFSFETs

operation of the $C–V_G$ characteristics originates from the ferroelectric polarization reversal or ion drift. The memory window width of the $C–V_G$ characteristic did not change, regardless of the scanning speed of the gate voltage. This implies the ferroelectric behavior of the PVDF film in the β-phase. The direction of the hysteretic loop corresponds to the polarization switching from the accumulation state to the inversion state. This denotes that the capacitance hysteresis behavior is due to the dipolar polarization caused by the PVDF layer. The memory window width of the 6 wt% MFSFETs is approximately 3.8 V for sweeping voltages from +5 to −5 V, which is sufficient for non-volatile memory device applications. By using the accumulation capacitance determined from the $C–V_G$ curve of the 6 wt% PVDF MFSFET with 40 nm² pad dimensions and a 180 nm thickness, the

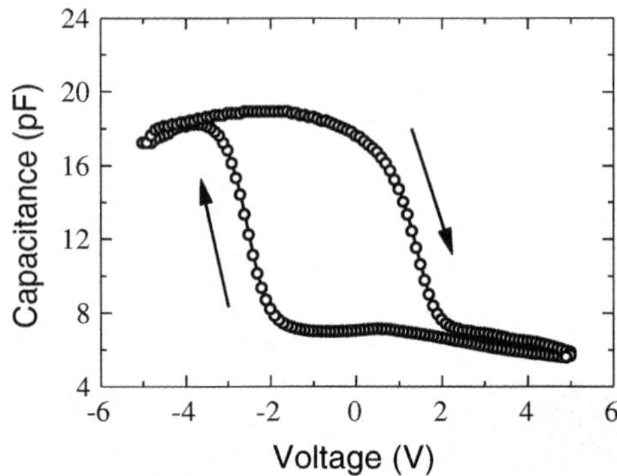

Fig. 7.8 Typical *C–V* characteristic of the Au/PVDF(6 wt%)/Si(100) structure in the fabricated MFSFETs

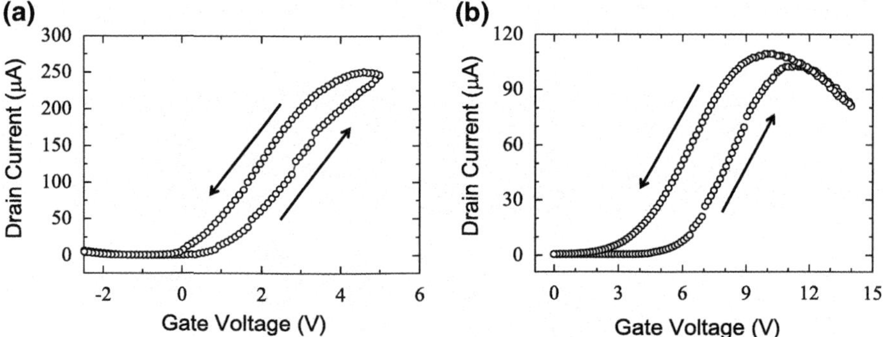

Fig. 7.9 **a** The drain current–gate voltage (I_D–V_G) characteristics of the MFSFETs with the 2 wt% PVDF film. **b** I_D–V_G characteristics of the MFSFETs with the 6 wt% PVDF film. The biased drain voltage was 2 V for both the 2 and 6 wt% MFSFETs

permittivity of the fabricated PVDF film is estimated to be approximately 10. Additionally, the same results were obtained in the case of the 2 wt% MFSFETs. However, the accumulation capacitances in some samples decreased at a high voltage. This phenomenon may be due to the fact that the current leakage may flow through the path between the crystalline and non-ferroelectric amorphous phases [13].

Additionally, the drain current–gate voltage (I_D–V_G) characteristics of the MFSFETs were also investigated. The I_D–V_G characteristics of the MFSFETs with the 2 and 6 wt% PVDF films are shown in Fig. 7.9. The biased constant drain voltage was 2 V for both the 2 and 6 wt% MFSFETs. Simultaneously, sweeping voltages from -2 to 5 V were applied to the 2 wt% MFSFETs. In the case of the 6 wt% MFSFETs, gate voltages from 0 to 14 V were delivered. As the thickness of the PVDF films increased according to the concentration of the PVDF solutions, the threshold voltage increased. The threshold voltages were 1 and 6.1 V for the 2 and 6 wt% MFSFETs, respectively. The saturated current of the 2 wt% MFSFETs is much larger than that of the 6 wt% MFSFETs because I_D is reciprocally proportional to the thickness. In terms of the direction, all of the I_D–V_G curves show hysteretic loops with a counterclockwise trace, which is the direction opposite to the C–V_G curves, as indicated by the arrows. Various scanning speeds of the gate voltage from 0.1 to 0.5 V/s were applied to the 2 and 6 wt% PVDF MFSFETs and no changes in the memory window width were observed. This also indicates that the ferroelectric nature results from the PVDF film in the β-phase. The memory window widths defined using the threshold voltage shift were approximately 1 V for the 2 wt% PVDF MFSFETs and 2.1 V for the 6 wt% PVDF MFSFETs. An observed difference between the memory window width of the I_D–V_G characteristics and that of the C–V_G characteristics of the 6 wt% PVDF MFSFETs is presented in Fig. 7.8, which was approximately 3.8 V. The cause of this difference in the memory window width between the C–V_G and I_D–V_G characteristics has not been clarified. A leakage current is a possible reason. In Fig. 7.9b, the I_D decreases above 10 V. This leakage current affects the threshold voltage of the MFSFETs. Thus, the memory window width of the I_D–V_G characteristics is less than that of the C–V_G characteristics by comparing Fig. 7.9a with Fig. 7.8. Incidentally, the transfer characteristic of the 6 wt% MFSFETs shown in Fig. 7.9b is adversely affected by a gate leakage current greater than 12 V. The cause of this leakage current is not yet clearly recognized. However, it is conjectured that it is ascribed to the carrier transport through the interface or gap states in the semiconductor [28]. It is likely due to process damage, which is a major cause.

Figure 7.10 shows the drain current–drain voltage (I_D–V_D) characteristics of the 2 and 6 wt% PVDF MFSFETs. Sweeping voltages from 0 to 5 V and step voltages of 1 V from 0 to 5 V were applied to the drain and gate electrodes of the 6 wt% MFSFETs. In the case of the 2 wt% PVDF MFSFETs, step voltages of 0.5 V from

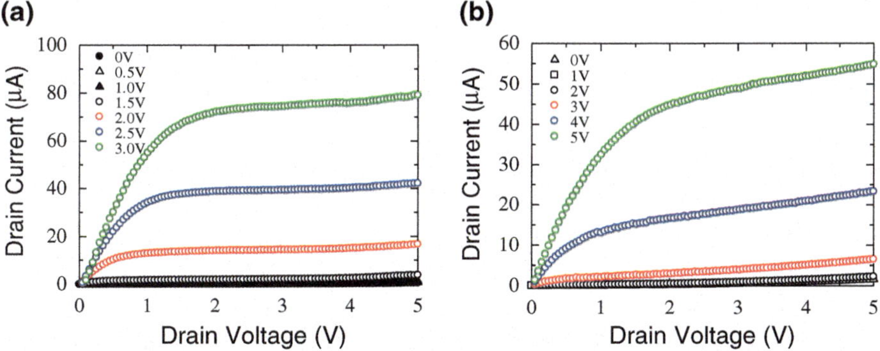

Fig. 7.10 a The drain current–drain voltage (I_D–V_D) characteristics of the 2 wt% PVDF MFSFETs. Sweeping voltages from 0 to 5 V and step voltages of 0.5 V from 0 to 3 V were applied. **b** The I_D–V_D characteristics of the 6 wt% PVDF MFSFETs. Sweeping voltages from 0 to 5 V and step voltages of 1 V from 0 to 5 V were applied

0 to 3 V were applied because the thickness was only approximately 25 nm. The 2 wt% PVDF MFSFET is relatively thinner than the 6 wt% PVDF MFSFET, and its current density was greater than 10^{-5} A/cm^2 at 5 V [13]. Therefore, in this MFSFET, breakdown easily occurred and the current density markedly increased to be greater than 3 V gate voltage. Because the ferroelectric layer is used as the gate dielectric layer in MFSFETs, the I_D–V_D characteristics show a typical operation of the metal–oxide–semiconductor field effect transistors (MOSFETs). Electrons are attracted to the surface by the polarization of the PVDF formed channel region and are permitted to flow from the source to the drain. The drain current I_D increased as the gate voltage increased, which implies that the PVDF thin films function as a gate dielectric layer. A comparison of the drain current magnitudes shows that the measured drain current of the 2 wt% PVDF MFSFETs is much larger than that of the 6 wt% PVDF MFSFETs at 3 V. The calculated gate electric fields are 1.2 MV/cm for the 2 wt% PVDF MFSFETs and 0.166 MV/cm for the 6 wt% PVDF MFSFETs at 3 V of gate voltage. This difference in electric field resulting from the difference between the thicknesses of the PVDF films affects the number of electrons that are induced on the surface. In Figs. 7.9 and 7.10, the I_D values in the I_D–V_G characteristics of the 2 wt% PVDF MFSFETs at V_G = 3 V and V_D = 2 V are approximately 130 (V_G increases) and 196 (V_G decreases) µA, respectively (Fig. 7.9a), but the I_D in the I_D–V_D characteristics of the 2 wt% PVDF MFSFETs at V_G = 3 V and V_D = 2 V is approximately 72 µA. Moreover, the I_D values in the I_D–V_G characteristics of the 6 wt% PVDF MFSFETs at V_G = 5 V and V_D = 2 V are approximately 14 (V_G increases) and 29 (V_G decreases) µA, respectively (Fig. 7.9b), but the I_D in the I_D–V_D characteristics of the 6 wt% PVDF MFSFETs at V_G = 5 V and V_D = 2 V is approximately 45 µA. Two possible reasons are considered for this phenomenon. One is the polarization direction; as voltage is delivered to the gate electrode, the direction of the polarization changes. Depending on the polarization direction of the PVDF film, the magnitude of the I_D characteristics will change. This might be the cause of the difference between the I_D–V_D and I_D–V_G characteristics at the same voltage. The other is the leakage current. The current density of the 2 wt% PVDF film in the MFS diode is approximately 10^{-5} A/cm^2 at 3 V, and the accumulation capacitance in the C–V characteristics decreases [13]. This leakage current may cause the difference in I_D between the I_D–V_D and I_D–V_G characteristics.

7.3.3 Electrical Properties of MFSFETs with PVDF-TrFE Thin Films

Figure 7.11 shows a typical capacitance-voltage characteristic for a Au/P(VDF-TrFE)/Si structure that was measured at 1 MHz. The thickness of the P(VDF-TrFE) film was approximately 100 nm. The P(VDF-TrFE) films showed clockwise hysteresis, as indicated by the arrows, and the value of the memory window width was approximately 1.4 V for the sweep range of ± 3 V. Additionally, the P(VDF-TrFE) film of the C–V curve was fully saturated in the

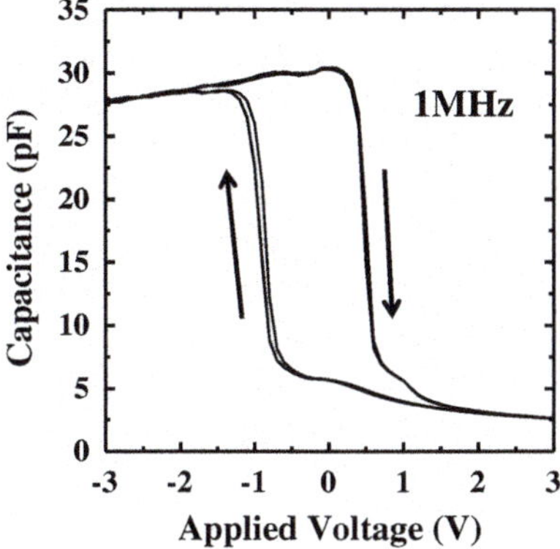

Fig. 7.11 Typical C–V characteristic for a Au/P(VDF-TrFE)/Si structure, which was measured at 1 MHz

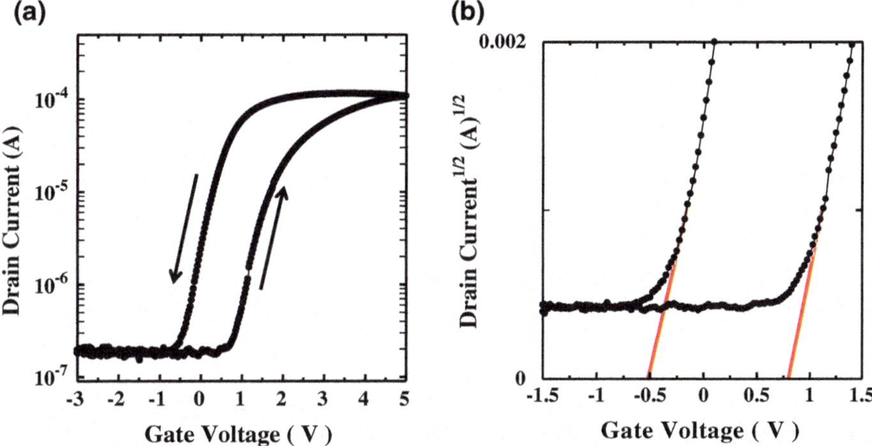

Fig. 7.12 **a** The drain current–gate voltage (I_D–V_G) characteristic curve of MFSFETs with P (VDF-TrFE) film. **b** The plot of the square root of the drain current–gate voltage ($I_D^{1/2}$–V_G)

accumulation region. In order to check the effect of mobile ionic charges on the hysteretic loop with the scanning speed of the bias voltage, the memory window width did not change within the scanning speed range investigated. These results suggest that the hysteresis characteristics are due to the ferroelectric property of the P(VDF-TrFE) film.

Figure 7.12a shows the I_D–V_G characteristic curve of MFSFETs with a P(VDF-TrFE) film. A hysteresis loop was obtained with a counterclockwise trace as indicated by the arrows. For measurement, gate voltage sweeps from −3 V to 5 V were performed while the drain voltage was fixed at 0.5 V. Additionally, the leakage current density and current on/off ratio were on the order of approximately 10^{-7} A and 10^3, respectively. The memory window width was defined by the threshold voltage shift, which also indicates that the threshold voltage (V_{th}) was obtained due to the shift of the turn-on voltage of the MFSFETs by the nature of the ferroelectric P(VDF-TrFE) film. Thus, the threshold voltage (V_{th}) values of the MFSFETs with P(VDF-TrFE) film were determined from the plot of the square root of the drain current versus gate voltage ($I_D^{1/2}$–V_G). As shown in Fig. 7.12b, the obtained memory window width (i.e., threshold voltage shift) of the MFSFETs with P(VDF-TrFE) film was approximately 1.3 V. This threshold voltage (V_{th}) value of the MFSFETs was nearly the same as that of the memory window width for the capacitor with the Au/P(VDF-TrFE)/Si structure shown in Fig. 7.11. Simply compared to the MFSFETs with PVDF film from the perspective of current on/off ratio, memory window width, and the shape of the hysteresis curve, the electrical properties of the MFSFETs with P(VDF-TrFE) are better than those of the PVDF MFSFETs. It is conceivable that a P(VDF-TrFE) film with ferroelectricity can be more easily formed than the PVDF film.

7.4 Conclusions

In summary, this study demonstrated a new and realizable possibility of next generation 1T-type ferroelectric RAM using MFS structures with ferroelectric polymer PVDF and P(VDF-TrFE) films for the first time. Hysteresis loops with a counterclockwise trace resulting from the ferroelectricity of PVDF with a β-phase were observed in the I_D–V_G characteristics. Moreover, the I_D–V_D characteristics of these MFSFETs also showed that PVDF thin films are functional as a gate dielectric layer. Additionally, MFSFETs can demonstrate a memory function due to the ferroelectric polarization of the P(VDF-TrFE) thin film. The threshold voltage of the MFSFETs with a P(VDF-TrFE) film shifted by approximately 1.3 V and the current on/off ratio was on the order of 10^3.

Without a buffer layer, the generation of the depolarization field was eliminated, which had been a major impediment in the development of a functioning device. Additionally, the structure of the memory cell can be simplified and fabrication process steps can be reduced. Furthermore, the proposed memory can still work, even with less than 1 V of operational voltage. Therefore, the proposed 1T-type MFSFETs with ferroelectric polymer PVDF and P(VDF-TrFE) films is a reliable method for achieving ultra-low cost, high-density, non-volatile random access memory with very minimal power consumption. This 1T-type polymer ferroelectric memory will expedite technical and industrial developments in the memory business.

References

1. J.F. Scott, C.A. Araujo, Science **246**, 1400–1405 (1989)
2. O. Auciello, J.F. Scott, R. Ramesh, Phys. Today **51**, 22–27 (1998)

3. Y. Arimoto, H. Ishiwara, MRS Bull. **29**, 823–828 (2004)
4. I.M. Ross, US Patent No. 2,791,760 (1957)
5. B.E. Park, H. Ishiwara, Appl. Phys. Lett. **79**, 806–808 (2001)
6. B.E. Park, K. Takahashi, H. Ishiwara, Appl. Phys. Lett. **85**, 4448–4450 (2004)
7. K.H. Kim, J.P. Han, S.W. Jung, T.P. Ma, IEEE Electron. Dev. Lett. **23**, 82–84 (2002)
8. K. Aizawa, B.E. Park, Y. Kawashima, K. Takahashi, H. Ishiwara, Appl. Phys. Lett. **85**, 3199–3201 (2004)
9. S. Sakai, R. Ilangovan, IEEE Electron. Dev. Lett. **25**, 369–371 (2004)
10. M. Takahashi, S. Sakai, Jpn. J. Appl. Phys. **44**, L800 (2005)
11. K. Takahashi, K. Aizawa, B.E. Park, H. Ishiwara, Jpn. J. Appl. Phys. **44**, 6218 (2005)
12. S.Y. Wu, IEEE Trans. Electron Devices **21**, 499 (1974)
13. J.H. Kim, D.W. Kim, H.S. Jeon, B.E. Park, Jpn. J. Appl. Phys. **46**, 6976 (2007)
14. J.H. Kim, J.N. Kim, B.E. Park, J. Korean Phys. Soc. **51**, 723 (2007)
15. T. Furukawa, Phase Transition **18**, 143 (1989)
16. D.W. Kim, G.G. Lee, B.E. Park, J. Korean Phys. Soc. **51**, 719 (2007)
17. R.C.G. Naber, J. Massolt, M. Spijkman, K. Asadi, P.W.M. Blom, Appl. Phys. Lett. **90**, 113509 (2007)
18. S. Fujisaki, Y. Fujisaki, H. Ishiwara, Appl. Phys. Lett. **90**, 162902 (2007)
19. S.J. Kang, Y.J. Park, J.W. Sung, P.S. Jo, C.M. Park, K.J. Kim, B.O. Cho, Appl. Phys. Lett. **92**, 012921 (2008)
20. T. Furukawa, Phase Trans. **18**, 143–211 (1989)
21. Q.M. Zhang, V. Bharti, X. Zhao, Science **280**, 2101–2104 (1998)
22. G.H. Gelinck et al., Nature Mater. **3**, 106–110 (2004)
23. A. Rose, Z. Zhu, C.F. Madigan, T.M. Swager, V. Bulovic, Nature **434**, 876–879 (2005)
24. J. Veres et al., Adv. Funct. Mater. **13**, 199–204 (2003)
25. B.E. Park, Korea Patent No. KR 10-0966301 (2010)
26. B. E. Park, Japan Patent No. JP 5,241,489 B2 (2013)
27. B.E. Park, US Patent No. US 8,120,082 B2 (2012)
28. M.W.J. Prins, S.E. Zinnemers, J.F.M. Cillessen, J.B. Giesbers, Appl. Phys. Lett. **70**, 458 (1997)

Chapter 8
Poly(Vinylidenefluoride-Trifluoroethylene) P(VDF-TrFE)/Semiconductor Structure Ferroelectric-Gate FETs

Yoshihisa Fujisaki

Abstract Ferroelectric field effect transistors (FeFETs) composed of P(VDF-TrFE) (Poly(Vinylidenefluoride-Tirfluoroethylene)) thin films and semiconductor substrates show excellent ferroelectric transistor characteristics. Since P(VDF-TrFE) has the ferroelectric polarization as large as those of oxide ferroelectric materials with much lower dielectric constant, it is an ideal material to build FeFETs with the combination to inorganic semiconductor materials. In addition, the process condition to form P(VDF-TrFE) is much milder to underlying semiconducting material compared to oxide ferroelectrics. Therefore, the improvement on the retention characteristics is expected by employing P(VDF-TrFE) ferroelectrics in FeFET instead of oxide ferroelectrics. The potential of P(VDF-TrFE) FeFET is discussed in this chapter.

8.1 Introduction

A ferroelectric gate transistor (FeFET; Ferroelectric Field Effect Transistor) is thought to be an ultimate non-volatile memory because of its low power and non-destructive read out characteristics. Therefore a FeFET is superior to a capacitor type ferroelectric memory in which stored information is destroyed during read out operation (destructive read out). However FeFET is not in production yet since it has crucial deficit of a short retention character. A number of demonstrations has been reported to overcome this inherent fault of FeFETs till now, but no one has succeeded to solve this difficult problem.

Y. Fujisaki (✉)
YourFriend, Hachioji, Tokyo 192-0916, Japan
e-mail: yoshihisa.fujisaki@yourfriend.tokyo

© Springer Science+Business Media Dordrecht 2016

B.-E. Park et al. (eds.), *Ferroelectric-Gate Field Effect Transistor Memories*, Topics in Applied Physics 131, DOI 10.1007/978-94-024-0841-6_8

8.1.1   Problem of Oxide/Silicon Based FeFETs

The short retention characteristics of FeFETs is originated from a low dialectic layer emerged between a ferroelectric layer and under lying semiconductor channel. In general, the material used for a ferroelectric layer is oxide material such as SBT ($SrBi_2Ta_2O_9$) or BLT ($Bi_{3.25}La_{0.75}Ti_3O_{12}$), and the material used for a semiconductor layer is Si. During fabrication process, Si channel is damaged by the highly oxidizing ambient of oxide ferroelectric layer formation. This damaged inter-layer has a lower dielectric constant compared to that of the oxide ferroelectric layer. As a result, it becomes difficult to apply enough voltage to a ferroelectric gate layer of a FeFET according to the capacitance coupling law. The capacitance coupling law is explained schematically in Fig. 8.1. Two capacitors are composed of a ferroelectric capacitor and a capacitor representing the damaged inter-layer. The thickness and dielectric constant of these two capacitors are d_1, ε_1 and d_2, ε_2, respectively. Because of the charge neutrality law, the amount of charge appear between two capacitors should be equal with different signs. If the voltage V is applied to this series capacitance, the resultant voltage applied to the ferroelectric capacitor is calculated as shown in (8.1).

$$V \times \left(\frac{V_1}{V_1 + V_2} \right) = \frac{V}{1 + \frac{V_2}{V_1}} = \frac{V}{1 + \frac{(\varepsilon_1/d_1)}{(\varepsilon_2/d_2)}} \tag{8.1}$$

For instance, consider that the gate of a FeFET is composed of a 100 nm thick ferroelectric layer having the dielectric constant of 200 and a 5 nm damaged inter-layer having the dielectric constant of 5. Substituting these parameters to the (8.1) gives the result shown in (8.2).

$$V \times \left(\frac{V_1}{V_1 + V_2} \right) = \frac{V}{1 + \frac{(\varepsilon_1/d_1)}{(\varepsilon_2/d_2)}} = \frac{V}{1 + \frac{(200/100)}{(5/5)}} = \frac{1}{3}V \tag{8.2}$$

Fig. 8.1 The capacitance coupling law

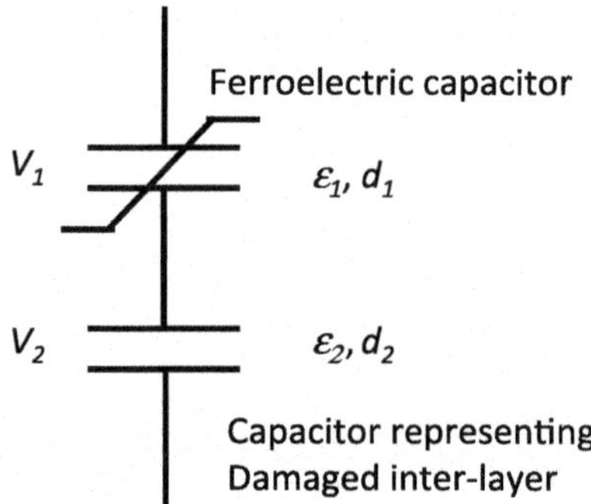

This means that only one third of applied voltage works effectively to operate a ferroelectric capacitor. To improve the voltage coupling of a ferroelectric capacitor, it is important to avoid the emergence of a damaged inter-layer. However, it is also important to deposit high quality ferroelectric layer on silicon under highly oxidizing ambience in order to achieve long retention characteristics. It is a big dilemma to fabricate high quality ferroelectric layer without forming a thick damaged inter-layer on a silicon substrate.

Therefore, the gate structure having a very thin diffusion barrier layer with high dielectric constant is frequently employed. This structure is called MFIS which stands for the stacked layers made of Metal, Ferroelectric, Insulator, and Semiconductor. Insulators used in this MFIS structure are such as Si_3N_4, HfO_x, and AlO_x having higher dielectric constants compared to SiO_2. This insulator should also work as a diffusion barrier of oxygen during the oxide ferroelectric layer formation. If this diffusion barrier works well, relatively longer retention characters can be achieved [1–4]. This is because dipole moments in a ferroelectric layer become stable if 100 % of moments are aligned in one direction by applying enough voltage to saturate the ferroelectricity. It is difficult to achieve this situation since electric field in the thin insulator in a MFIS structure easily reaches the breakdown field with the low voltage application.

It is also effective to employ a thin ferroelectric material with low dielectric constant in order to improve the above mentioned voltage coupling. However, oxide ferroelectrics usually have dielectric constants around 100 with thickness larger than several tens nm. Therefore, the insulating layer in a MFIS structure should be as thin as a few nm and also should have dielectric constant larger than 10 at the same time. Exploring the insulator having such physical parameters is most concern of the researchers developing FeFETs with excellent retention characteristics.

Organic ferroelectric materials have lower dielectric constants compared to those of oxide ferroelectrics. In addition, organic ferroelectric films can be deposited on silicon substrates under much milder condition compared to oxide ferroelectrics do. Therefore, it is one of solutions to employ organic ferroelectrics in a MFIS structure to achieve high performance FeFET with excellent retention characteristics.

8.1.2 Property of Organic Ferroelectric Material; P(VDF-TrFE)

An organic ferroelectric material P(VDF-TrFE) is a random co-polymer composed of vinylidenefluoride; VDF ($-CF_2-CH_2-$) and trifluoroethylene; TrFE ($-CF_2-CHF-$). The molecule model of a P(VDF-TrFE) oligomer is shown in Fig. 8.2. In a P(VDF-TrFE) molecule, VDF and TrFE are aligned like a chain in which TrFE is randomly arranged as shown in Fig. 8.2a. In Fig. 8.2b, electron clouds are schematically drawn. The maldistribution of electrons like in Fig. 8.2b is the origin of polarization in an oligomer molecule. It is easily understood that ferroelelcricity

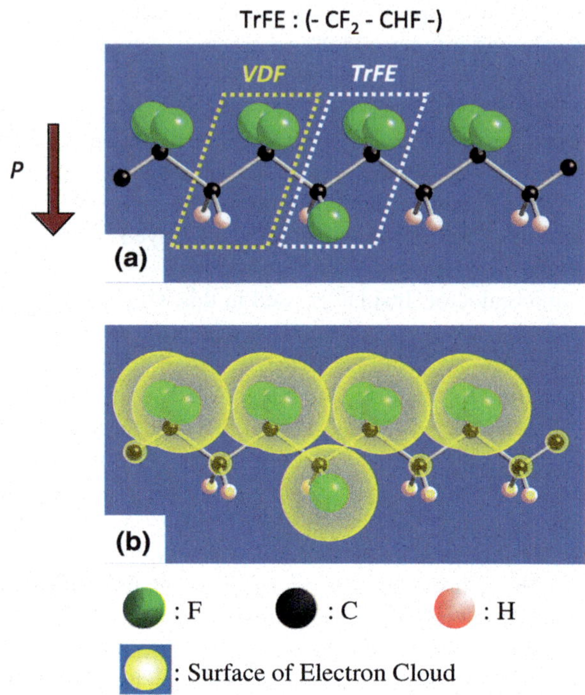

Fig. 8.2 The molecule model of a P(VDF-TrFE) oligomer

of P(VDF-TrFE) is originated from VDF that was found by DUPON as a family of tefron in 1938. The ferroelectricity of VDF was found by Kawai [5]. In order to understand the ferroelectricity of P(VDF-TrFE), it is worthy to understand the ferroelectric property of VDF. PVDF is a crystalline polymer that is expressed by the chemical formula $(-CF_2-CH_2-)_n$. It is known that PVDF has four crystalline phases called α, β, γ and δ. The most stable phase is α and has a TGTG′ structure. (T; Trans, G; Gauche, G′; anti-Gauche, Trans and Gauche are French words that mean straight and left, respectively.) The crystal structure of α phase is shown in Figs. 8.3 and 8.4 [6]. The drawings in Fig. 8.3 show a unit cell of α phase crystal. If the direction of a polymer chain is determined to be c-axis, lattice parameters of a $(-CF_2-CH_2-)$ monomer become a = 0.496 nm, b = 0.964 nm and c = 0.462 nm, respectively. The structure of α phase crystal with plural unit cells observed from a-axis is shown in Fig. 8.4. As shown in this figure, zigzag chains composed of carbon atoms aligned in c-axis make the main structure of α phase crystal. This zigzag alignment of carbon atoms is expressed by TGTG′ in chemistry. This zigzag structure is suitable for relaxing the interaction of fluorine atoms in the neighboring chains, since a fluorine atom has a large ionic radius. If this structure is glanced from the direction parallel to c-axis, it is clear from Fig. 8.3 that fluorine atoms in the neighboring zigzag chains are arranged in the other side of carbon chains with each other. This is also effective to relax the interaction of fluorine atoms in the

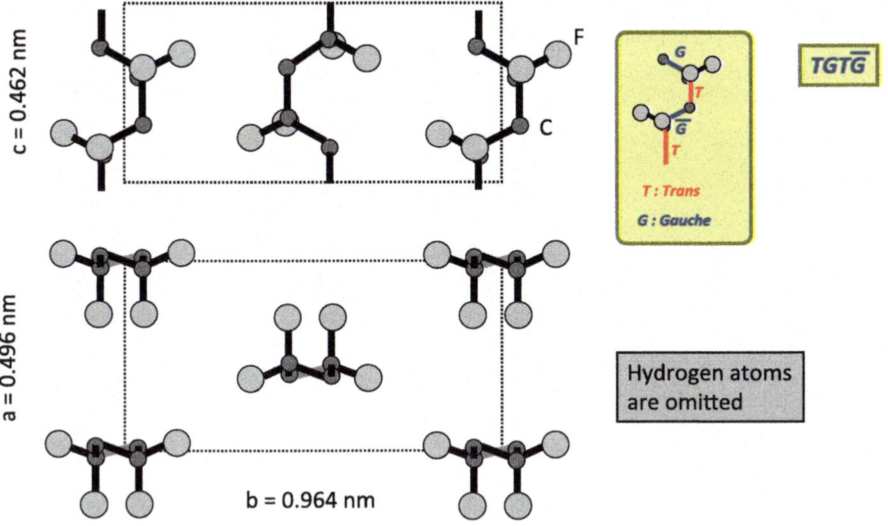

Fig. 8.3 Unit cell of α phase crystal

neighboring zigzag chains. Therefore, α phase is the most stable structure among four phases of a PVDF crystal. In this configuration, dipole moments emerge across neighboring zigzag carbon chains and point to the counter directions with each other. Therefore, polarization is not created in α phase since dipole moments in the neighboring chains contradict with each other.

The crystal structure of β phase PVDF is shown in Figs. 8.5 and 8.6. In β phase, fluorine and hydrogen atoms are arranged in the 180° counter side of a carbon chain. This structure is more shown in Fig. 8.5 in which ab-plain is perpendicular to the c-axis. In this figure, dipole moments directed from fluorine to hydrogen atoms are shown by arrows. In this configuration, it is clear that dipole moments are aligned in one direction. Therefore, β phase crystal of PVDF is ferroelectric. It is known that β phase crystal of PVDF shows the maximum ferroelectricity, piezo-electricity and pyroelectric property among various organic materials. Therefore this material is the most important one in organics for wide practical applications. The lattice parameters of β phase crystal are a = 0.496 nm, b = 0.858 nm and c = 0.256 nm. The β phase crystal structure observed from the direction parallel to the a-axis is shown in Fig. 8.6. As is clear from Fig. 8.6, β phase has a structure in which carbon chains are aligned in more straight fashion compared to that in α phase having the zigzag carbon chain configuration. This β phase configuration is called an all-trans structure. This all-trans configuration is composed of a structure in which fluorine atoms having high electron negativity are located in one side of a carbon chain. Therefore, this configuration cannot exist with a single polymer chain since the internal energy of a chain becomes too high. This configuration becomes stable by composing β phase crystal structure. This β phase structure with all-trans configuration is the key issue in controlling the ferroelectricity of PVDF.

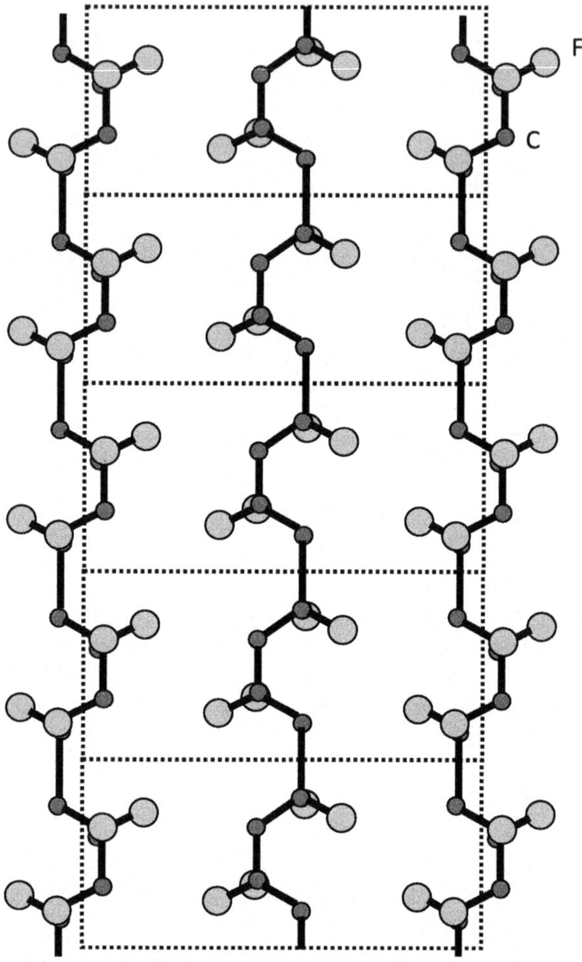

Fig. 8.4 The structure of α phase crystal with plaural unit cells observed from a-axis

PVDF has other crystal phases called γ and δ. In Fig. 8.7, crystal structures of γ and δ phases are shown. The γ phase structure has T_3GT_3G' configuration. The δ phase structure has the same chain structure but the direction of dipole moments of neighboring chains is different from that in α phase. In these figures, dipole moments originated from a chain in a VDF structure are shown by arrows. Since arrows point in one direction, both γ and δ phase crystals are ferroelectric. However, the density of dipole moments is much smaller compared to that of β phase, thus γ and δ phase crystal structures are not so important in the practical application.

It is known that four phases of PVDF crystals mentioned above can be transformed from one another by applying energy to the crystal. In Fig. 8.8, phase transfer diagram among four phases of PVDF crystals is shown [7]. By applying tension to the most stable thin film in α phase, the crystal can be transformed to β

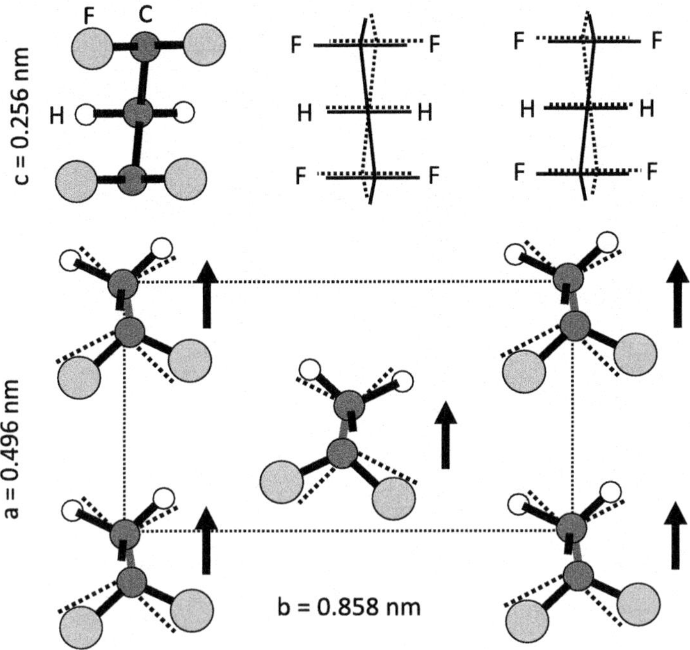

Fig. 8.5 This structure is more explicitly

phase. If one directional tension is applied to PVDF film in α phase, the film is converted to 100 % β phase. If two dimensional tension is applied to the film, it is converted to the structure composed of 50 % β phase and 50 % α phase. It is expected that PVDF molecular chain in a TGTG′ zigzag form shown in Fig. 8.4 is stretched to a straight all-trans form shown in Fig. 8.6 by the tension applied to α phase film. Since pyroelectric α phase is most stable, it is difficult to make PVDF crystal in single ferroelectric β phase. However, the measured remnant polarization shows almost maximum value as if the film is in the single β phase. This is caused by the phase transformation by electric field as shown in Fig. 8.8. Under the very high electric field, α phase can be transformed directly to β phase. Beside the direct path, Fig. 8.8 shows that α phase can be transformed to δ phase and then δ phase can be transformed to β phase only by high electric field. By comparing Figs. 8.3 and 8.7, it can be understood that α phase is transformed to δ phase by the 180° rotation of one zigzag carbon chain. In other words, TGTG′ configuration is transformed to TG′TG configuration. Such phase transformation can be achieved by applying high electric field. However, Lovinger showed that transformation from TGTG′ to TG′TG can be achieved by local shift of atoms in a zigzag chain [3].

Since the Curie temperature Tc of PVDF is higher than the melting point Tm, the phase transformation from α to β was thought to be impossible [4]. However, it was experimentally demonstrated that remnant polarization of 10 μC/cm^2 was achieved by annealing α phase film. The discussion is not over yet.

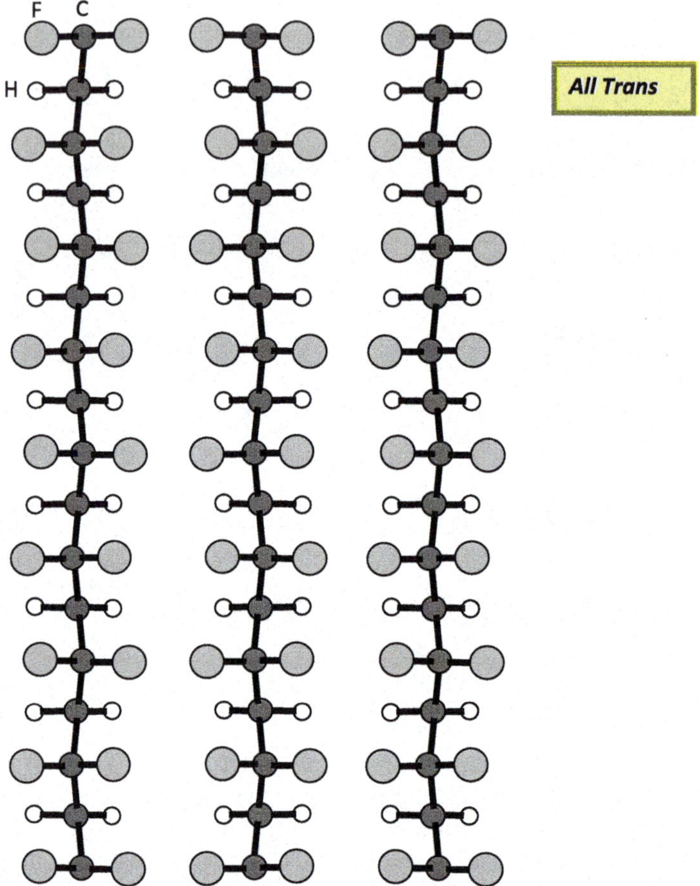

Fig. 8.6 The β phase crystal structure observed from parallel to the a-axis

Commercially available PVDF has the molecular weight around 4×10^6 that is composed of about 60,000 VDF monomers in average. Polymer chains have $(-CF_2-CH_2-)_n$ structure but commercially available ones have defects with the density of 4–5 %. There are two types of defects called "head to head; HH" and "tail to tail; TT". HH and TT have the structures of $(-CF_2-CF_2-)$ and $(-CH_2-CH_2-)$, respectively. By controlling the synthesis condition, defect density can be changed from about 3.5–6 % [5, 6].

There is a strong relation between the defect density of polymer chain and the crystal structure. It was theoretically expected that α phase is most stable if the defect density is less than 10 % and that β phase is most stable if the defect density is larger than 10 %. Farmer investigated theoretically the effect of HHTT defect

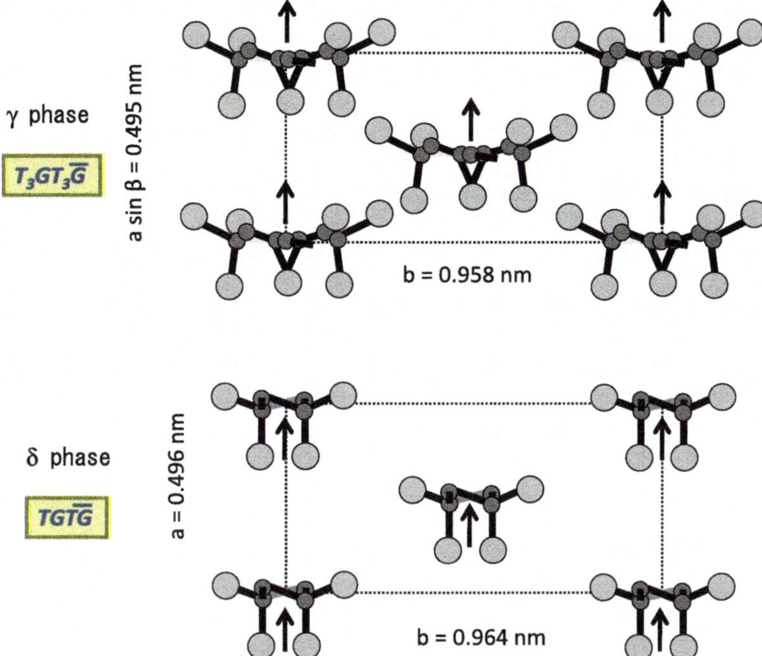

Fig. 8.7 Crystal structures of γ and δ phases

density on the energy of β phase (all-trans) and α phase (TGTG′) [7]. In the calculation, both steric and electrostatic potential were taking into account using the potential energy function proposed by DeSantis [8]. The calculated energy of two chain configurations as a function of defect density is shown in Fig. 8.9. If the defect density is less than 10 %, the energy of TGTG′ configuration becomes smallest and α phase is expected to be most stable. On the other hand, β phase is most stable if defect density is larger than 12 % since the energy of all-trans configuration becomes smallest. This expectation was experimentally proved.

The effect of HHTT defect density on crystal structure is understood as follows. In Fig. 8.10, TGTG′ configuration is shown in which HH and TT defects are introduced. In the part of neighboring chains starting from TT defect to HH defect, the steric hindrance of fluorine atoms (colored in red) takes place between two neighboring polymer chains. Since fluorine atoms have large ionic radii of 0.15 nm and also have large electro-negativity, the energy of the crystal with the defect configuration shown in Fig. 8.10 becomes too large. To the contrary, the energy increase in the all-trans β phase configuration by the introduction of HHTT defects in carbon chains is negligible since steric hindrance is not so serious as is obvious from Fig. 8.11. Therefore, β phase configuration is more tolerable for the intro-duction of defects and α phase is not tolerable to the defect introduction. In other words, ferroelectric β phase becomes stable if more than 10 % of defects are

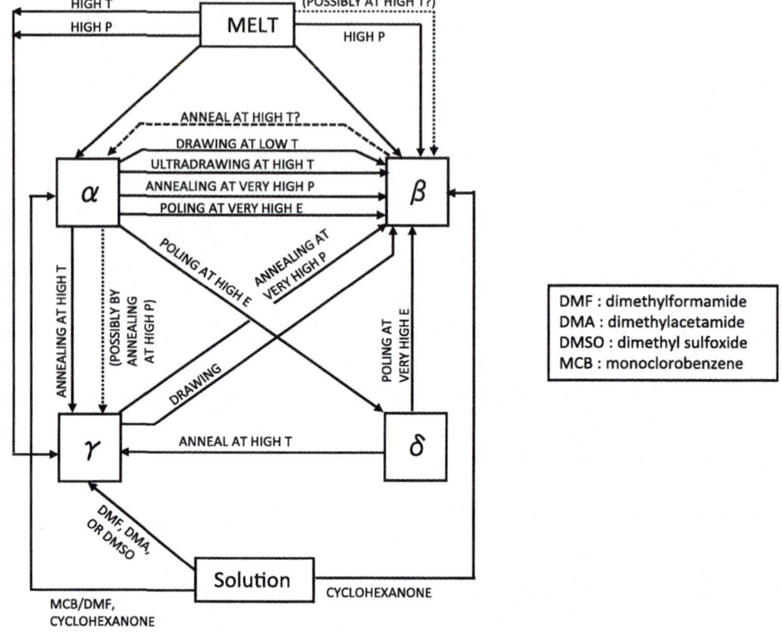

Fig. 8.8 The phase transformation by electric field

Fig. 8.9 The calculated energy of two chain configurations as a function of defect density

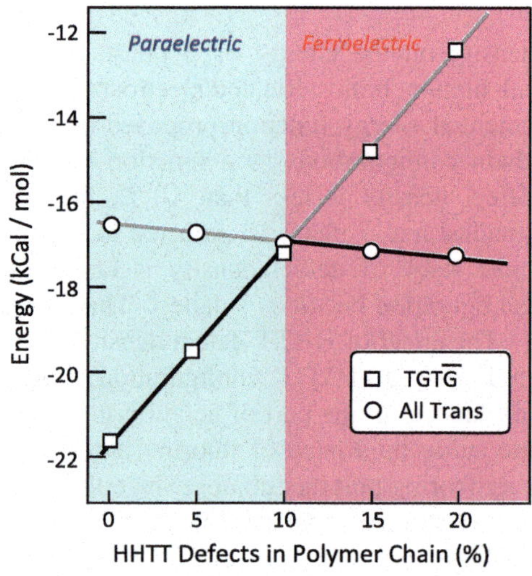

introduced in the crystal. As shown in Fig. 8.9, the energy of TGTG′ configuration increases monotonically as a function of defect density. This dependence can be understood as a result of the steric hindrance of fluorine atoms in neighboring carbon chains. Since steric hindrance is not serious in β phase crystal as shown in

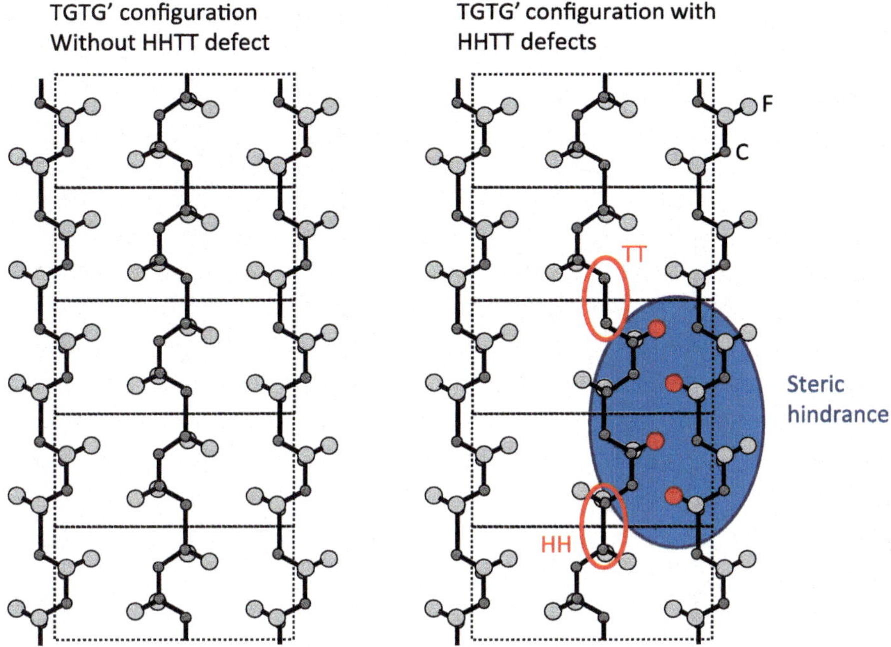

Fig. 8.10 The energy of the crystal with the defect configuration

Fig. 8.11, the energy of all-trans configuration does not much depend on defect density as shown in Fig. 8.9.

Because HH and TT defects in PVDF are created statistically, it is difficult to control their density precisely only by modifying the synthesis conditions. Introduction of TrFE in PVDF makes similar steric hindrance just as HHTT defects do in the polymer chains. In Fig. 8.12, the α phase configuration of the copolymer made of VDF and TrFE is shown. Clear from the figure, steric hindrance of fluorine atoms in the neighboring TGTG' chains takes place at the position where TrFE is introduced. This is the similar configuration as the α phase PVDF with HH and TT defects as shown in Fig. 8.10. This means that the introduction of TrFE to PVDF works as if HH and TT defects are made in the polymer chains. To the contrary, no serious steric hindrance is observed in the case of β phase crystal as shown in Fig. 8.13. Therefore, phase control becomes possible by doping TrFE to PVDF based on the calculation shown in Fig. 8.9. In other words, β phase ferroelectric PVDF can be obtained by doping some amount of TrFE to PVDF. Since HH and TT defects are always introduced in a carbon chain in a pair, the critical amount of TrFE with which α to β phase transition occurs should be twice of the HHTT defect density in Fig. 8.9. Namely, by doping 20–24 mol% of TrFE to PVDF, β phase can be obtained with high controllability.

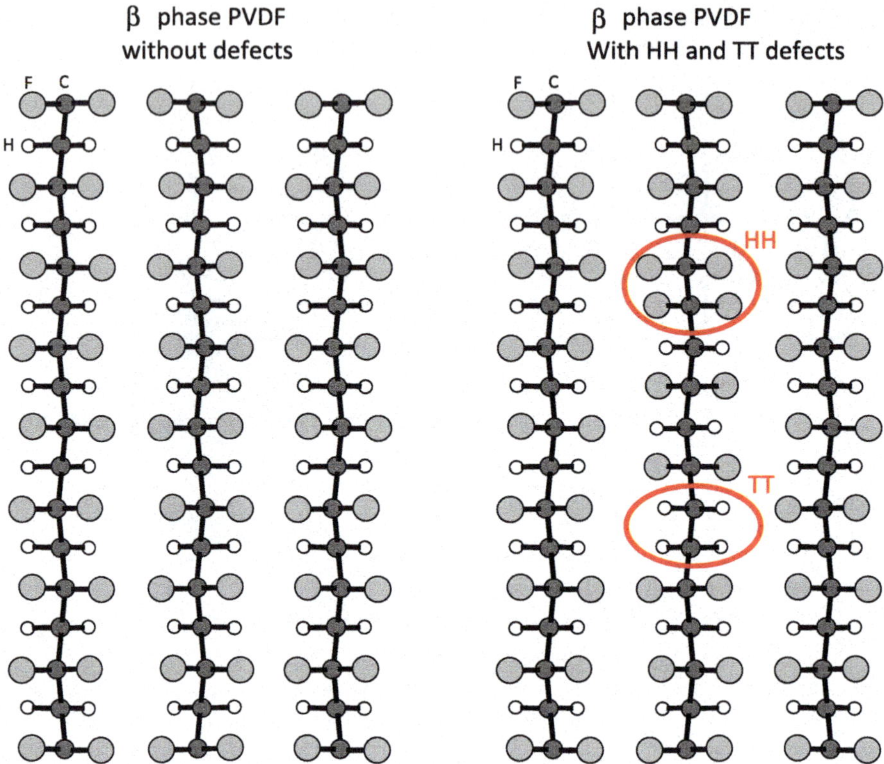

Fig. 8.11 Steric hindrance is no serious in β phase crystal

8.2 Ferroelectric Properties of VDF Based Polymers

8.2.1 Ferroelectricity of PVDF

It was confirmed experimentally that the β phase PVDF is ferroelectric. It was proved by X-ray diffraction that polarization reversal of β phase PVDF is performed by the 60° rotation of polymer chains. It is explained by Figs. 8.14 and 8.15. In Fig. 8.14, X-ray diffraction spectra of α and β phase PVDF are shown [9]. It is obvious that the peaks of XRD spectra shown in Fig. 8.14 correspond to the lattice parameters shown in Figs. 8.3 and 8.5. The β phase specimen used in taking the spectrum in Fig. 8.14 was prepared by rewinding a hot PVDF film. During the rewind process, the film was stretched and cooled rapidly. This rewind process applies one-dimensional tension to a PVDF film during cooling process. As a result of this process, PVDF becomes 100 % ferroelectric β phase. Before taking X-ray diffraction, the PVDF sheet was poled by applying strong electric field to make dipole moments in the specimen aligned in one direction. By keeping the 110 and 200 diffraction conditions of β phase ($2\theta \sim 20°$), sample was rotated by 360°.

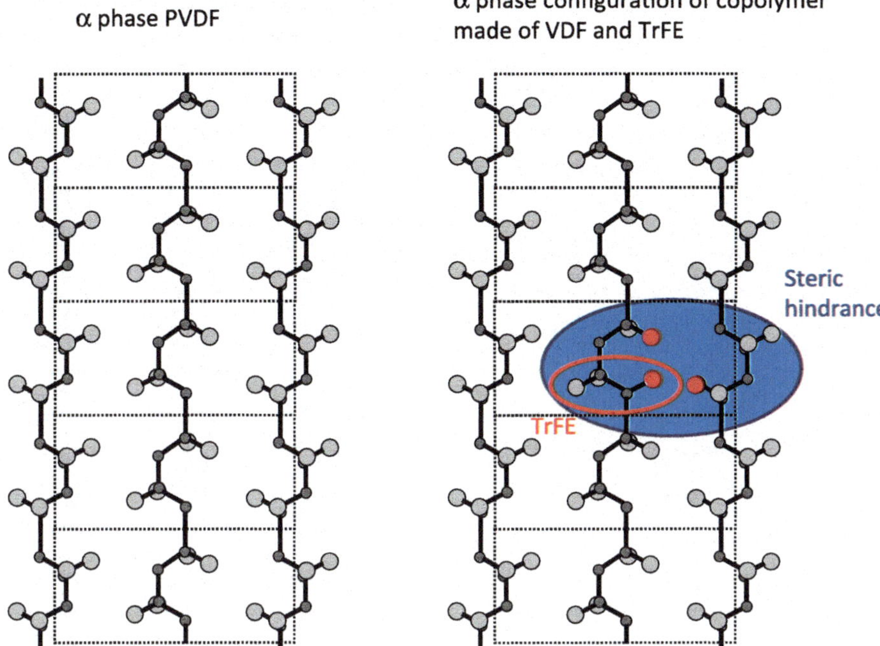

Fig. 8.12 The α phase configuration of the copolymer made of VDF and TrFE

The result is shown in the bottom chart of Fig. 8.15 [10]. Six peaks were observed at every 30° while the sample was rotated by 360°. The dots in the chart were the experimental data and the curve in the chart was the calculated one from the model assuming three arrangements drawn above the chart. The three arrangements were (a) 90° domain that was not poled, (b) the domain lent 30° to the left, and (c) the domain lent 30° to the right. It is obvious from the chart that the experimental result agrees well with the calculated curve. Therefore, it was proved that polarization reversal of β phase PVDF takes place by rotating the domain by 60°. It should be noted that 180° rotation of the domain in PVDF will not take place.

Dvey-Aharon calculated the energy needed to rotate polymer chains in β phase PVDF [11]. Essence of the calculation is shown in Fig. 8.16. The energy required to rotate an all-trans chain was calculated by taking the following three factors into account; interaction between chains, electric field and the stiffness of the chain itself against the rotation. As a result, the stable angle of rotation was found to be 30° and 60°. These molecular chains form domains to minimize the internal energy of PVDF crystal. Therefore, polarization switching of ferroelectric PVDF was found to take place by moving a domain boundary which is made of two domains having 60° angle with each other. A domain wall moves with the one by one rotation of a chain by 60°. For instance, consider the rotation of a chain hatched by a circle. This polymer chain is surrounded by six neighboring chains, three of which have the

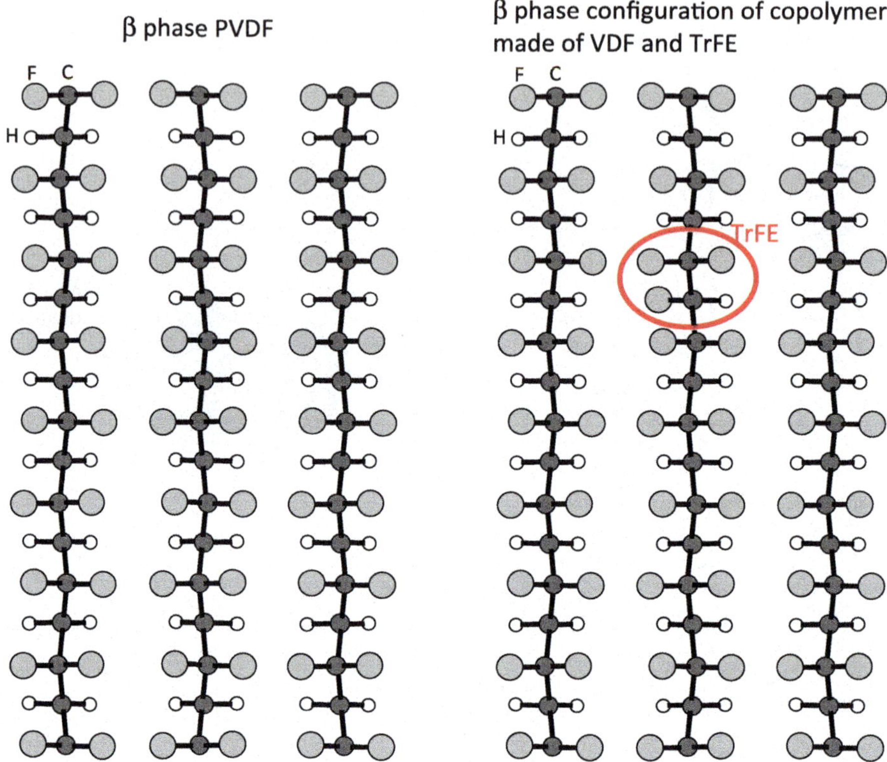

Fig. 8.13 To the contrary, no serious steric hindrance is not observed in the case of β phase crystal

Fig. 8.14 X-ray diffraction spectra of α and β phase PVDF

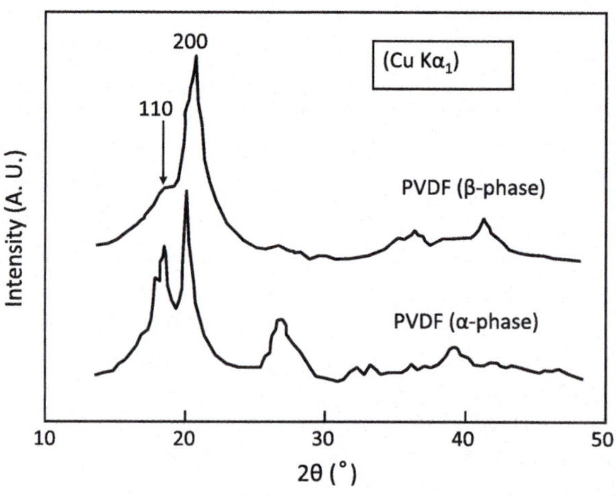

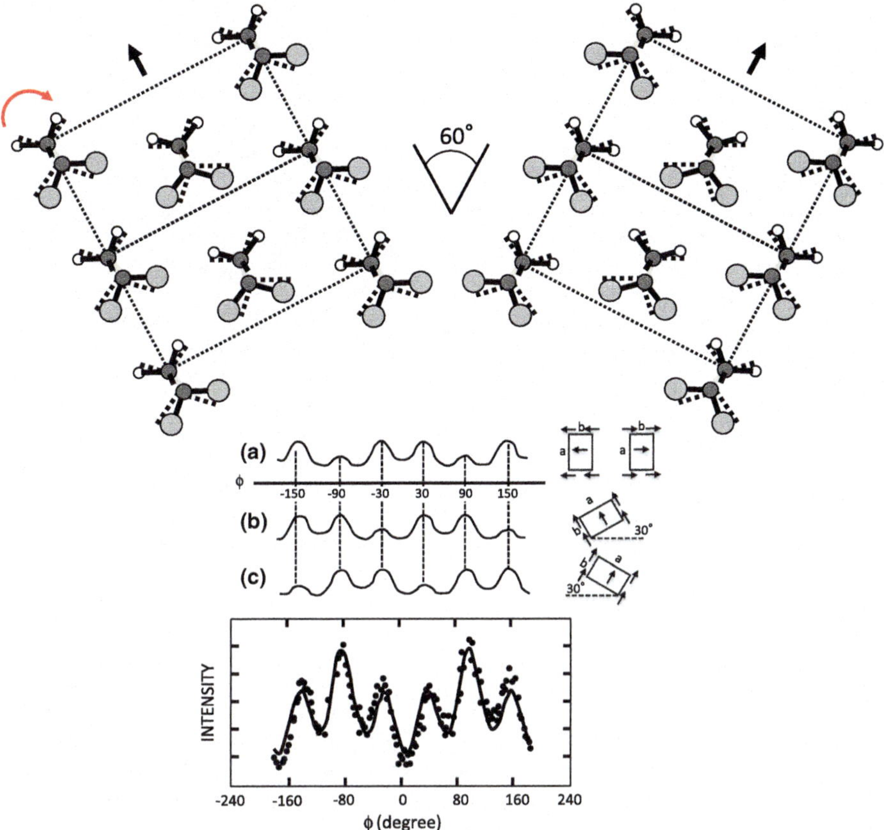

Fig. 8.15 X-ray diffraction that polarization reversal of β phase PVDF is performed by the 60°
rotation of polymer chains

aligned dipole moments making 60° angle with the aligned dipole moments of the
other three polymer chains. Therefore, the polymer chain in the circle feels the
almost neutral electric field balanced by surrounding six polymer chains. As a
result, this polymer chain can easily rotate by the electric field applied from the
electrodes attached to the sample. Since a polymer chain is quite flexible for 60°
rotation, domain walls move at high velocity.

By incorrect understanding on the electrical response of organic materials,
polarization reversal of PVDF was thought to be slow. However, the actual
response is very quick since the domain wall movement is performed according to
the scheme shown in Fig. 8.16. It was investigated both theoretically and experimentally how the polarization reversal depends on temperature and electric field.
Clark and Taylor estimated the velocity of polarization reversal based on the 60°
domain model shown in Fig. 8.16 [12]. Result is summarize in Fig. 8.17. It is
expected that the velocity of polarization reversal depends both on temperature and

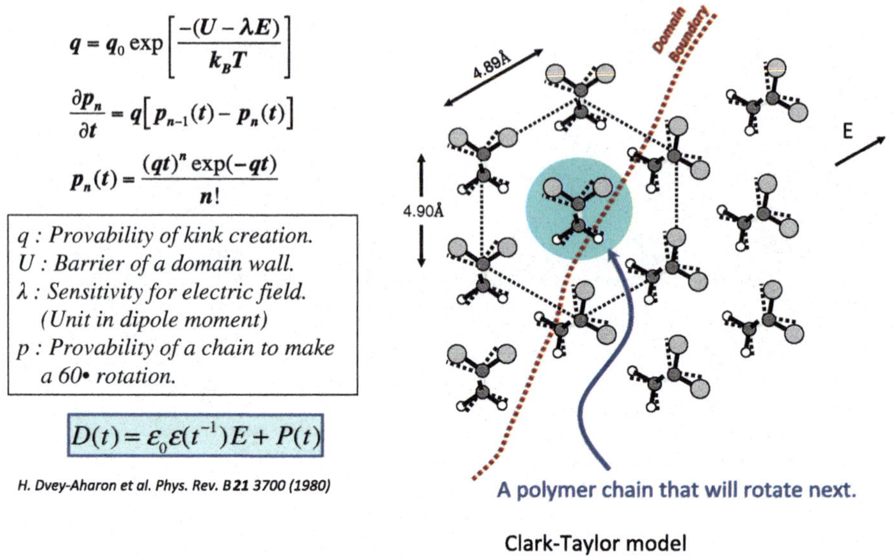

$$q = q_0 \exp\left[\frac{-(U - \lambda E)}{k_B T}\right]$$

$$\frac{\partial p_n}{\partial t} = q\left[p_{n-1}(t) - p_n(t)\right]$$

$$p_n(t) = \frac{(qt)^n \exp(-qt)}{n!}$$

q : Provability of kink creation.
U : Barrier of a domain wall.
λ : Sensitivity for electric field.
  (Unit in dipole moment)
p : Provability of a chain to make
  a 60° rotation.

$$D(t) = \varepsilon_0 \varepsilon(t^{-1})E + P(t)$$

H. Dvey-Aharon et al. Phys. Rev. B **21** 3700 (1980)

4.89Å
4.90Å
Domain Boundary
E
A polymer chain that will rotate next.

Clark-Taylor model

Fig. 8.16 Essence of the calculation

electric field. It is also expected that the polarization can be performed during a few ns duration at room temperature. These were proved by Furukawa experimentally. The result is summarized in Fig. 8.18 [11]. Dots in the figure were data taken by Furukawa and the curves were calculated based on the 60° model by Clark and Taylor shown in Fig. 8.17. Experiments and calculation match very well as shown in the figure. However, there is a little quantitative incoincidence between the calculation and experimental data in the dependence at low electric field less than 80 MV/m. To improve the prediction, phenomenological correction has to be introduced to the theory just like in the case of oxide ferroelectric materials. It was proved that the velocity of polarization reversal is very high. However, there seems to be a problem that the velocity of polarization reversal depends strongly on electric field. To operate actual devices in an integrated circuit, this type of characteristics is not desirable.

The density of remnant polarization of PVDF can be estimated by the simple calculation assuming that origin of a unit dipole moment is made by a H-C-F arrangement in a polymer chain. If the density of H-C-F in PVDF is simply taken into account, the polarization density becomes 13.1 μC/cm^2. If the inner electric field made by polarization is taken into account, the polarization density becomes 12.7 μC/cm^2 [13–20]. The experimental values vary from 6 to 10 μC/cm^2 since the actual material is not 100 % crystallized.

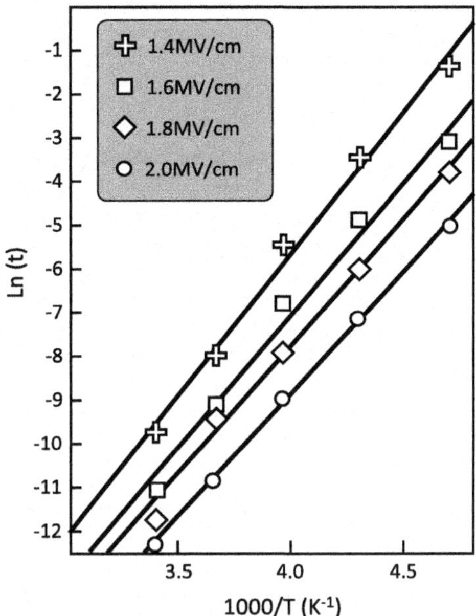

Fig. 8.17 Result is summarize

8.2.2   *Issues that Should Be Solved Before Application*

It becomes clear that β phase ferroelectric film can be obtained by adding 20–24 mol% TrFE to PVDF with high reproducibility. It is also confirmed that random copolymer P(VDF-TrFE) can be used as a material to make a high performance FeFET. This is because that P(VDF-TrFE) can be deposited on silicon at low temperature without degrading the channel of FET located at the surface of silicon. The maximum temperature of ferroelectric P(VDF-TrFE) film deposition is around 150 °C that is much lower than those of oxide ferroelectrics formation. Therefore, it is expected that the emergence of damaged inter-layer in silicon channel can be suppressed by using P(VDF-TrFE) instead of oxide ferroelectrics. Another important point is the low dielectric constant of P(VDF-TrFE). There are variations in the reported values but the average value for the dielectric constant is about 15 that is smaller by one order of magnitude than those of oxide ferroelectrics.

The merit of the introduction of P(VFE-TrFE) ferroelectrics instead of oxide materials in a FeFET can be confirmed easily. For instance, the voltage coupling of a FeFET is calculated as follows. Since the process temperature is very low, the damaged inter-layer emerged on the surface of silicon is limited. Assume that the dielectric constant and the thickness of P(VDF-TrFE) are 15 and 100 nm, respectively and the dielectric constant and the thickness of damaged inter-layer on silicon are 5 and 1 nm, respectively. The voltage coupling calculated by the (1) becomes

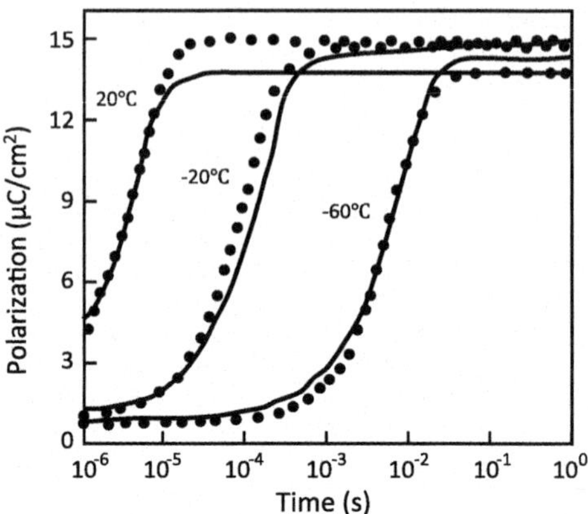

Fig. 8.18 The result was summarized

$$V \times \left(\frac{V_1}{V_1 + V_2} \right) = \frac{V}{1 + \frac{(\varepsilon_1/d_1)}{(\varepsilon_2/d_2)}} = \frac{V}{1 + \frac{(15/100)}{(5/1)}} = \frac{1}{1.03} \times V \qquad (8.3)$$

Compared with the calculated result in the case of oxide materials as shown in (8.2), the replacement of oxide ferroelectrics with P(VDF-TrFE) in a FeFET improves the voltage coupling from 1/3 to 1/1.03. The achieved value obtained by adopting P(VDF-TrFE) makes the loss of electric field almost negligible in a FeTET structure. This is the most important advantage of employing ferroelelctic P(VDF-TrFE) in a FeFET device.

8.2.3 Ferroelectricity of Random Copolymer P(VDF-TrFE)

It was disclosed that β phase PVDF can be obtained quite easily by adding 20–24 mol% TrFE to VDF based on calculation shown in Fig. 8.9. The ferroelectricity of random copolymer P(VDF-TrFE) is discussed in this section. Since Curie temperature of P(VDF-TrFE) depends on the mole percentage of TrFE in P(VDF-TrFE), ferroelectricity of copolymer can be controlled by the amount of TrFE added to PVDF [21]. It is expected that the maximum ferroelectricity can be obtained by the minimum amount of TrFE added to PVDF because the dipole moment originated from TrFE monomer is smaller than that of VDF. Therefore, P(VDF-TrFE) with 23 mol% TrFE is used for FeRAMs in general. Ferroelectric property of P(VDF-TrFE) thin films has been investigated by MFM (Metal-Ferroelectric-Metal) capacitors. In earlier studies, aluminum was used as electrode material for the MFM capacitors but enough ferroelectricity was not obtained compared to the expected value from the theory [22–25]. This is because chemically active fluorine atoms in P(VDF-TrFE) molecules degrade aluminum electrodes and make inter-layers with poor conductivity between P(VDF-TrFE) and aluminum electrodes. If P(VDF-TrFE) films are thin, the voltage loss at these

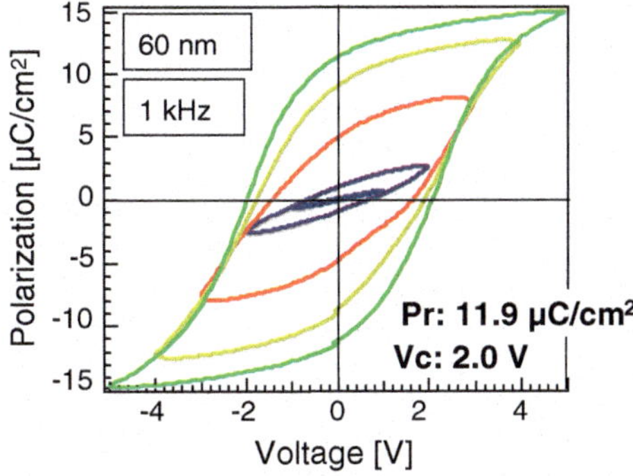

Fig. 8.19 Ferroelectric hysteresis loops and saturation property of polarizations

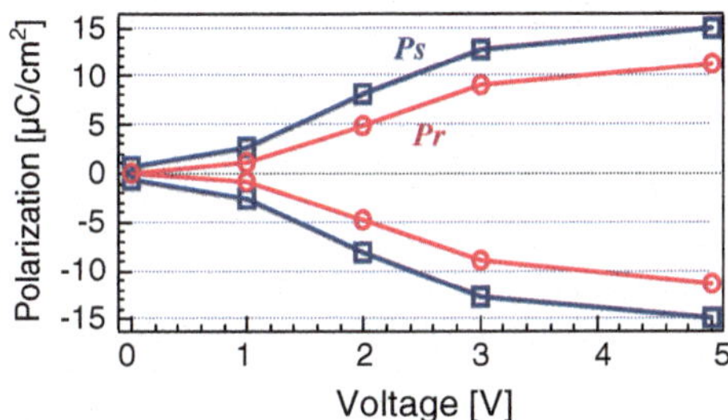

Fig. 8.20 Ferroelectric hysteresis loops and saturation property of polarizations

degraded inter-layers becomes serious. Therefore, relatively thick P(VDF-TrFE) films were used in these earlier reports. As a result, the operating voltage of MFM capacitors became as large as a few tens or hundreds volts.

Fujisaki et al. succeeded in making MFM capacitors with thin $P(VDF_{0.77}\text{-}TrFE_{0.23})$ films by using electrodes made of noble metals [26]. The thickness of P(VDF-TrFE) was less than 60 nm and the metal used for electrodes were gold and platinum. With the use of noble metals, it was possible to avoid the emergence of degraded inter-layer between P(VDF-TrFE) and electrodes. Ferroelectric hysteresis loops and saturation property of polarizations are shown in Figs. 8.19 and 8.20. Obvious from Fig. 8.20, the operating voltage was made less than 4 V that is applicable to Si LSIs. The obtained remnant polarization was 11.9 $\mu C/cm^2$ that is as large as the expected value from theory. As shown in Figs. 8.21 and 8.22, the more advanced

Table 8.1 List of reported physical parameters for P (VDF-TrFE) random co-polymer

P(VDF-TrFE) Copolymer Properties	
Physical Form	Pellets
Composition	VDF 77~56 mol%
Melting Temp.	154.5 °C (VDF=70%)
Cryst. Temp.	129 °C (VDF=70%)
Curie Temp.	106 °C (VDF=70%) 126 °C (VDF=77%)[*]
(50 °C, 1kHz)	~13 (VDF=70%) ~14 (VDF=77%)

[*] Extrapolated from 56~70 mole % materials

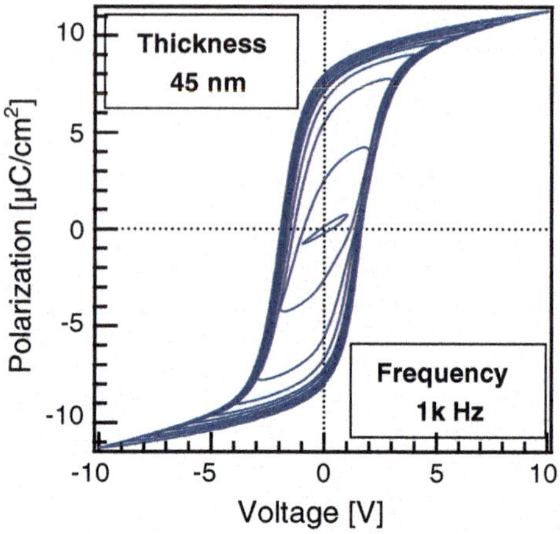

Fig. 8.21 The obtained
remnant polarization

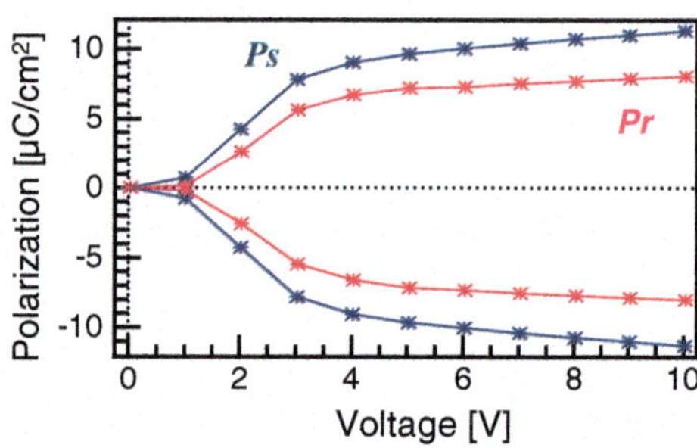

Fig. 8.22 The more
advanced result with 45 nm P
(VDF$_{0.77}$-TrFE$_{0.23}$)

result with 45 nm P(VDF$_{0.77}$-TrFE$_{0.23}$) was also reported in which remnant
polarization and coercive voltage were 8.1 μC/cm^2 and 1.5 V, respectively [27].
The physical parameters reported till now are summarized in Table 8.1.

Random copolymer P(VDF-TrFE) is usually deposited on a substrate by spin
coating since polymer pellets can be easily solved in organic solvent such as MEK
(Methyl Ethyl Ketone). The spun coated solution is then annealed at the temperature
above Curie temperature and below melting temperature. Because coercive field Ec of
P(VDF-TrFE) is as large as several hundreds kV/cm, films should be as thin as a few
tens nm in order to operate the device with a few volt. It is not easy to eliminate
pin-holes from such thin films. Therefore, it is important to evaluate the current
leakage characteristics of a capacitor. These characteristics are shown in Fig. 8.23
which were taken using the same capacitor having the ferroelectric characteristics
shown in Fig. 8.21. Obvious from the figure, the film has sufficient insulating property
up to 5 V. Frequency dependence of a capacitor operation is another important feature
to apply organic films to electronics. It was demonstrated that P(VDF-TrFE) capacitors can well operate up to 100 kHz if the film are thin enough [28] (Fig. 8.24).

It is concluded from discussions in this sub-section, random copolymer
P(VDF-TrFE) has enough potential to be employed in FeFETs for non-volatile
memory applications.

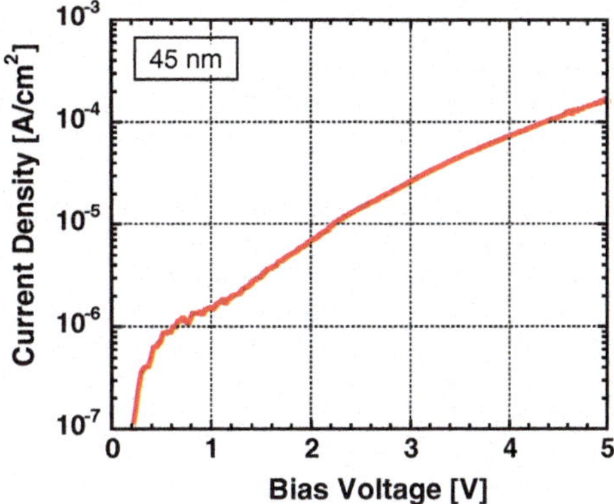

Fig. 8.23 Evaluate the current leakage characteristics of a capacitor

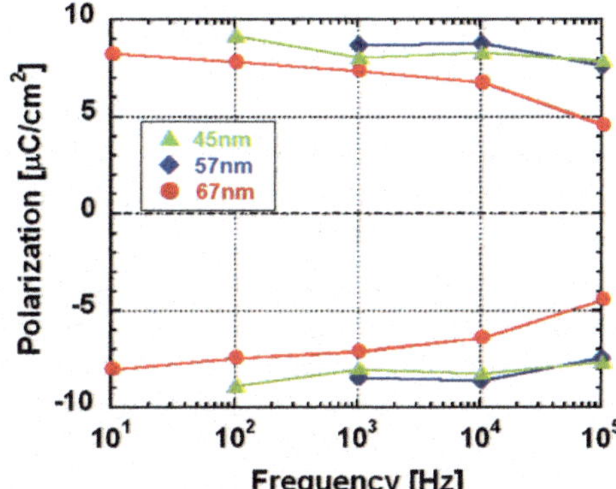

Fig. 8.24 Frequency dependence of polarization as a function of P(VDF-TrFE) film thickness

8.3 Ferroelectric P(VDF-TrFE) Gate FeFETs

8.3.1 FeFETs Using Inorganic Semiconductors

As discussed in the introduction, it is a matter of great interest to make a FeFET by adopting P(VDF-TrFE) to a ferroelectric gate layer. This is because the challenge of replacing oxide ferroelectrics by P(VDF-TrFE) in a FeFET structure can solve a lot of issues that prevent realizing reliable ferroelectric gate transistors. The most serious problem in realizing FeFET using oxide ferroelectrics is how to avoid the emergence of damaged inter-layer between oxide ferroelectrics and Si semiconductor channel as discussed in the Sect. 1.1. By employing organic ferroelectrics P(VDF-TrFE) in a FeFET structure, the largest problem can be solved as discussed in the Sect. 1.4. Therefore a number of papers have been reported using this scheme.

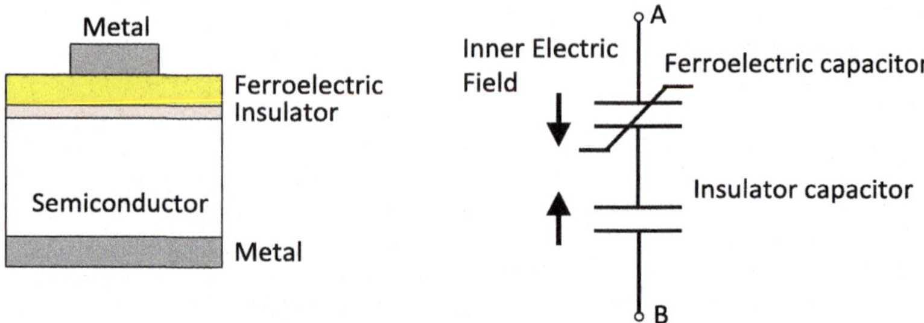

Fig. 8.25 The circuit model of a MFIS structure

There have been a lot of demonstrations of FeFETs using oxide ferroelectrics. However, retention characteristics have to be much improved for the practical applications. The degradation of retention characteristics is attributable to the damaged inter-layer that emerges between semiconductor channel and oxide ferroelectrics during the deposition process of the ferroelectric layer. To avoid the emergence of damaged inter-layer, MFIS (Metal Ferroelectrics Insulator Semiconductor) structure is employed in general. The insulator film sandwiched between semiconductor and ferroelectrics works to prevent semiconductor layer from being oxidized during ferroelectric layer formation. The circuit model of a MFIS structure is shown in Fig. 8.25. A ferroelectric capacitor and a capacitor composed of the I-layer of a MFIS structure are connected in series. If a ferroelectric capacitor is poled in one direction, the same amount of charge as that at the ferroelectric capacitor will appear at the dielectric capacitor made of the I-layer in a MFIS structure. As a result, the electric field with the same amplitude as that in a ferroelectric capacitor will appear in the capacitor made of the I-layer. However, the directions of the electric fields in a ferroelectric capacitor and in the capacitor made of the I-layer have 180° inverse direction. Therefore, the ferroelectric capacitor will feel the field that cancels the field in itself if the MFIS is in a retention state, i.e., the terminals A and B are short circuited. This field in the dielectric capacitor is called depolarization field since the polarization in a ferroelectric capacitor will be made depolarized by this electric field. The existence of this depolarization field is the most serious problem of devices having a MFIS structure in making highly reliable memory cells with long retention characteristics. In other words, the device with a MFIS structure has native and inevitable problem in retention characteristics. This is the main reason why FeFET devices have not been in the practical application yet. To the contrary, this serious retention problem can be solved by eliminating the I-layer from a MFIS structure. This structure is called MFS (Metal Ferroelectric Semiconductor) that is most efficient way to make FeFETs highly reliable. The I-layer in a MFIS structure works as a buffer layer to protect underlying semiconductor channel from being oxidized if oxide ferroelectrics are employed to make a MFIS structure. Therefore, it is almost impossible to fabricate a MFS structure if oxide ferroelectrics are used. To the contrary, there is a possibility to fabricate a MFS structure if polymer ferroelectrics are used since the formation process of a

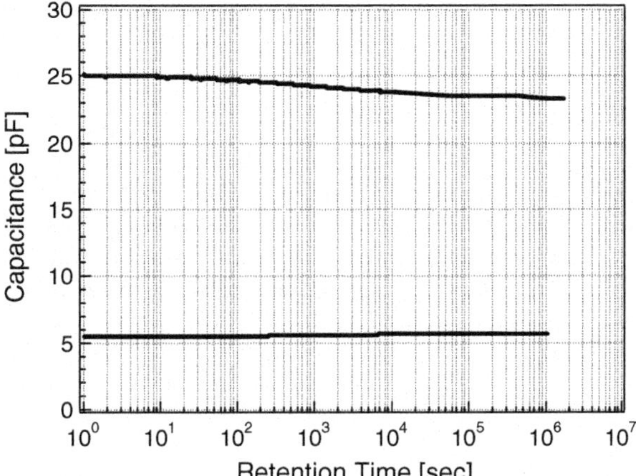

Fig. 8.26 They achieved the retention time longer than at least 10^6 s (12 days)

polymer film is much less damaging for a semiconductor channel compared to that of an oxide film. On the remaining part of this section, the actual performance of a FeFET with P(VDF-TrFE) ferroelectrics will be introduced and discussed with the comparison of those with oxide ferroelectrics.

Gerber et al. fabricated a MFIS FeFET using 36 nm thick P(VDF-TrFE) ferroelectric layer and 10 nm thick SiO_2 insulator on p-type Si channel [29]. However, retention performance is not acceptable for practical applications because of the existence of depolarization field made by 10 nm thick SiO_2 capacitor.

The challenge to make a MFS FeFET is motivated from the situation mentioned above. A MFS FeFET was demonstrated using the PVDF ferroelectric film on Si semiconductor [30]. Non-volatile transistor characteristics, such as hysteresis loop in drain current vs. gate voltage (I_d-V_g) curve, were confirmed to be originated from the ferroelectricity of PVDF. Although the advantage of MFS structure on retention performance over the MFIS structure was discussed, no data related to retention was disclosed. It is expected that the improvement of retention performance could not be achieved in this scheme.

PVDF or P(VDF-TrFE) film formation process on a semiconductor channel is still damaging because fluorine atoms in PVDF or P(VDF-TrFE) is also oxidizing active for semiconductor materials. Therefore, MFIS structure is employed in most of reports on FeFET made of P(VDF-TrFE) and inorganic semiconductor material [31–33].

Fujisaki et al. reported MFIS structure with a very thin I-layer [28]. They used 3 nm thick Ta_2O_5 on Si as an I-layer. This very thin layer works as a buffer to protect Si during P(VDF-TrFE) formation because the crystallization temperature of P(VDF-TrFE) is as low as 140 °C that is not very damaging for Si material. In addition, the capacitance made by Ta_2O_5 does not work to depolarize the dipole moments in a ferroelectric film since Ta_2O_5 of this thickness is in the direct tunneling regime. If insulator has the thickness in this regime, charges separated on both sides of insulator tunnel easily with the application of very small electrical field. Since charge density originated from dipole moments in P(VDF-TrFE) is as large as 10 μC/cm^2, almost all charges appear on both sides of Ta$_2$O$_5$ recombine promptly by huge electric field caused by polarization in P(VDF-TrFE). It can be said that MFIS structure with very thin I-layer works like MFS in the equivalent circuit model. They demonstrated retention performance longer than 1800 s. Their group reported further improvement of retention character by increasing insulating property of P(VDF-TrFE) doped with PMMA (Poly(methyl methacrylate)) [34]. They achieved the retention time longer than at least 10^6 s (12 days) by blending 1.65 wt% PMMA to P(VDF$_{0.77}$-TrFE$_{0.23}$) (Fig. 8.26). The test was terminated just because it was time consuming although the further retention was expected. Such long retention performance is hard to be obtained under the existence of depolarization field.

The idea of replacing semiconductor materials such as Si by oxide semiconductors is efficient way to eliminate I-layer because oxide material is much more robust against oxidation ambient. This scheme is also advantageous to fabricate transparent TFTs (Thin Film Transistors) on flexible substrates since both P (VDF-TrFE) and oxide semiconductors are transparent. Lee et al. fabricated transparent TFTs on a flexible polyethylene-naphthalate (PEN) substrate [35]. Amorphous indium gallium zinc oxide (a-IGZO) was deposited on a PEN substrate first at low temperature, then P(VDF-TrFE) ferroelectric gate was spun coated and crystallized at 140 °C. Since all the processes were carried out at low temperature, polymer PEN substrate can endure the thermal budget. Sophisticated transistor performance was demonstrated that endures hard bending tests. Memory window of 8.4 V in I_d-V_g characteristics and sub-threshold swing value of 400 mV/dec were achieved at the same time. Judging from these performances, P(VDF-TrFE) formation process did make only negligible damage on underlying a-IGZO channel. The oxide channel TFT is one of the most suitable applications for P(VDF-TrFE) because it can be deposited at low temperature, thus at low damaging ambient. If reliable process is build up, these devices can be used as non-volatile pixel memory for display applications. Similar reports were found with ZnO oxide channels but in which MFIS structure was employed because of the degradation of crystalline ZnO semiconductor during P(VDF-TrFE) formation process [36, 37]. As a result, FeFET made of P(VEF-TrFE) and ZnO channel did not show good retention performances.

8.3.2 FeFETs Using Organic Semiconductors

The interface problem, that causes degradation of retention performance, can be also avoided by using organic semiconductors. As a result, this type of FeFETs becomes all organic transistors. Naber et al. and other groups demonstrated a MFS FeFET made of P(VDF-TrFE) and organic semiconductors [38–41]. They observed a hysteresis loop in I_d-V_g characteristics originated from the ferroelectricity of a P(VDF-TrFE) gate. The retention performance up to 10^4 min. (7 days) was achieved. All organic MFS FeFETs, in which P(VDF-TrFE) ferroelectric gates were employed, were demonstrated with the combination of various organic semiconductor materials. In [38–41], P3HT,[1] MEH-PPV,[2] OC_1OC_{10}-PPV[3] and PCBM[4] were used as semiconductors. In addition to these materials, the more popular organic semiconductors such as Pentacene and F8T2[5] were also tried to be employed in a MFS structure [42–44]. Since depolarization field does not exist in these structures, relatively good retention performances, as long as, few hours were reported by these demonstrations. However, operating voltage of the reported devices are as high as a few tens volts, and these are too high for practical applications. Since one origin of high operating voltage comes from low mobility of organic semiconductor, it is difficult to decrease operation voltage lower than 10 V.

Instead, all organic FeFETs are suitable for flexible memory applications since fabrication temperature is much lower than those of inorganic material formation. It is crucial to find practical application with these schemes.

8.4 Summary

FeFETs with P(VDF-TrFE) ferroelectric gates have promising potential because of excellent ferroelectricity of P(VDF-TrFE) and low process temperature to form it. The application in the field of highly integrated non-volatile memories, it is not realistic since formation of P(VDF-TrFE) film with dry process, such as chemical vapor deposition, is difficult. This means that uniform fabrication of FeFET type memory cells composed of FeFET in a large diameter substrate is difficult if P(VDF-TrFE) is used as a ferroelectric gate. The most important advantage of the employment of P(VDF-TrFE) is low process temperature with which highly reliable MFS can be fabricated. System on chip (SOC) applications for wireless cards and smart cards with relatively smaller scale memory integration are suitable for FeEFTs with a P(VDF-TrFE) ferroelectric gate.

As discussed in the previous section, non-volatile TFTs are another important application since MFS structure can be used if P(VDF-TrFE) is combined with oxide or organic semiconductors. The size of devices for these applications is also suitable for organic ferroelectrics because requirement for uniformity is relatively tolerant when the device size is large compared to that in highly integrated applications.

---

[1]poly(3-hexyltiophene).

[2]poly[2-methoxy, 5-(2′-ethyl-hexyloxy)-p-phenylene-vinylene].

[3]poly(2-methoxy-5-(3′,7′-dimethylocthyloxy)-phenylenevinylene).

[4][6,6]-phenyl-C61-butyricacidmethylester.

[5]p(fluorene-bithiophene).

References

1. Y. Fujisaki, T. Kijima, H. Ishiwara, Appl. Phys. Lett. **78**, 1285 (2001)
2. Y. Tabuchi, B. Park, K. Aizawa, Y. Kawashima, K. Takahashi, K. Kato, Y. Arimoto, H. Ishiwara, Integr. Ferroelectr. **65**, 125 (2004)
3. S. Sakai, M. Takahashi, R. Ilangovan, Technical Digest International Electron Device Meeting (2004), p. 915
4. E. Tokumitsu, G. Fujii, H. Ishiwara, Appl. Phys. Lett. **75**, 575 (1999)
5. H. Kawai, Jpn. J. Appl. Phys. **18**, 976 (1969)
6. R. Hasegawa, Y. Takahashi, Y. Chatani, H. Tadokoro, Polymer J. **3**, 600 (1972)
7. A.J. Lovinger, *Development in Crystalline Polymers-1* (Applied Science Publishers, London, New Jersey, 1982), p. 195
8. P. DeSantis, E. Giglio, A.M. Liquori, A. Ripamonti, J. Polymer Sci. A **1**, 1383 (1963)
9. H. Horibe, M. Taniyama, J. Electrochem. Soc. **153**, G119 (2006)
10. A.J. Bur, J.D. Barnes, K.J. Wahlstrand, J. Appl. Phys. **59**, 2345 (1986)
11. H. Dvey-Aharon, T.J. Slucjin, P.L. Taylor, A.J. Hopfinger, Phys. Rev. B **21**, 3700 (1980)
12. J.D. Clark, P.L. Taylor, Phys. Rev. Lett. **49**, 1532 (1928)
13. F. Mospik, M.G. Broadhurst, J. Appl. Phys. **46**, 4992 (1978)
14. H. Kakutani, J. Polymer Sic. **8**, 1177 (1970)
15. M.G. Groadhurst, G.T. Davis, *Topics in modern physics-electronics*, 2nd edn. (Springer, Berlin, 1987), p. 285
16. C.K. Purvis, P.L. Taylor, Phys. Rev. B **26**, 4547 (1982)
17. C.K. Purvis, P.L. Taylor, J. Appl. Phys. **54**, 1021 (1983)
18. R. Al-Jishi, P.L. Taylor, J. Appl. Phys. **57**, 897 (1985)
19. R. Al-Jishi, P.L. Taylor, J. Appl. Phys. **57**, 902 (1985)
20. H. Ogura, A. Chiba, Ferroelectrics **74**, 247 (1987)
21. H. Tanaka, H. Yukawa, T. Nishi, Macromolecules **21**, 2469 (1988)
22. T.R. Dargaville, J.M. Elliott, M. Celina, J. Polym. Sci. B **44**, 3253 (2006)
23. T.R. Dargaville, M. Celina, P.M. Chaplya, J. Polym. Sci. B **43**, 1310 (2005)
24. H. Xu, J. Appl. Polymer Sci. **80**, 2259 (2001)
25. M.P. Wenger, P.L. Almeida, P. Blanas, R.J. Shuford, D.K. Das-Gupta, Polym. Eng. Sci. **39**, 483 (1999)
26. S. Fujisaki, H. Ishiwara, Y. Fujisaki, Appl. Phys. Lett. **90**, 162902 (2007)
27. S. Fujisaki, *Research on the application of organic ferroelectric P(VDF-TrFE) to non-volatile memories* (Doctorial theses in Japanese, Japan, 2008)
28. S. Fujisaki, Y. Fujisaki, H. Ishiwara, I.E.E.E. Trans, IEEE Trans. Ultrason. Ferroelectric. Freq Control **54**, 2592 (2007)
29. A. Gerber, H. Kohlstedt, M. Fitsilis, R. Waser, T.J. Reece, S. Ducharme, E. Rije, J. Appl. Phys. **100**, 024110 (2006)
30. J.H. Kim, B.E. Park, H. Ishiwara, Jpn. J. Appl. Phys. **47**, 8472 (2008)
31. G.A. Salvatore, D. Bouvet, A.M. Ionescu, In *Proceedings of IEEE International Electronic Device Meeting* (2008), p. 1
32. T.J. Reece, S. Ducharme, A.V. Sorokin, M. Poulsen, App. Phys. Lett. **82**, 142 (2003)
33. S.H. Lim, A.C. Rastogi, S.B. Desu, J. Appl. Phys. **96**, 5673 (2004)
34. S. Fujisaki, Y. Fujiski, H. Ishiwara, Appl. Phys. Express **1**, 081801 (2008)
35. G.-G. Lee, E. Tokumitsu, S.-M. Yoon, Y. Fujisaki, J.-W. Yoon, H. Ishiwara, Appl. Phys. Lett. **99**, 012901 (2011)
36. C.H. Park, G. Lee, K.H. Lee, S. Im, B.H. Lee, M.M. Sung, Appl. Phys. Lett. **95**, 153502 (2009)
37. S.-M. Yoon, S.-H. Yang, S.-H. Ko Park, S.-W. Jung, D.-H. Cho, C.-E. Byun, S.-Y. Kang, C.-S. Hwang, B.-G. Yu, J Phys. D **42**, 245101 (2009)
38. R.C.G. Naber, J. Massolt, M. Spijkman, K. Asadi, P.W.M. Blom, D.M. de Leeuw, Appl. Phys. Lett. **90**, 113509 (2007)

39. R.C.G. Naber, C. Tanase, P.W.M. Blom, G.H. Gelinck, A.W. Marsman, F.J. Touwslager, S. Sepas, D.M. de Leeuw, Nature Mater. **4**, 243 (2005)
40. G.H. Gelinck, A.W. Marsman, F.J. Touwslager, S. Setaesh, M. de Leeuw, R.C.G. Naber, P.W.M. Blom, Appl. Phys. Lett. **87**, 092903 (2005)
41. K. Müller, I. Paloumpa, K. Henkel, D. Schmeisser, J. Appl. Phys. **98**, 056104 (2005)
42. K.N. Narayanan, Unni, Remi de Bettignies, Sylvie Dabos-Seignon and Jean-Michel Nunzi. Appl. Phys. Lett. **85**, 1823 (2004)
43. Raoul Schroeder, Leszek A. Majewski, Martin Grell, Adv. Mater. **16**, 633 (2004)
44. J. Karasawa, in *Applications of Ferroelectrics 2007, International Symposium Application Ferroelectrics* (2007), p. 41. doi:10.1109/ISAF.2007.4393161

Part V
Practical Characteristics of Organic Ferroelectric-Gate FETs: Thin Film-Based Ferroelectric-Gate Field Effect Transistors

Chapter 9
P(VDF-TeFE)/Organic Semiconductor Structure Ferroelectric-Gate FETs

Takeshi Kanashima and Masanori Okuyama

Abstract Organic ferroelectric-gate field-effect transistor (FET) memories were fabricated using pentacene and rubrene thin films as the semiconductors and a poly (vinylidene fluoride-tetrafluoroethylene) [P(VDF-TeFE)] thin film as the ferroelectric gate. The P(VDF-TeFE) film was prepared by spin coating and annealing at 170 °C for 2.5 h, and the pentacene was prepared by vacuum evaporation. In contrast, the rubrene thin film sheet was grown by physical vapor transport and placed onto a spin-coated P(VDF-TeFE) thin film layer. The polarization-electric field hysteresis of the P(VDF-TeFE) thin film was observed, and the obtained remanent polarization of 3.9 μC/cm^2 was sufficient for controlling the surface potential of pentacene or rubrene. A hysteresis loop was clearly observed in the drain current-gate voltage behavior of the ferroelectric-gate FET. In the case of the ferroelectric-gate FET with P(VDF-TeFE)/pentacene, the ON/OFF ratio of drain current was 830, and the carrier mobility was 0.11 cm^2/Vs. On the other hand, the maximum drain current of the FET with P(VDF-TeFE)/rubrene was 1.6×10^{-6} A, which is about two orders of magnitude larger than that of the P(VDF-TeFE)-gate FET using the pentacene thin film. The mobility of the organic ferroelectric-gate FET using the rubrene thin film was 0.71 cm^2/Vs, which is 6.5 times larger than that of the FET with pentacene thin film.

T. Kanashima (✉)
Graduate School of Engineering Science, Osaka University,
Machikaneyama 1-3, Toyonaka, Osaka 560-8531, Japan
e-mail: kanashima@ee.es.osaka-u.ac.jp

M. Okuyama
Institute for NanoScience Design, Osaka University,
Machikaneyama 1-3, Toyonaka, Osaka 560-8531, Japan
e-mail: okuyama@insd.osaka-u.ac.jp

© Springer Science+Business Media Dordrecht 2016

B.-E. Park et al. (eds.), *Ferroelectric-Gate Field Effect Transistor Memories*,
Topics in Applied Physics 131, DOI 10.1007/978-94-024-0841-6_9

9.1 Introduction

9.1.1 Organic Ferroelectric-Gate FETs

Recently, much attention has been paid to organic electronics because devices based on organic materials have several unique advantages such as flexibility, light weight, low preparation cost, nontoxicity, and transparency. In addition, because organic electronic materials can dissolve in solvents, simple solution processes can be used for to fabrication such devices. Thus, the cost of fabrication is expected to be low, due to the possibility of using low-temperature processes on cheap and abundant substrates such as glass, organic plates, and metal films. Moreover, a printing technique without any vacuum process can be used to prepare these devices, allowing for their cost-effective mass production.

If a nonvolatile memory function is added to organic devices, their use could be expanded to include applications such as smart cards, display panel devices, and telecommunication. Moreover, power consumption can be decreased using non-volatile memories, making these devices important components in ubiquitous computing. New types of memories such as a spin-torque-transfer magnetic random access memory (STT-MRAM), phase change RAM (PCRAM), resistive RAM (RRAM), and ferroelectric RAM (FeRAM) are currently under development. Among these, FeRAM has the advantages of high-speed access, good endurance, and low power consumption. Moreover, device scalability can be improved in the case of ferroelectric-gate field-effect transistor (FET) memory (1T-Type FeRAM; Fig. 9.1), an emerging ferroelectric type memory reported in the International Technology Roadmap for Semiconductors (ITRS). FET-type memories that have metal-ferroelectric-semiconductor (MFS) structures fabricated using organic ferro-electric thin films tend to be more suitable for organic devices. Moreover, this type of memory has low fabrication cost, which is advantageous.

In this memory, ferroelectric materials are used as gate insulators. Ferroelectric-gate FETs using inorganic ferroelectric materials generally require insulating buffer layers between the semiconductors and ferroelectric materials to prevent semiconductor-ferroelectric interface degradation; buffer layers are necessary because the preparation of an inorganic ferroelectric material requires high-temperature processing above approximately 600 °C to achieve crystallization

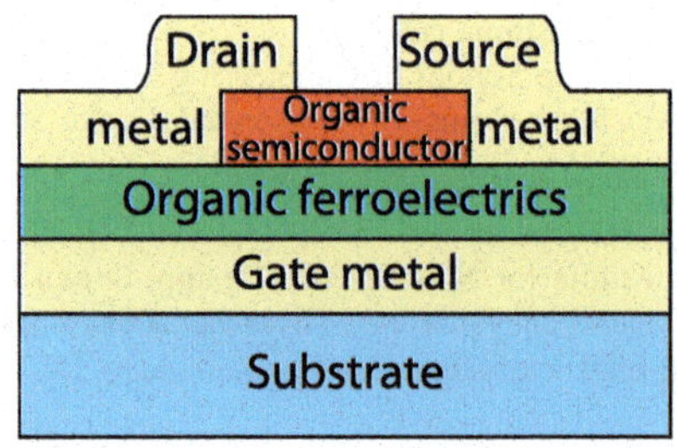

Fig. 9.1 Illustration of the cross-sectional structure of an organic 1T-type FeRAM in which ferroelectric material is used as a gate insulator

[1, 2]. On the other hand, the preparation temperatures for organic ferroelectric and semiconducting materials are not as high. Thus, insulating buffer layers are not required in organic ferroelectric-gate FETs.

9.1.2   PVDF-TeFE Organic Ferroelectrics

Organic ferroelectric materials are very important to realize organic ferroelectric-gate FET-type memories. Poly(vinylidene fluoride-trifluoroethylene) P[(VDF-TrFE)] copolymer and poly(vinylidene fluoride-tetrafluoroethylene) P[(VDF-TeFE)] copolymer are well-known ferroelectric materials for memory applications [3, 4]. Figure 9.2 illustrates the structure and bonding configurations of poly(vinylidene fluoride) (PVDF), P(VDF-TrFE), and P(VDF-TeFE). Combination of these materials and the organic semiconductor pentacene has been applied in many types of ferroelectric-gate FETs; for example, P(VDF-TrFE) and pentacene [5, 6], P(VDF-TeFE) and pentacene [7], and PVDF and pentacene [8]. In comparison with P(VDF-TrFE) and P(VDF-TeFE), P(VDF-TrFE) is more widely used as an organic ferroelectric material because it shows good polarization hysteresis, piezoelectricity, and pyroelectricity. However, the polymerization reaction of VDF-TrFE is more difficult to produce, and a large amount of P(VDF-TrFE) is not easily supplied [9–11]. Figure 9.3 shows PVDF, P(VDF-TrFE), and P(VDF-TeFE) molecular models for the dipole moment calculations. The arrows indicate the calculated dipole moments. The

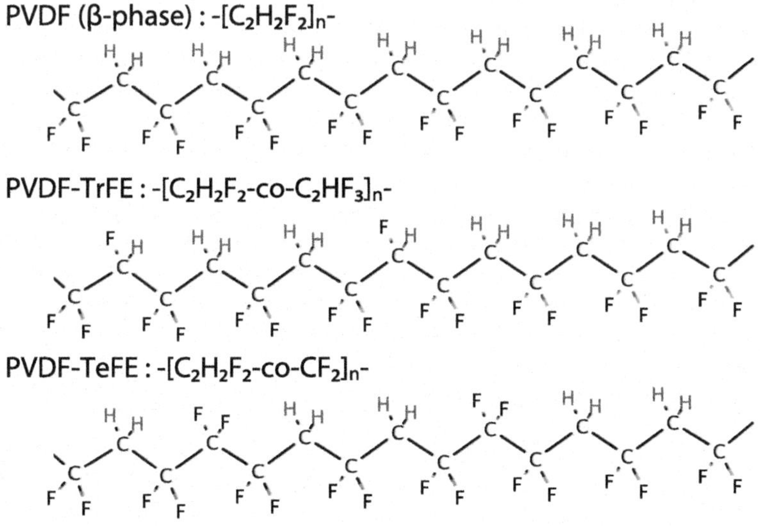

PVDF (β-phase) : -[$C_2H_2F_2$]$_n$-

PVDF-TrFE : -[$C_2H_2F_2$-co-C_2HF_3]$_n$-

PVDF-TeFE : -[$C_2H_2F_2$-co-CF_2]$_n$-

Fig. 9.2 Illustrations of the chemical bonds of PVDF, P(VDF-TrFE), and P(VDF-TeFE)

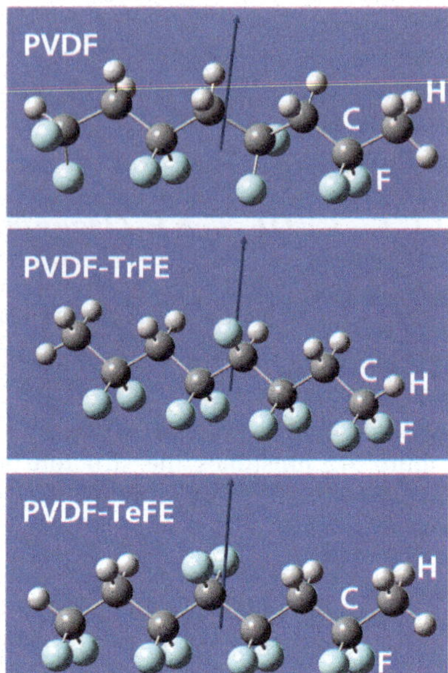

Fig. 9.3 PVDF, P(VDF-TrFE), and P(VDF-TeFE) molecular models for dipole moment calculations. Structures were optimized using Gaussian software. *Arrows* show dipole moments of molecules

geometries of these molecules were optimized using molecular orbital calculation software (Gaussian 09 [12]). The 3-21G basis function and density functional theory method were used to optimize the geometry. The resulting calculated dipole moments for PVDF, PVDF-TrFE, and PVDF-TeFE were 9.0 D, 8.0 D, and 6.8 D, respectively, when the geometries were not optimized and fixed to the same to PVDF structure. The dipole moment decreased with increasing number of F atoms when the geometries were not optimized and fixed to the same to PVDF structure. On the other hand, when the geometries were optimized, the calculated dipole moments for PVDF, PVDF-TrFE, and PVDF-TeFE were 7.1 D, 4.7 D, and 5.9 D, respectively. Thus, the dipole moment of P(VDF-TeFE) becomes larger than that of P(VDF-TrFE) when the molecular structure is stable. Moreover, the TeFE molar ratio used for ferroelectrics is typically 20 % in the case of P(VDF-TeFE); in comparison, the TrFE molar ratio is about 30 % in the case of P(VDF-TrFE), which indicates that the ratio of VDF that has a large dipole moment is large in P(VDF-TeFE). Thus, it is worthwhile to investigate alternative organic ferroelectric materials such as P(VDF-TeFE). The experimental remnant polarization of P(VDF-TeFE) is similar to that of P(VDF-TrFE) [13]. Furthermore, P(VDF-TeFE) can be produced easily and abundantly. Thus, it is expected to find applications in nanoelectronics, and the characteristics of an FET with P(VDF-TeFE) as the gate insulator have been reported [10]. Thus, we chose to focus on P(VDF-TeFE), which is more easily obtainable than P(VDF-TrFE).

9.1.3 Organic Semiconductors

Many types of organic semiconductors such as acenes (e.g., naphthalene, pentacene, and tetracene), oligomer-based materials (e.g., rubrene and quinoidal oligoth-iophene derivative (QQT(CN)4)), and phthalocyanines have been reported [14–18]. Among organic semiconductor materials, in this study we focused on a pentacene thin film and a rubrene single crystal with a relatively high mobility [19]. Figure 9.4 shows the structures of pentacene ($C_{22}H_{14}$) and rubrene ($C_{42}H_{28}$) [20]. Pentacene is commercially available and is widely used in organic FETs. On the other hand, single crystals of rubrene can be prepared separately by physical vapor transport for which the optimum conditions can be set, independent of those of P(VDF-TeFE) and the substrate. This is the advantage of organic semiconductor single crystals over materials such as amorphous pentacene and oxide semiconductors. Hence, it is expected that drain currents can be increased in ferroelectric-gate FETs using rubrene single crystals.

FETs using a PVDF-based ferroelectric polymer on Si have been reported [21–26]. The mobility of Si is high; however, Si is not transparent and flexible (an advantage held by organic devices) and does not form well on plastic or glass substrates. In addition, memory FETs with ferroelectric-copolymer-gate insulators and oxide semiconductor channels have also been reported [27]. In these FETs, zinc oxide (ZnO) [28, 29], amorphous In-Ga-Zn oxide (a-IGZO) [30, 31], Al-Zn-Sn oxide (AZTO) [32], Zn-Sn oxide (ZTO) [33], and tin oxide (SnO) [34] were used as the channel materials. These semiconductors can be formed on glass substrates and are transparent. Moreover, a relatively high mobility of approximately 1 cm^2 V^{-1} s^{-1} has been reported [30]. However, the growth temperature is not so low.

In the present chapter, we focus on ferroelectric-gate FETs using P(VDF-TeFE) thin films as an organic ferroelectric material and pentacene and rubrene organic semiconductors [35–37].

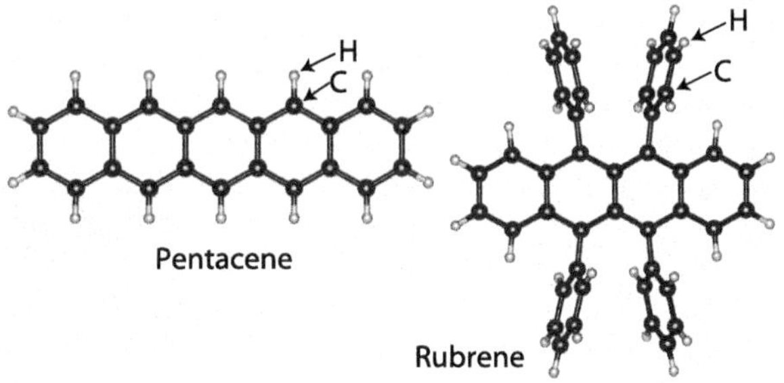

Fig. 9.4 Structures of pentacene and rubrene

9.2 Experimental Procedure

9.2.1 Preparation of P(VDF-TeFE) Thin Films

Figure 9.5 shows the schematic diagram of the ferroelectric-gate FET. A Pt (200 nm) gate electrode and a Ti (15 nm) thin film were deposited successively on a SiO_2/Si substrate as gate electrodes by RF sputtering. P(VDF-TeFE) thin films were deposited on a Pt/Ti/SiO_2/Si substrate using a spin-coating method. An 80/20 mol% P(VDF-TeFE) dissolves well in methyl ethyl ketone (MEK), and a 5 wt% of P(VDF-TeFE) solution in MEK was prepared. The P(VDF-TeFE) solution was spin-coated onto the Pt/Ti/SiO_2/Si substrate at 1500 rpm for 10 s and dried at 170 °C for 10 min. The thickness of the film depends on the concentration of P(VDF-TeFE), rotation speed, and the spin-coating time. Diethyl carbonate was also used as the solvent instead of MEK to prepare the P(VDF-TeFE) solution. The Curie transition temperature of P(VDF-TeFE) is close to its melting point of approximately 130–140 °C [38–42]. Thus, the annealing temperatures should be over the Curie temperature of 140 °C. Accordingly, the spin-coated P(VDF-TeFE) thin film was annealed at 170 °C for 2 h in air and then cooled quickly. The typical deposited P(VDF-TeFE) film thickness was 500 nm.

9.2.2 Preparation of Organic Semiconductors

The organic semiconductors used were pentacene and rubrene. The organic semiconductor film of pentacene was prepared on the ferroelectric-gate insulator of P(VDF-TeFE) by vacuum evaporation at a rate of 50 nm/h. The pentacene film was patterned by vacuum evaporation with a shadow mask. The pentacene source material was purchased from Sigma-Aldrich (P1802, assay 99 %). The organic semiconductor rubrene thin film was separately grown at 225 °C for 7 h by horizontal physical vapor transport in a stream of inert gas (Ar) to obtain single crystals [43].

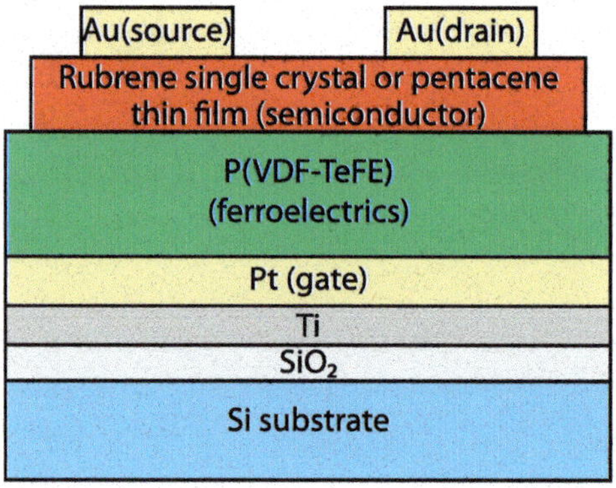

Fig. 9.5 Illustration of cross-sectional structure of ferroelectric-gate FET with rubrene or pentacene semiconductor thin film and P(VDF-TeFE) ferroelectric thin film

The temperature of the source zone was approximately 280–290 °C and the flow rate was approximately 100 ccm. The rubrene source material was purchased from Sigma-Aldrich (554073, assay ≥98 %). The prepared rubrene thin film (thickness = 130–350 nm) was carefully placed on a P(VDF-TeFE) thin film.

9.2.3 Preparation and Characterization Methods for Ferroelectric-Gate FETs

After the deposition of P(VDF-TeFE) on the $Pt/Ti/SiO_2/Si$ substrate, a semiconductor layer was deposited or placed, and Au source and drain electrodes (50 nm thick) were formed by vacuum evaporation with a shadow mask. The Pt layer was used as a gate electrode. The channel length and width of the fabricated device were both 100 µm. The thin films and FET samples were characterized by X-ray diffraction (XRD; RIGAKU RINT-2000), atomic force microscopy (AFM; Keyence VN-8010), and the examination of electrical properties such as current-voltage (I-V) and polarization-electric field (P-E) curves. The hysteresis loops in the P-E curves were evaluated using a ferroelectric characterization evaluation system (TOYO Co., FCE). The I-V and other FET characteristics were evaluated using a semiconductor parameter analyzer (Agilent, HP4155C). All measurements were performed at room temperature in air.

9.3 Results and Discussion

9.3.1 Basic Properties of Fabricated P(VDF-TeFE) Thin Films

The memory retention property and ON/OFF ratio are important characteristics in ferroelectric FET memory and depend on the quality of the ferroelectric thin film. Therefore, it is important to prepare a high-quality ferroelectric thin films to improve these characteristics. The effect of an electric field during the thermal treatment of P(VDF-TeFE) has been reported to increase the remnant polarization [44, 45], and the slow cooling process is thought to improve the crystallization and ferroelectric properties. However, organic ferroelectric thin films with rough surfaces were obtained when the sample was slowly cooled from 140 to 110 °C at a cooling rate of 15 °C/h. Typically, organic semiconductors are prepared on P(VDF-TeFE) thin films by methods other than solution growth. Because P(VDF-TeFE) was dissolved in an organic solvent for spin-coating, and the organic solvent damages organic semiconductors in MFS structures. When pentacene is used for the semiconductor layer, its thickness is approximately 50 nm [46], which makes it difficult to prepare a high-quality thin film on a rough surface. Thus, the

Fig. 9.6 AFM image of
P(VDF-TeFE) thin film.
Measured area is 100 μm^2.
Lower figure shows
cross-sectional profile

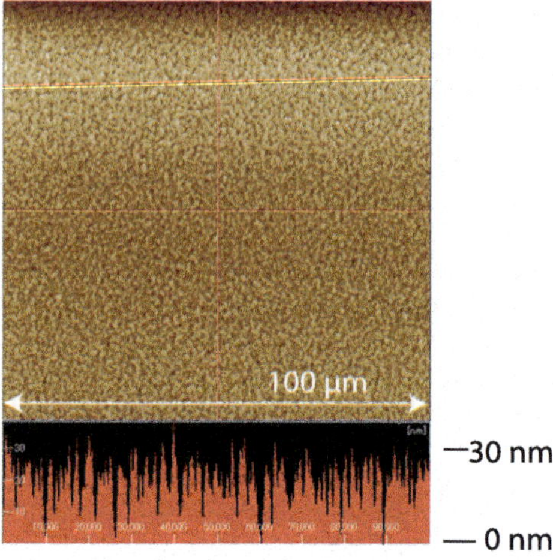

qualities of the semiconductor and the interface between the ferroelectric thin film
and the semiconductor tend to decrease if the surface of the ferroelectric thin films
is rough; therefore, very flat ferroelectric thin films are desirable. Moreover, if
organic ferroelectric thin films with very flat surfaces are realized, FET memory can
be fabricated by laminating the organic semiconductor crystal. To obtain very flat
ferroelectric thin films, quick cooling and flattening processes can be used [36]. In
this study, quick cooling was employed because of its versatility and simplicity.

Figure 9.6 shows the surface morphologies and cross-sectional profile of the
P(VDF-TeFE) thin film observed by AFM. The film thickness was 500 nm. The
lamellar structures that are generally observed in slowly cooled samples are not
apparent, and the average surface roughness of the sample was 7.4 nm. A flat
surface was obtained because this film was quickly cooled after annealing, and
grain growth was suppressed [36].

The ferroelectric properties of the P(VDF-TeFE) thin films were characterized
based on their metal-ferroelectric-metal structures. Figure 9.7 shows the typical
P-E hysteresis loops of the P(VDF-TeFE) thin film. These *P-E* curves were

Fig. 9.7 *P-E* hysteresis loops
of P(VDF-TeFE) thin film
with thickness of 500 nm.
Measured frequency is
100 Hz

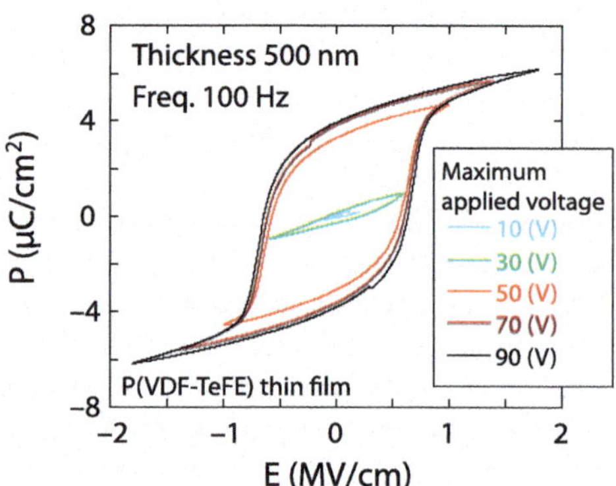

measured at 100 Hz. The applied voltage was increased from 10 to 90 V, and the step voltage was set at 20 V. The remnant polarization quickly increases above 30 V. The hysteresis curves are well saturated above 70 V (1.4 MV/cm), and the parallelogram shape of the hysteresis loop indicates the high quality of the ferroelectric thin film. The remanent polarization (P_r) is 3.9 μC/cm^2, and the coercive force (E_c) is 0.64 MV/cm. These values agree reasonably well with the previously reported results [13, 47].

9.3.2 Basic Properties of Pentacene and Rubrene Semiconductors

The crystallinity of the pentacene thin film was characterized by XRD analysis. Figure 9.8 shows the XRD pattern of the pentacene thin film. Pentacene was deposited on P(VDF-TeFE), and the film thickness was 80 nm. Only a very small peak at 6°, attributed to pentacene (001), is observable in the case of this thick sample, and the pentacene (002) peak is not apparent. Thus, the crystallinity of the deposited pentacene thin film was very poor, and the phase of the thin film was mainly amorphous.

The crystalline structure of the rubrene thin film was also characterized by XRD (Fig. 9.9). The inset shows a unit cell of single crystalline rubrene [20]. Only narrow (00l) peaks with FWHMs of approximately 0.05° are observable. The peaks marked by asterisks originated from the sample holder because the size of the obtained rubrene thin film was smaller than the XRD beam size. The lattice constant was estimated from the Nelson-Riley plots under the assumptions that the film was oriented along the c-axis and that the crystal structure was orthorhombic [48]. Each diffraction peak was deconvoluted by applying a pseudo-Voigt function, and the lattice constant was calculated to be 2.70 nm by extrapolating the measured peak positions. This value agrees well with the c-axis lattice constant of the rubrene crystal [49, 50]. Figure 9.10 shows the surface of the rubrene thin film observed by

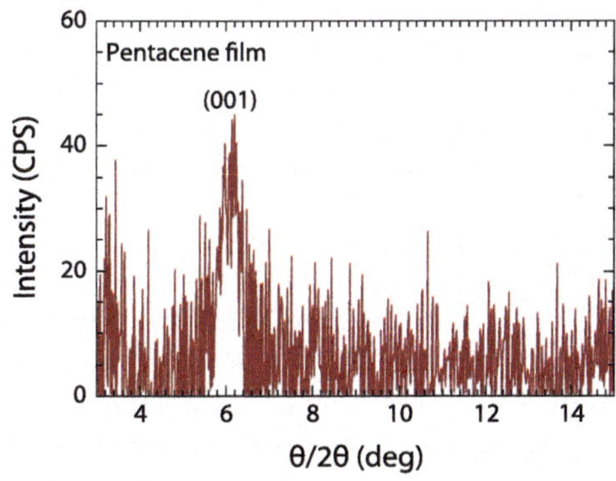

Fig. 9.8 XRD pattern of pentacene thin film

Fig. 9.9 XRD pattern of grown rubrene thin film. *Inset* shows unit cell of single-crystalline rubrene [20]

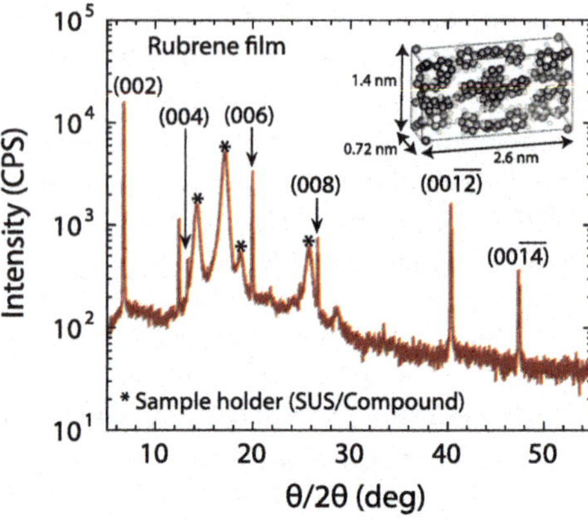

Fig. 9.10 AFM image of rubrene thin film. Measured area is 50 μm^2. *Lower* figure shows cross-sectional profile

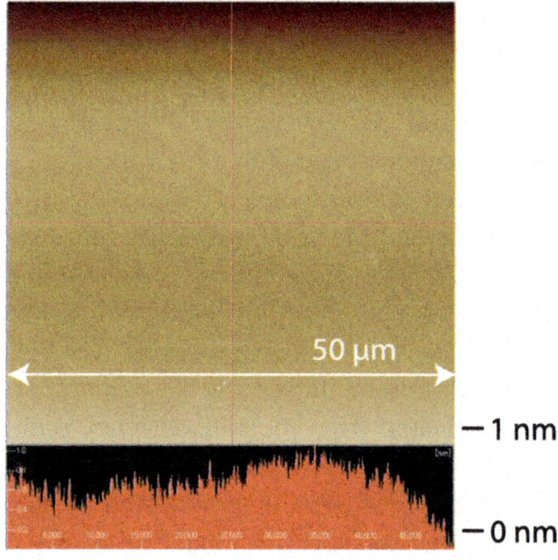

AFM. It is evident that the surface is very smooth, and the height difference between the highest and lowest pixels is <1 nm in an area of 50 μm^2. Moreover, no grain boundaries are observable. Based on these findings, we deduced that the obtained rubrene was either highly oriented or a single crystal.

9.3.3 P(VDF-TeFE)/Pentacene Ferroelectric-Gate FETs

Figure 9.11 shows the drain current-drain voltage (I_D-V_D) characteristics of the FET with P(VDF-TeFE) and pentacene thin films. The gate width (W) and length (L) were 100 and 100 μm, respectively. The gate voltage (V_G) was swept from +10 to −40 V. The drain current is linearly proportional to drain voltage in the low-voltage region, very similar to the behavior of a p-channel

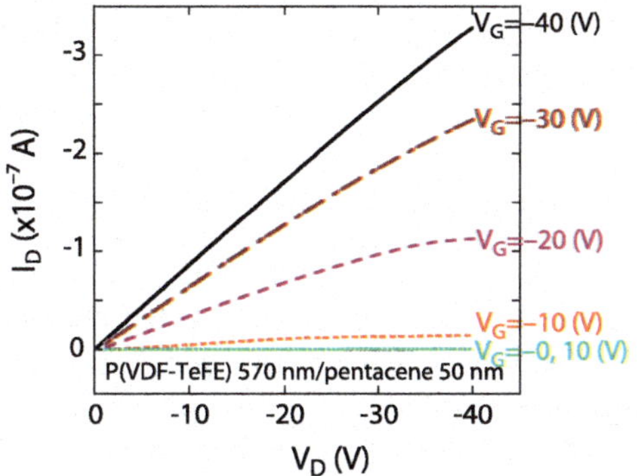

Fig. 9.11 I_D-V_D characteristics of ferroelectric-gate FET with P(VDF-TeFE) and pentacene thin films. Thicknesses of P(VDF-TeFE) and pentacene thin films are 570 and 50 nm, respectively

metal-insulator-semiconductor field-effect transistor (MISFET) fabricated using pentacene. Figure 9.12 shows the drain (I_D), source (I_S), and gate (I_G) currents as functions of V_G. The gate leakage current is below 2 nA and is very small in the entire gate voltage region. The gate current has a peak at near $V_G = -60$ V, which is considered noise because I_D(40 nA) + I_G(17 nA) does not equal I_S(43 nA). The small peaks near $V_G = -35$ and $+30$ V represent polarization reversal current. However, the peak at 30 V is not clearly observable in I_G because of leakage current. The drain and source currents in the OFF state are very small, and the drain current is almost equal to the source current. The memory window width was about 40 V. This FET memory can realize ON and OFF states at $V_G = 0$ corresponding to P(VDF-TeFE) polarization as the I_D-V_G characteristics are clockwise. The center of the threshold voltages of the I_D-V_G hysteresis curves is around 0 V, and the I_D-V_G characteristics have good symmetry. The ON/OFF ratio is 830, and the calculated mobility is 0.11 cm^2/Vs. The mobility of the organic semiconductor is not very high, and that of the semiconductor prepared on the organic ferroelectric thin film is even lower than those of the films themselves. As a result, the drain current is low in the organic ferroelectric-gate FET with the organic semiconductor. Thus, a wide gate is required to obtain a high drain current; consequently, the device must be

Fig. 9.12 I_D, I_S, and I_G as functions of V_G in ferroelectric-gate FET with P(VDF-TeFE) and pentacene thin films. Drain voltage is -5 V

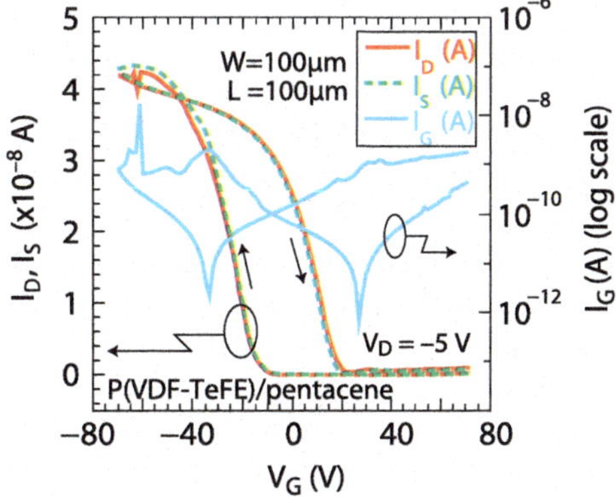

large. Therefore, an organic semiconductor with high mobility is required to be prepared on an organic ferroelectric layer.

9.3.4 P(VDF-TeFE)/Rubrene Ferroelectric-Gate FETs

Figure 9.13 shows the I_D-V_D characteristics of the ferroelectric-gate FET with P(VDF-TeFE) and rubrene thin films. A highly (001)-oriented rubrene thin film with high mobility was used as the semiconductor. Because I_D is almost linearly proportional to V_D in the low-voltage region, this behavior is similar to that of the p-channel MISFET fabricated using pentacene (Fig. 9.11). The gate voltage was swept from +4 to −24 V in steps of 4 V. The drain current becomes saturated in the high-drain-voltage region. Moreover, the drain current is much higher than that in the FET prepared using the pentacene thin film, because of the high mobility of the rubrene thin film and the reasonably good interface between the rubrene layer and the P(VDF-TeFE) layer resulting from direct bonding. Figure 9.14 shows I_D, I_S, and I_G as functions of V_G for the ferroelectric-gate FET. The rubrene thickness was 130 nm, the drain voltage was set to −5 V, and V_G was varied from −60 to +60 V. The drain and source currents in the OFF state were below 0.1 nA, which is very low. The gate leakage current was below 2.5 nA, which is also very low. Moreover, the drain current was almost equal to the source current. Thus, the gate leakage current of this FET is very low, and a good insulation property for P(VDF-TeFE) is obtained.

A clockwise hysteresis loop is clearly observable in Fig. 9.14, and this device acts as a ferroelectric-gate FET. The memory window width resulting from the ferroelectricity of P(VDF-TeFE) was 9.6 V, and the ON/OFF ratio at $V_G = 0$ was 2.39. The maximum drain current was 1.6×10^{-6} A, two orders of magnitude greater than that of the ferroelectric-gate FET with pentacene. The calculated mobility of this FET with rubrene is 0.71 cm^2 V^{-1} s^{-1}, which is 6.5 times larger than that of the FET fabricated using the pentacene thin film. This mobility is the

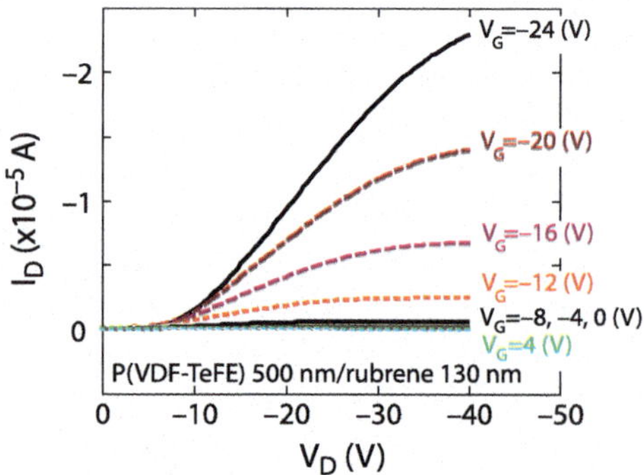

Fig. 9.13 I_D-V_D characteristics of ferroelectric-gate FETs with P(VDF-TeFE) and rubrene thin films. Thicknesses of P(VDF-TeFE) and rubrene thin films are 500 and 130 nm, respectively

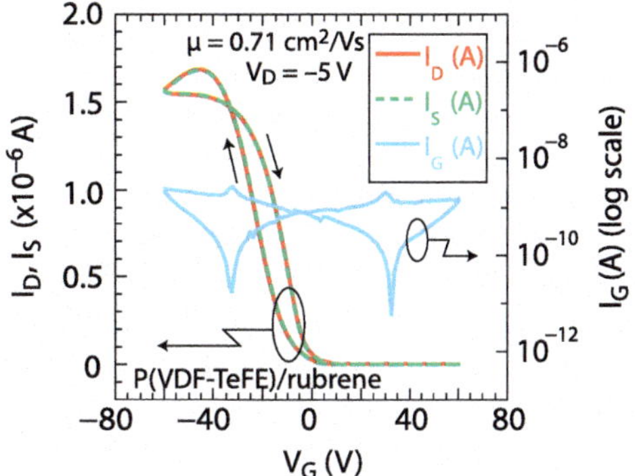

Fig. 9.14 I_D, I_S, and I_G as functions of V_G in ferroelectric-gate FETs with P(VDF-TeFE) and rubrene thin films. Drain voltage is −5 V. I_D is almost equal to I_S; thus, their lines overlap

largest among the values reported for ferro-electric-gate FETs using PVDF-related materials and organic semiconductors [5, 35, 51–63].

9.4 Conclusions

In the present study, we fabricated ferroelectric-gate FETs using pentacene and highly (001)-oriented rubrene thin films as semiconductors and P(VDF-TeFE) thin films as a ferroelectric material. A flat surface and saturated *P-E* curves were obtained in the P(VDF-TeFE) thin film. The pentacene and highly (001)-oriented rubrene thin films were deposited and placed on the flat P(VDF-TeFE) thin film, and the ferroelectric-gate FET was subsequently fabricated by metallization. Typical I_D-V_D characteristics were observed, and a clockwise hysteresis in the I_D-V_G characteristics originating from ferroelectricity was obtained in the P(VDF-TeFE)/pentacene and P(VDF-TeFE)/rubrene ferroelectric-gate FETs. Moreover, a high drain current was obtained in the organic ferroelectric-gate FET with the highly (001)-oriented rubrene and P(VDF-TeFE) thin films; its mobility was 0.71 cm² V⁻¹ s⁻¹, which was 6.5 times larger than that of pentacene and P(VDF-TeFE) thin films.

References

1. Y. Matsui, M. Okuyama, M. Noda, Y. Hamakawa, Appl. Phys. A **28**, 161 (1982)
2. T. Hirai, Y. Fujisaki, K. Nagashima, H. Koike, Y. Tarui, Jpn. J. Appl. Phys. **36**, 5908 (1997)
3. D.Y. Kusuma, C.A. Nguyen, P.S. Lee, J. Phys. Chem. B **114**, 13289 (2010)
4. S.-M. Yoon, S.-W. Jung, S.-H. Yang, S.-Y. Kang, C.-S. Hwang and B.-G. Yu, Jpn. J. Appl. Phys. **48**, 09KA20 (2009)
5. K.H. Lee, G. Lee, K. Lee, M.S. Oh, S. Im, Appl. Phys. Lett. **94**, 093304 (2009)
6. B. Kam, X. Li, C. Cristoferi, E. Smits, A. Mityashin, S. Schols, J. Genoe, G. Gelinck, P. Heremans, Appl. Phys. Lett. **101**, 033304 (2012)

7. R. Tamura, S. Yoshita, E. Lim, T. Manaka, M. Iwamoto, Jpn. J. Appl. Phys. **48**, 021501 (2009)
8. S. Kang, I. Bae, J.-H. Choi, Y. Park, P. Jo, Y. Kim, K.-J. Kim, J.-M. Myoung, E. Kim, C. Park, J. Mater. Chem. **21**, 3619 (2011)
9. E. Fukada, Kobayashi Institute of Physical Research News in Japanese. **63**, 5 (1999)
10. J. Hulburt, A. Feiring, Chem. Eng. News **75**, 6 (1997)
11. A.E. Feiring, E.R. Wonchoba, Macromolecules **31**, 7103 (1998)
12. M.J. Frisch, G.W. Trucks, H.B. Schlegel, G.E. Scuseria, M.A. Robb, J.R. Cheeseman, G. Scalmani, V. Barone, B. Mennucci, G.A. Petersson, H. Nakatsuji, M. Caricato, X. Li, H.P. Hratchian, A.F. Izmaylov, J. Bloino, G. Zheng, J.L. Sonnenberg, M. Hada, M. Ehara, K. Toyota, R. Fukuda, J. Hasegawa, M. Ishida, T. Nakajima, Y. Honda, O. Kitao, H. Nakai, T. Vreven, J.A. Montgomery Jr., J.E. Peralta, F. Ogliaro, M. Bearpark, J.J. Heyd, E. Brothers, K.N. Kudin, V.N. Staroverov, R. Kobayashi, J. Normand, K. Raghavachari, A. Rendell, J.C. Burant, S.S. Iyengar, J. Tomasi, M. Cossi, N. Rega, J.M. Millam, M. Klene, J.E. Knox, J.B. Cross, V. Bakken, C. Adamo, J. Jaramillo, R. Gomperts, R.E. Stratmann, O. Yazyev, A.J. Austin, R. Cammi, C. Pomelli, J.W. Ochterski, R.L. Martin, K. Morokuma, V.G. Zakrzewski, G.A. Voth, P. Salvador, J.J. Dannenberg, S. Dapprich, A.D. Daniels, O. Farkas, J.B. Foresman, J.V. Ortiz, J. Cioslowski, D.J. Fox, Gaussian 09, Revision E.01, Gaussian Inc., Wallingford CT (2009)
13. S. Tasaka, S. Miyata, J. Appl. Phys. **57**, 906 (1985)
14. S.K. Gupta, P. Jha, A. Singh, M.M. Chehimi, D.K. Aswal, J. Mater. Chem. C **3**, 8468 (2015)
15. A. Yassar, Polym. Sci. Ser. C **56**, 4 (2014)
16. C. Wang, H. Dong, W. Hu, Y. Liu, D. Zhu, Chem. Rev. **112**, 2208 (2012)
17. A. Facchetti, Mater. Today **10**, 28 (2007)
18. J. Mei, Y. Diao, A.L. Appleton, L. Fang, Z. Bao, J. Am. Chem. Soc. **135**, 6724 (2013)
19. J. Takeya, M. Yamagishi, Y. Tominari, R. Hirahara, Y. Nakazawa, T. Nishikawa, T. Kawase, T. Shimoda, S. Ogawa, Appl. Phys. Lett. **90**, 102120 (2007)
20. O.D. Jurchescu, A. Meetsma, T.T.M. Palstra, Acta Crystallogr. B **62**, 330 (2006)
21. S.-H. Lim, A. Rastogi, S. Desu, J. Appl. Phys. **96**, 5673 (2004)
22. S. Fujisaki, H. Ishiwara, Y. Fujisaki, Appl. Phys. Lett. **90**, 162902 (2007)
23. Y. Park, I.-S. Bae, S. Ju Kang, J. Chang, C. Park, IEEE Trans. Dielectr Electr. Insul. **17**, 1135 (2010)
24. S. Das, J. Appenzeller, Nano Lett. **11**, 4003 (2011)
25. Y. Park, S. Kang, Y. Shin, R.-H. Kim, I. Bae, C. Park, Curr. Appl. Phys. **11**, e30 (2011)
26. H. Miyashita, T. Watanabe, T. Kanashima, M. Okuyama, Ext Abstr (69th Autumn Meeti, 2008); Japan Society of Applied Physics [in Japanese] (2008) 4a-K-8
27. S.-M. Yoon, S. Yang, C.-W. Byun, S.-W. Jung, M.-K. Ryu, S.-H.K. Park, B. Kim, H. Oh, C.-S. Hwang, B.-G. Yu, Semicond. Sci. Tech. **26**, 034007 (2011)
28. C.H. Park, S. Im, J. Yun, G.H. Lee, B.H. Lee, M.M. Sung, Appl. Phys. Lett. **95**, 223506 (2009)
29. S.-M. Yoon, S.-H. Yang, C.-W. Byun, S.-H. Ko Park, S.-W. Jung, D.-H. Cho, S.-Y. Kang, C.-S. Hwang, H. Ishiwara, Jpn. J. Appl. Phys. **49**, 04DJ06 (2010)
30. G.-G. Lee, E. Tokumitsu, S.-M. Yoon, Y. Fujisaki, J.-W. Yoon, H. Ishiwara, Appl. Phys. Lett. **99**, 012901 (2011)
31. A. Van Breemen, B. Kam, B. Cobb, F.G. Rodriguez, G. Van Heck, K. Myny, A. Marrani, V. Vinciguerra, G. Gelinck, Org. Electron. **14**, 1966 (2013)
32. S.-M. Yoon, S. Yang, C. Byun, S.-H.K. Park, D.-H. Cho, S.-W. Jung, O.-S. Kwon, C.-S. Hwang, Adv. Funct. Mater. **20**, 921 (2010)
33. S.-M. Yoon, S.-W. Jung, S.-H. Yang, C.-W. Byun, C.-S. Hwang, S.-H. Ko Park, H. Ishiwara, Org. Electron. **11**, 1746 (2010)
34. J.A. Caraveo-Frescas, P.K. Nayak, H.A. Al-Jawhari, D.B. Granato, U. Schwingenschloegl, H.N. Alshareeft, ACS Nano **7**, 5160 (2013)
35. T. Watanabe, H. Miyashita, T. Kanashima, M. Okuyama, Jpn. J. Appl. Phys. **49**, 04DD14 (2010)

36. T. Kanashima, K. Yabe, M. Okuyama, Jpn. J. Appl. Phys. **51**, 02BK06 (2012)
37. T. Kanashima, T. Watanabe, M. Okuyama, in International Conference on Advanced Electromaterials (ICAE 2011) (2011) FM859
38. A.J. Lovinger, Macromolecules **16**, 1529 (1983)
39. R. Khalfin, A. Bezprozvannykh, V. Poddubnyi, Polym. Sci. U.S.S.R. **30**, 2138 (1988)
40. K. Tashiro, H. Kaito, Polymer, 2915 (1992)
41. H. Ohigashi, Jpn. J. Appl. Phys. Suppl. **24S2**, 23 (1985)
42. Y. Murata, Polym. J. **19**, 337 (1987)
43. R.A. Laudise, C. Kloc, P.G. Simpkins, T. Siegrist, J. Cryst. Growth **187**, 449 (1998)
44. J.-H. Jeong, C. Kimura, H. Aoki, T. Sugino, Jpn. J. Appl. Phys. **49**, 04DK23 (2010)
45. J.-H. Jeong, D. Terashima, C. Kimura, H. Aoki, Jpn. J. Appl. Phys. **50**, 04DK09 (2011)
46. L. Mariucci, D. Simeone, S. Cipolloni, L. Maijo, A. Pecora, G. Fortunato, S. Brotherton, Solid State Electron. **52**, 412 (2008)
47. J.C. Hicks, T.E. Jones, J.C. Logan, J. Appl. Phys. **49**, 6092 (1978)
48. Y. Waseda, E. Matsubara, K. Shinoda, *X-Ray Diffraction Crystallography: Introduction, Examples and Solved Problems* (Springer, Berlin Heidelberg, 2011)
49. D.E. Henn, W.G. William, D.J. Gibbons, J. Appl. Crystallogr. **4**, 256 (1971)
50. X. Zeng, D. Zhang, L. Duan, L. Wang, G. Dong, Y. Qiu, Appl. Surf. Sci. **253**, 6047 (2007)
51. W.-H. Kim, M.-H. Kim, C.-M. Keum, J. Park, J. Appl. Phys. **109**, 024508 (2011)
52. R.C.G. Naber, K. Asadi, P.W.M. Blom, D.M. De Leeuw, B. De Boer, Adv. Mater. **22**, 933 (2010)
53. C. Nguyen, S. Mhaisalkar, J. Ma, P. Lee, Org. Electron. **9**, 1087 (2008)
54. K.N. Narayanan Unni, S. Dabos-Seignon, J.-M. Nunzi, J. Phys. D-Appl. Phys. **38**, 1148 (2005)
55. T.N. Ng, B. Russo, A.C. Arias, J. Appl. Phys. **106**, 094504 (2009)
56. S.J. Kang, I. Bae, Y.J. Park, T.H. Park, J. Sung, S.C. Yoon, K.H. Kim, D.H. Choi, C. Park, Adv. Funct. Mater. **19**, 1609 (2009)
57. R. Naber, C. Tanase, P. Blom, G. Gelinck, A. Marsman, F. Touwslager, S. Setayesh, D. De Leeuw, Nat. Mater. **4**, 243 (2005)
58. K. Mueller, K. Henkel, I. Paloumpa, D. Schmeiber, Thin Solid Films **515**, 7683 (2007)
59. S. Kang, Y. Park, I. Bae, K.-J. Kim, H.-C. Kim, S. Bauer, E. Thomas, C. Park, Adv. Funct. Mater. **19**, 2812 (2009)
60. K.N.N. Unni, R. de Bettignies, S. Dabos-Seignon, J.-M. Nunzi, Mater. Lett. **59**, 1165 (2005)
61. R. Naber, M. Mulder, B. De Boer, P. Blom, D. De Leeuw, Org. Electron. **7**, 132 (2006)
62. J. Chang, C. Shin, Y. Park, S. Kang, H. Jeong, K.-J. Kim, C. Hawker, T. Russell, D. Ryu, C. Park, Org. Electron. **10**, 849 (2009)
63. W. Choi, S.H. Noh, D.K. Hwang, J.-M. Choi, S. Jang, E. Kim, S. Im, Electrochem. Solid-State Lett. **11**, H47 (2008)

Chapter 10
Nonvolatile Ferroelectric Memory Thin-Film Transistors Using a Poly(Vinylidene Fluoride Trifluoroethylene) Gate Insulator and an Oxide Semiconductor Active Channel

Sung-Min Yoon

Abstract Nonvolatile memory thin-film transistor using an organic ferroelectric gate insulator and oxide semiconductor active channel is proposed as a promising memory elements embedded onto the next-generation flexible and transparent electronic systems. In this chapter, some important technical issues for this device, such as device structure, process optimization and memory array integration, are extensively discussed. Feasible applications and remaining technological issues to be solved for practical applications are also reviewed.

10.1 Introduction

Large-area electronics provided new chance and opportunity in future consumer electronics. Various devices are researched and developed to enhance the function and design performance of the large-area electronic systems implemented on glass and plastic substrates. Among them, nonvolatile memory is one of the most important components. Unlike silicon-based electronic industries demanding high performance and aggressive device scaling, the specifications for the nonvolatile memory devices integrated into the large-area electronic system are quite different. First, the memory devices employed in this field should be fabricated with a low cost. Lower power consumption and higher device reliability are required with a simpler process at lower temperature. Mechanical flexibility and optical transparency will be great benefits. Various methodologies and operation origins have been proposed to realize the most appropriate nonvolatile memory element for this

S.-M. Yoon (✉)
Department of Advanced Materials Engineering for Information and Electronics,
Kyung Hee University, Yongin, Gyeonggi-Do 446-701, Republic of Korea
e-mail: sungmin@khu.ac.kr

© Springer Science+Business Media Dordrecht 2016
B.-E. Park et al. (eds.), *Ferroelectric-Gate Field Effect Transistor Memories*,
Topics in Applied Physics 131, DOI 10.1007/978-94-024-0841-6_10

application. The ferroelectric field-effect thin-film transistor (TFT) is a very promising candidate. The combination of an organic ferroelectric gate insulator and an oxide semiconductor active layers is the best selection for high performance nonvolatile memory transistors embeddable into the various large-area electronic systems because the device can be reproducibly operated with a definitely designable principle and be fabricated with a very simple process.

However, there are few studies on the memory TFTs (MTFTs) with the organic ferroelectric-oxide semiconductor (OrFOx) gate stack. For the practical applications of the OrFOx-MTFTs, (1) lower program voltage, (2) higher program speed, (3) longer data retention time, and (4) easier integration with peripheral driving circuit should be totally reflected into the device design schemes. In this chapter, we provide a focused overview on the technical issues for the OrFOx-MTFTs. In Sect. 10.2, the features and requirements of employed materials are briefly introduced. Section 10.3 explains the device design schemes for optimizing the memory operations. In Sect. 10.4, some important issues related to the device fabrication process are discussed. Promising practical applications of the OrFOx-MTFTs are introduced in Sect. 10.5. Then, the integration issues, such as process compatibility and disturb-free operation for the memory cells, are proposed in Sect. 10.6. Section 10.7 deals with the remaining technical issues for further improvements in memory performance and commercial applications. Finally, future perspectives and outlooks for the proposed OrFOx-MTFTs are summarized in Sect. 10.8.

10.2 Choice of Materials

10.2.1 Organic Ferroelectric Gate Insulators

Poly(vinylidene fluoride) (PVDF, $(CH_2CF_2)_n$) and a copolymer with tri-fluoroethylene (P(VDF-TrFE), $(CH_2CF_2)_n$-$(CHFCF_2)_m$) are the most typical polymer ferroelectric materials [1–3]. Various organic ferroelectric materials such as cynopolymer derivatives, polyurea, polythiourea, odd-nylon, and croconic acid have also been introduced [1, 4]. However, PVDF-based ferroelectrics exhibit superior properties of a relatively large remnant polarization, a short switching time and a good thermal stability. The ferroelectric nature of the PDVF results from the dipole moments in the molecule which can be aligned with the applied electric field by a rotation of the polymer chain. The strongly electronegative fluorine atoms work as dipole moments [2, 3]. The material properties of the melting temperature, Curie temperature, and crystallization temperature are varied with the relative composition of PVDF to TrFE. For the 70/30 composition, those are known as 155, 106, and 129 °C, respectively. The dielectric constant of P(VDF-TrFE) is in the range from 12 to 25 depending on the composition [1]. There have been encouraging results on nonvolatile ferroelectric memory device behaviors of various device structures using the P(VDF-TrFE) thin films [5–18]. P(VDF-TrFE) thin film can be easily formed by a solution spin-coating method and be crystallized at a lower temperature of about 140 °C. The solution was prescribed by melting the powder source of P(VDF-TrFE) into suitable solvents such as dimethylformamide, methylethylketone, cyclohexanone, and diethyl carbonate. For the proposed nonvolatile ferroelectric thin-film transistors, P(VDF-TrFE) thin film is employed as gate insulator.

10.2.2　Oxide Semiconductor Active Channels

For large-area electronics, the oxide semiconductor channel can be a powerful solution for enhancing the device performance for the TFTs. Recently, oxide channel TFTs have attracted much interest as backplane devices for flat-panel displays. Oxide semiconductors exhibit such beneficial features such as high carrier mobility, excellent uniformity, and robust device stability, compared to conventional amorphous silicon and low-temperature polycrystalline. Amorphous indium-gallium-zinc oxide (IGZO) is a typical composition of oxide semiconductor channel material. Other compositions have been employed as active channels for the oxide TFTs [19–25]. Although many works on the fabrication and characterization for the nonvolatile memory devices employing the P(VDF-TrFE) have mainly been focused for realizing the all-organic memory TFTs with organic semiconductors, the weakness of a low carrier mobility, an unsatisfactory ambient stability, and a difficult device integration seriously restrict the practical applications. Therefore, the features of oxide semiconductors are desirable for the ferroelectric nonvolatile memory TFTs. Consequently, the combination of an organic ferroelectric gate insulator and an oxide semiconductor channel will be the best choice for the high performance memory TFTs embeddable into various electronic systems implemented on large-area glass or plastic substrates.

10.3　Design of Device Structures

10.3.1　Design Schemes for Device Operations [26]

A typical top-gate bottom-contact structure of the OrFOx-MTFTs is shown in Fig. 10.1. Because the P(VDF-TrFE) is vulnerable to the plasma-induced process for the preparation of oxide channel layer, it is difficult to fabricate the bottom-gate structure with an excellent interface between P(VDF-TrFE) and oxide channel layers. The post annealing process at a temperature of higher than 200 °C is sometimes performed to enhance the TFT characteristics after the deposition of the oxide channel. However, for the bottom-gate structure, an available process temperature after the formation gate insulator is restricted to below 150 °C because of the melting temperature of P(VDF-TrFE). An interface buffer layer is introduced

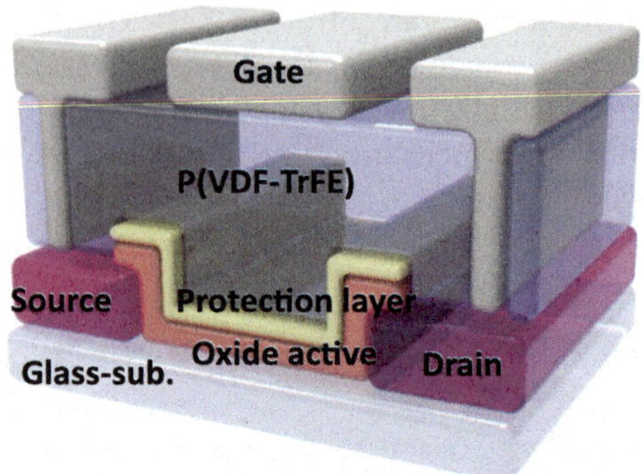

Fig. 10.1 Typical schematic cross-sectional diagram of the OrFOx-MTFT

between the P(VDF-TrFE) and oxide channel layer. The buffer layer is effective for protecting the oxide channel during the spin-coating and etching processes of P(VDF-TrFE). The chemical solvent of P(VDF-TrFE) and oxygen plasma employed for the etching might seriously degrade the electrical natures of the oxide semiconductor channels. In order to obtain the sound behaviors of the OrFOx-MTFTs having the device structure shown in Fig. 10.1, it is very important to optimize such parameters as thicknesses of interface buffer and oxide channel layers.

When compared to the organic ferroelectric MTFTs prepared on Si substrate, the device operations for the OrFOx-MTFTs are totally different owing to the difference in channel material. While the Si channel device operates between inversion and accumulation mode, for the OrFOx-MTFT, the device shows the switching between depletion and accumulation. Therefore, we have to note that the oxide channel can be fully depleted and remain as an insulating layer when negative gate voltage is applied because a thin layer below 50 nm is generally employed for the oxide semiconductor channel. Furthermore, the inversion layer is very difficult to be formed because of wide band gap. However, the formation of a full-depletion layer is definitely unfavorable in reducing the operation voltage owing to the capacitance coupling of the gate stack. The load-line analysis is helpful in estimating and designing the program voltages and depolarization field during the retention period. Figure 10.2a schematically shows the load-line in induced charge density-applied voltage (Q-V) plane for the OrFOx-MTFT. The gate stack is composed of serially-connected ferroelectric and insulator capacitors formed by P(VDF-TrFE) gate insulator and interface buffer layers. Total Q is expressed by $Q = Q_F = Q_O = C_O(V-V_F)$, where Q_F and Q_O are induced charge densities across the ferroelectric and insulator capacitors. A load-line formed by the insulator capacitor is described in V_F axis by red-straight line shown in Fig. 10.2a. When the negative gate voltage is applied, C_O is given by the geometric average of C_{OX} and C_{Dep}, which are capacitance per unit area of the insulator and channel depletion capacitors. Therefore, the slope of the load-line show different values between the positive and negative sides in voltage axis. The operating points at on- and off-program events are determined by the interceptions of ferroelectric $Q-V_F$ hysteresis and the load-line. As can be seen in Fig. 10.3b, different voltage amplitudes are required to

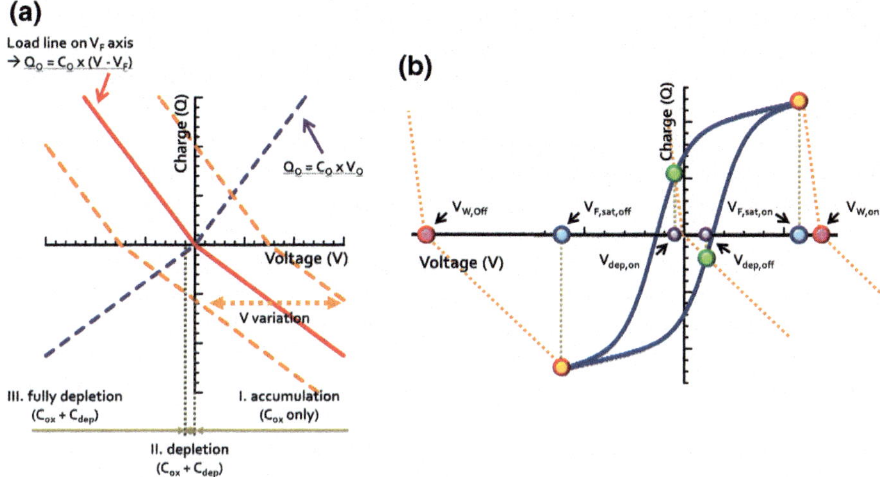

Fig. 10.2 **a** Load-line analysis in Q-V plane to determine the operating points of the OrFOx-MTFT [26]. The *dotted line* in the first and the third quadrants shows the variation in total induced charge density as a function of applied voltage. The load-line determined in the second and the forth quadrants moves on the V_F with the change in total applied voltage as shown in *dashed lines*. C_{OX} and C_{dep} correspond to the capacitance per unit area, which are induced in interface buffer layer and the depletion layer of oxide semiconductor channel, respectively. **b** The programming voltages for the on ($V_{W,on}$) and off ($V_{W,off}$) operations are determined by the interceptions of ferroelectric Q-V_F hysteresis curve and the load-line deduced in (**a**), in which it is assumed that the $V_{F,sat,on}$ and $V_{F,sat,off}$ are applied to the ferroelectric capacitor at on and off operations, respectively. During the data retention periods for on and off states, the depolarization fields of $V_{dep,on}$ and $V_{dep,off}$ are generated, respectively

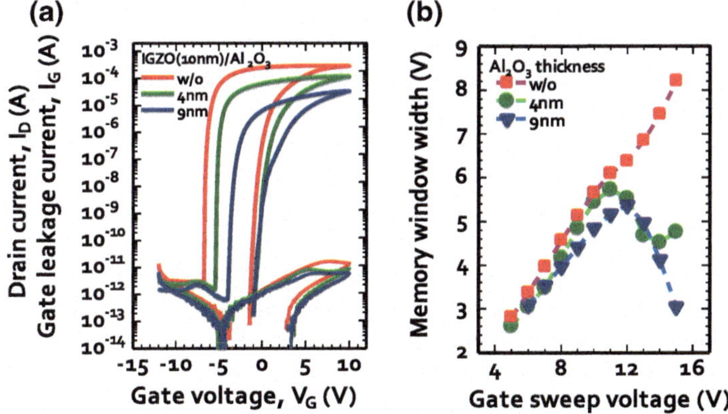

Fig. 10.3 **a** I_D-V_G transfer characteristics and gate leakage currents of the OrFOx-MTFTs using P(VDF-TrFE)/Al$_2$O$_3$/IGZO structures with the variation in Al$_2$O$_3$ thickness. **b** Variations in the programmed I_D values in on and off states were compared among the fabricated OrFOx-MTFTs with different thickness of Al$_2$O$_3$ buffer layer with the lapse of time [29]

completely program both on- and off-memory states. The program voltages for the on- (V_{ON}) and off-operations (V_{OFF}) of the OrFOx-MTFT can be estimated as (1) and (2), assuming that the ferroelectric polarization in the P(VDF-TrFE) film should be fully saturated for each program event. The use of a saturated ferroelectric hysteresis loop is absolutely desirable to obtain stable memory operations. Minor loops having partial polarization show a very weak immunity against the depolarization field during the data retention period.

$$V_{ON} = \frac{P_s \cdot d_{ox}}{\epsilon_0 \cdot \epsilon_{ox}} + V_{F(sat)} \tag{1}$$

$$V_{OFF} = \frac{P_s}{C_{ox}} \left(1 + \frac{\epsilon_{ox} \cdot d_s}{\epsilon_s \cdot d_{ox}} \right) + V_{F(sat)}, \tag{2}$$

where P_s and $V_{F(sat)}$ are saturated polarization charge per unit area of the P(VDF-TrFE) and corresponding applied voltage across the capacitor. ϵ_{ox}, ϵ_s, d_{ox}, and d_s are dielectric constants and film thicknesses of the buffer insulator and oxide channel layers, respectively. The V_{ON} can be reduced by decreasing the d_{ox}. However, we cannot minimize the V_{OFF} only with the reduction in buffer insulator thickness. The increase in d_s linearly increase V_{OFF}. Consequently, the film thicknesses of buffer insulator and oxide channel layers should be reduced for lowering the program voltages of the OrFOx-MTFTs. Alternatively, data retention time for the ferroelectric field-effect-based memory transistor is very sensitively affected by the depolarization field induced during the retention period. The depolarization field is defined as the electric field applied to the ferroelectric layer when the gate voltage is returned to 0 V [27], as shown in Fig. 10.2b. It is noteworthy that the depolarization field for the off-state is decided to be larger than that for the on-state. As results, the film thickness of the buffer insulator and oxide channel layers should also be reduced for minimizing the depolarization field and enhancing the memory retention time for the OrFOx-MTFTs.

10.3.2 Typical Device Structure and Characteristics

The effects of thickness variations in interface buffer and oxide channel layers on the memory device characteristics were investigated [28], in which atomic-layer-deposited Al_2O_3 and ZnO thin films were used as buffer and oxide channel layers for the OrFOx-MTFTs. The design schemes for the device structure can be varied with oxide channel materials. When the IGZO was chosen as the channel layer, excellent memory device characteristics were obtained [29]. Figure 10.3a shows the drain current–gate voltage (I_D-V_G) characteristics and gate leakage currents of the OrFOx-MTFTs employing IGZO channels when the thickness of Al_2O_3 buffer layer was varied to be 0, 4, and 9 nm. The memory window (MW) decreased with the increase in the Al_2O_3 thickness, and the smallest value of subthreshold swing (SS) was obtained for the device without Al_2O_3 layer. The field effect mobility at linear region, SS, MW (at ±12 V program) of the OrFOx-MTFT without Al_2O_3 layer were estimated to be approximately 61 cm^2/Vs, 120 mV/dec, and 6.4 V, respectively. However, there was a trade-off in introducing the Al_2O_3 buffer layer when the memory retention properties were examined. Figure 10.3b shows the variations in programmed values of I_D of the OrFOx-MTFTs with different Al_2O_3 thickness as a function of retention time. The memory retention time was remarkably enhanced by introducing the Al_2O_3 buffer layer with a thickness of 4 nm, and the on/off ratio of more than 70 could be confirmed after the lapse of 10^4 s. These results suggest that the device structures of the OrFOx-MTFTs, such as the film thicknesses of interface buffer and active channel layers, should be carefully investigated to compromise both memory TFT performance and retention time. The choice of oxide semiconductor channel material is also one of the important factors for optimizing the device structure of the OrFOx-MTFTs.

10.4 Process Optimization

In this section, the issues related to the fabrication process for OrFOx-MTFTs are discussed. The device characteristics are sensitively influenced by the employed process conditions.

10.4.1 Lithography Compatible Patterning Process

There have been many researches on micro- and nano-patterning process for the PVDF-based ferroelectric copolymers for realizing the high-density nonvolatile memory array. Although some approaches such as nanoimprint lithography [30–32] and direct writing method by piezoresponse force microscopy [33–35] have presented useful techniques for preparing the nanopatterns of ferroelectric thin films, they do not provide practical solutions for the conventional lithography-based device fabrication procedures using various chemicals. In order to guarantee the sound ferroelectric properties of the P(VDF-TrFE) after various fabrication processes routinely performed for device manufacturing, the chemical effects on the electrical and physical properties of the P(VDF-TrFE) should be investigated to confirm the process compatibility. It was proposed that the use of photoresist stripper during the photolithography process was related to the degradation of P(VDF-TrFE) [36]. Figure 10.4a, b compare the atomic-force microscopy observations and X-ray diffraction patterns of the P(VDF-TrFE) films treated by some developer and strippers with those of untreated films. The degradation in surface morphologies and crystallinities were detected for the film treated by stripper A, which is one of the typical commercial strippers. Undesirable increases in coercive field and leakage current were also observed as shown in Fig. 10.4c, d. Contrary, the suitable choice of stripper (stripper B) was very effective to minimize the degradations in physical and electrical properties of the P(VDF-TrFE). The ferroelectric copolymer can be easily removed by using an O_2 plasma in a conventional dry etching system. Figure 10.5a, b show scanning electron microscopy images of the P(VDF-TrFE) patterns. It was also found that the devices with a smaller size did not a harder degradation owing to the etch damage to larger edge areas. Thus, if the used chemicals are suitably chosen, the photolithography process for the P(VDF-TrFE) can be designed in a very compatible manner to full fabrication processes.

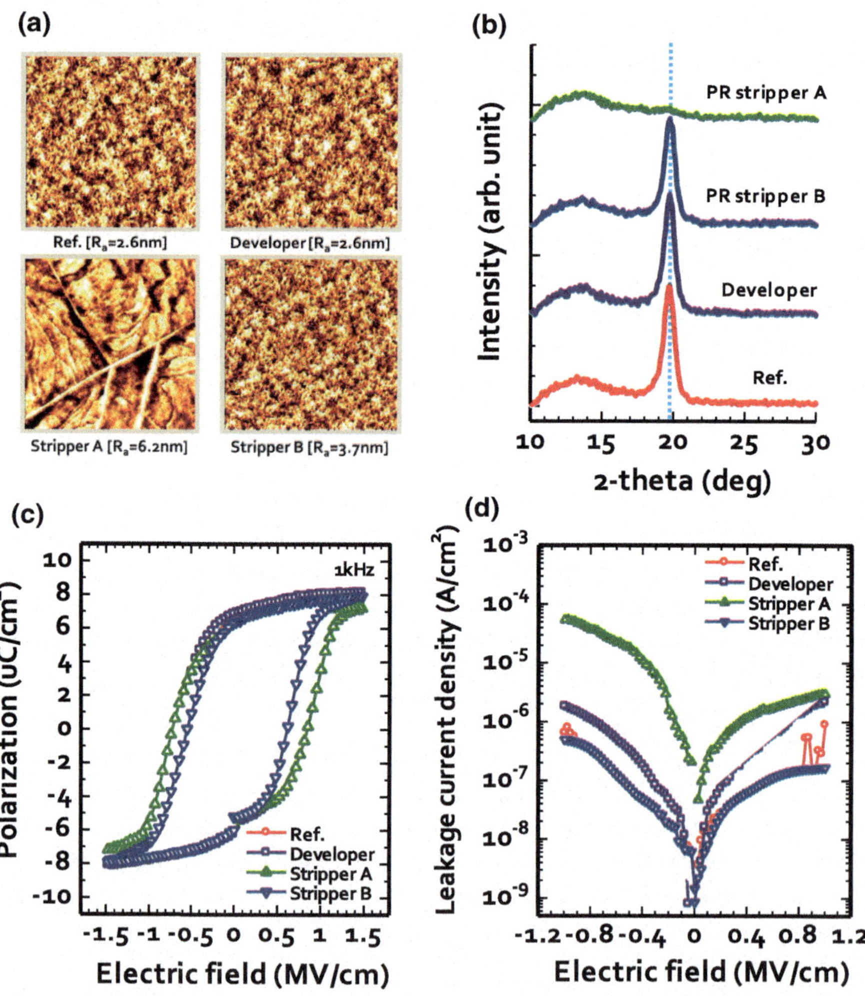

Fig. 10.4 a AFM images of surface morphology in the size of $5 \times 5\ \mu m^2$ and **b** XRD patterns for the P(VDF-TrFE) films treated with developer and photo-resist strippers. Single peaks at 19.7° correspond to the ferroelectric β phase of the P(VDF-TrFE) [36]

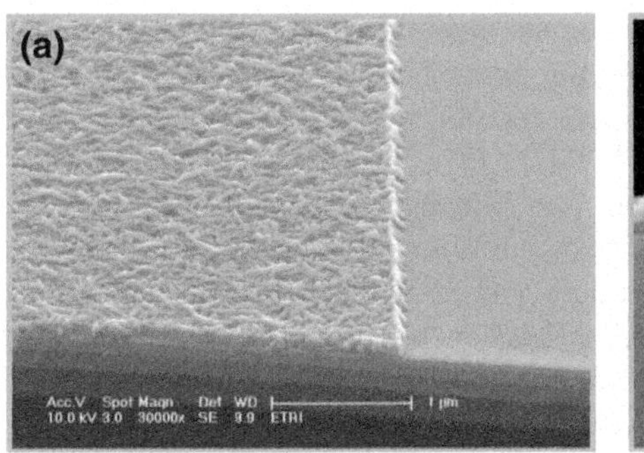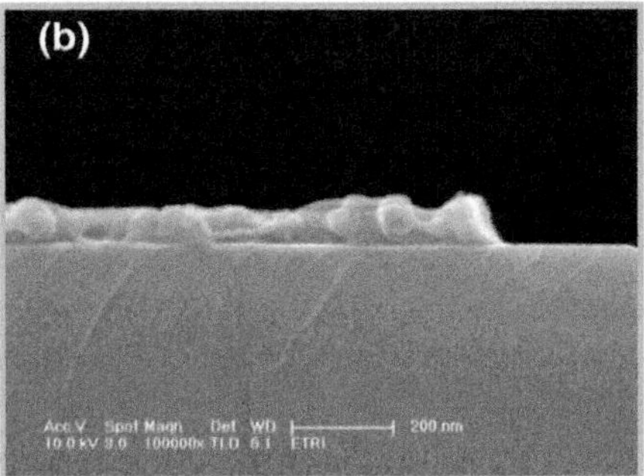

Fig. 10.5 SEM images of **a** a plane view and **b** a cross-section for the patterned P(VDF-TrFE) film [36]

10.4.2 Interface Protection Layer

For the proposed OrFOx-MTFTs, the main role of the interface is to effectively protect the surface of oxide channel layer during the fabrication procedures and to guarantee the interface quality. The device characteristics and reliabilities of the oxide TFTs were found to be significantly improved by the appropriate introduction of an interface controlling layer, which was sometimes called the 'protection layer' (PL) [37]. A successful fabrication of transparent and photo-stable ZnO TFT by employing the Al_2O_3 PL layer was reported, in which the Al_2O_3 PL was deposited by the ALD method using water vapor as oxygen precursor [38]. It was also confirmed that the ALD process conditions had critical impacts on the interfacial properties [39, 40]. As discussed in Sect. 10.3.2, in order to optimize the non-volatile memory characteristics of the OrFOx-MTFTs, the formation methods and thickness conditions for the interface controlling buffer layer between the P(VDF-TrFE) and oxide channel layers should be carefully controlled.

10.4.3 Oxide Channel Solution Process

The solution-based process for the oxide channel layer can be an alternative promising methods for the future large-area electronics, because it provides a low-cost process, a simple process design, and a high throughput [41–44]. Samsung Electronics demonstrated their first prototype active-matrix organic light-emitting diode display panel using solution-processed indium-zinc oxide (IZO) TFT [45]. It is also a very attractive approach for the proposed OrFOx-MTFTs, in which the oxide semiconducting channel and organic ferroelectric gate insulator can be suc-cessively formed in a simpler manner than the conventionally employed vacuum-deposition methods. Successful combinations of the spin-coated P(VDF-TrFE) with solution-processed oxide channel layers have been attempted as one of the challenging process issues for the OrFOx-MTFTs [46–48]. Various oxide semiconductors such as IZO, zinc-tin oxide (ZTO), indium-zinc-titanium oxide (IZTiO), and indium-zinc-silicon oxide (IZSiO) have been employed for the active channel materials for the OrFOx-MTFTs. Especially, it was suggested that the control of carrier concentration and channel conductance for the solution-processed oxide channel layer was very important for improving the nonvolatile memory characteristics including larger memory margin and longer retention time [49, 50]. One of the critical issues for the OrFOx-MTFTs fabricated by the solution process is that the post-annealing temperature is typically required to be higher than 400 °C to obtain sufficient device performance, even though the low-temperature solution process for oxide TFTs have been actively demonstrated [51–53]. Consequently, in order to employ the low-cost solution process for the large-area electronics implemented on glass or plastic substrates, new materials and process techniques should be urgently investigated to reduce the thermal budget as future works.

10.5 Promising Applications

In developing the OrFOx-MTFTs, it is important to define appropriate applications in advance. Because different specifications are required for each applications, device performance should be suitably developed to meet them. In this section, feasible and promising application fields for the embeddable OrFOx-MTFTs are introduced.

10.5.1 Nonvolatile Flexible Memory

Recently, various interesting approaches have been researched and developed in the fields of flexible electronics, which is featured to be prepared on a bendable or rollable thin glass and plastic substrates. Flexible sensor arrays [54–56], flexible displays [57, 58], radio-frequency identification tags [59–61], flexible electronic circuits [62–64], and sheet-type communication system [65] correspond to the feasible application examples. These devices necessarily demand embeddable nonvolatile memory elements. Several approaches have been tried to realize the nonvolatile memory devices on the flexible plastic substrates [67, 68]. The proposed OrFOx-MTFT can also be a good candidate. Because the oxide semiconductor channels are patterned into only small gate areas on the substrate, a relatively brittle nature of the oxide thin film is no longer be a fatal problem for the flexible memory applications. The sound characteristics of the P(VDF-TrFE) capacitors prepared on the polyethylene naphthalate (PEN) substrate were well verified and

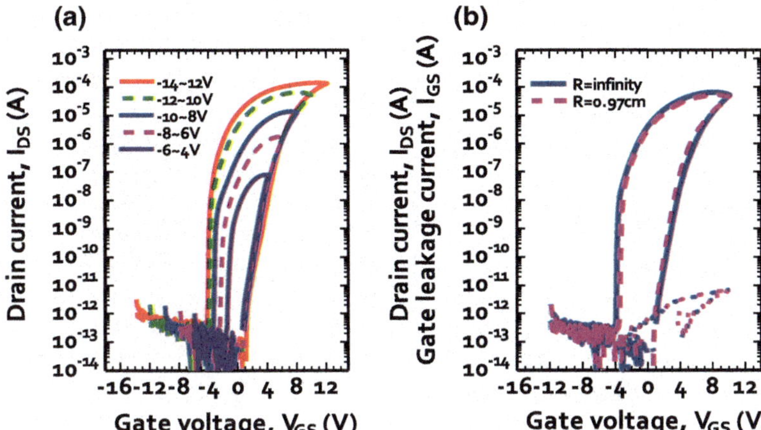

Fig. 10.6 **a** Sets of I_D-V_G transfer curves of the flexible OrFOx-MTFTs with the gate stack structure of P(VDF-TrFE)/Al$_2$O$_3$/ZnO when the V_G sweep ranges were varied. **b** Variations in transfer characteristics and memory behaviors under a substrate bending situation with a curvature radius of 0.97 cm [70]

the ferroelectric properties were not so changed with variations in curvature radius under the bending situations [69]. Figure 10.6a shows the transfer characteristics of the flexible OrFOx-MTFT with the gate stack structure of P(VDF-TrFE)/Al$_2$O$_3$/ ZnO [70]. All the processes were performed below 150 °C on a PEN substrate. The memory window and the on/off ratio of the flexible OrFOx-MTFT were 7.8 V and 10^8, respectively. These behaviors did not show any marked degradations at bending situations with a curvature radius of 9.7 mm, as shown in Fig. 10.6b. These results suggest that the OrFOx-MTFTs have great potential for the use as flexible nonvolatile devices in large-area electronics. However, the oxide channel materials should be further optimized so as to obtain better performances even at a temperature of lower than 200 °C.

10.5.2 Nonvolatile Transparent Memory

One of the new design concepts of 'consumer electronics' is 'see-through' electronic devices feature to be transparent to the visible light. These transparent electronic systems provide us with a new experience and specified functions. In realizing highly functional transparent devices and systems, embeddable transparent memories are quite demanding. Although some types of transparent memory devices have been demonstrated [71–77], several technical problems should be

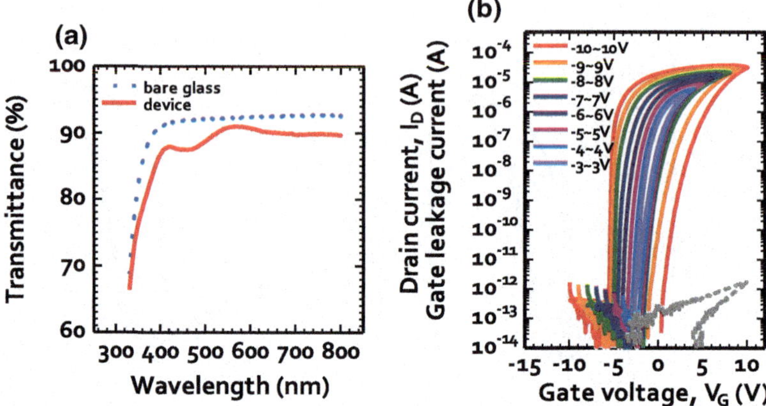

Fig. 10.7 **a** Transmittance of the transparent OrFOx-MTFT fabricated on the glass substrate in the visible range. The *inset* shows the optical image of the transparent glass substrate on which the transparent OrFOx-MTFTs were fabricated. The substrate size was 2×2 cm^2. **b** Sets of I_D-V_G transfer curves of the transparent OrFOx-MTFT when the V_G sweep ranges increased from ± 3 to ± 10 V [78]

solved for practical applications. The proposed OrFOx-MTFTs can be a promising solution for the nonvolatile memory devices embedded into the transparent electronic systems, in which the transparent conducting oxide (TCO) is utilized as electrodes and interconnections. Fully transparent OrFOx-MTFT was demonstrated for the first time [78]. Optical transmittance spectra of the OrFOx-MTFT prepared on the glass substrate exhibited a transmittance of approximately 90 % at a wavelength of 550 nm, as shown in Fig. 10.7a. Aluminum-zinc-tin oxide (AZTO) and ITO were chosen as active channel and TCO electrodes, respectively. Figure 10.7b shows sets of transfer curves of the transparent OrFOx-MTFT when the V_G sweep ranges increased from ± 3 to ± 10 V. The field effect mobility at the linear region, SS, MW at the V_G sweep of ± 10 V, and on/off ratio were successfully obtained to be 32.2 cm^2/Vs, 0.45 V/dec, 7.5 V, and 10^8, respectively. The MW increased in almost a symmetrical way toward both directions with the V_G sweeps. This suggest that there was not any marked charge-trapping events at the interfaces. The photo-response of the transparent OrFOx-MTFT should be investigated, because the memory behaviors may be significantly affected by the visible light owing to the transparency.

10.5.3 *Low-Power Backplane Device for Display Panel*

In the flat-panel display (FPD) field, the low-power consumption is one of the most urgent technical issues. A higher resolution and a larger panel size have

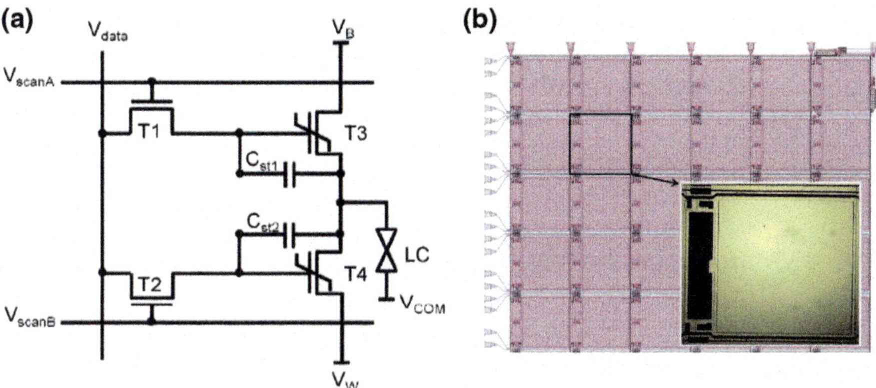

Fig. 10.8 **a** A unit pixel circuit of the proposed low-power reflective LCD. **b** A prototype low-power LCD with 5 × 6 pixels [81]

aggressively been pursued in the major FPD industries. Furthermore, the mobile information displays with an ultra-low power are strongly required for modern 'consumer electronics'. The 'memory-in-pixel' technology can be a promising approach to reduce the power consumption, in which nonvolatile memory cell is integrated into pixel of display panel [79, 80]. However, for these previously proposed techniques, the memory cell is composed of many TFTs occupying a large area and need refresh operations owing to their volatility in memory states. Therefore, the power consumption of the display panel can be effectively reduced only when a nonvolatile memory cell with a simple structure can be integrated into the pixels and/or driver circuits. Considering that the oxide TFT backplane device is very promising for future FPD panels, the proposed OrFOx-MTFT is one of the good candidates for realizing an ultra-low power display panel. A new pixel architecture and driving scheme for a low-power reflective LCD with a low refresh rate was successfully demonstrated [81]. The pixel architecture is composed of two oxide TFTs, two OrFOx-MTFTs, and two capacitors, as shown in Fig. 10.8a. Figure 10.8b shows a prototype LCD with 5 × 6 pixels, and the fabricated LCD operates at a 0.5-Hz refresh rate without flicker after programming the OrFOx-MTFTs.

10.5.4   Other Feasible Applications

Apart from the previously discussed main applications, the OrFOx-MTFTs can be utilized as nonvolatile memory components for other applications in oxide electronics. First, they can give a nonvolatile characteristics to the conventional logic elements so that the system power of the integrated circuits employing the oxide TFTs can be greatly reduced. The logic-in-memory concepts have been very popular for realizing lower-power VLSI in Si-based electronics. Second, the OrFOx-MTFTs can also be integrated with functional sensor arrays to store the sensing information.

10.6 Memory Array Integration

The OrFOx-MTFTs should be constructed in an array configuration and be integrated with drive and logic circuitry for practical applications. Therefore, the establishment of the integration process for the memory cell structure composed of the OrFOx-MTFTs and switching devices is very important. Furthermore, the memory cell should also be designed so as to perform the random access functions without any crosstalk problem.

10.6.1 Memory Cell Integration Process

As an good example of memory cell integration process, two transistor-type non-volatile memory cell was demonstrated [82], in which the integration process of the memory cell composed of an OrFOx-MTFT an oxide TFT were described. The temperature of the overall process was designed to below 200 °C. Figure 10.9a, b show a schematic circuit diagram and a microscopic photo image of the fabricated memory cell, respectively. The program and read-out operations were examined as shown in Fig. 10.9c. The memory on- and off-states were initially programmed by applying voltage pulses of 20 and −20 V, respectively. The programmed current could be modulated when the clock voltages were switched from −5 to 15 V. Nondestructive read-out operations were also confirmed for the 40 times repetitive operations, as shown in Fig. 10.9d.

10.6.2 Disturb-Free Memory Cell Array

Although the prototype memory cell and its nonvolatile/nondestructive operations were demonstrated in Sect. 10.6.1, the cell structure should be modified so that the data stored in the selected cell are not changed while accessing the neighboring cells. A new cell structure composed of two access oxide TFTs and one OrFOx-MTFT was proposed, as shown in Fig. 10.10a. In the actually fabricated memory cell shown in Fig. 10.10b, the program and read-out operations were confirmed when the program voltage pulse of ±10 V and 1 s were applied.

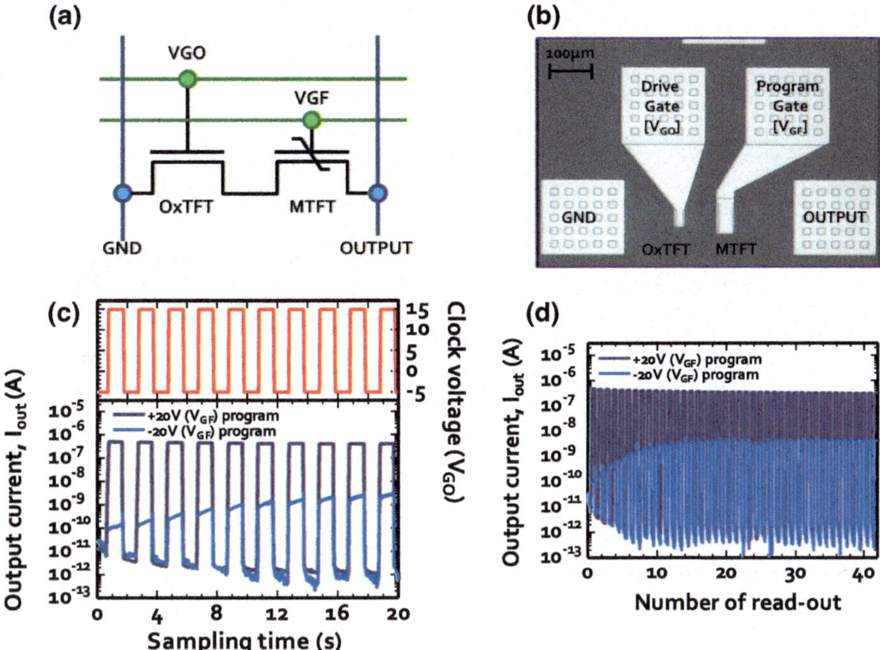

Fig. 10.9 **a** Schematic circuit diagram of the proposed 2T-type memory cell composed of one OrFOx-MTFT and one OxTFT. V_{GO} and V_{GF} correspond to drive gate of the OxTFT and program gate of the MTFT, respectively. **b** Photograph of the fabricated 2T-type nonvolatile memory cell. **c** Variations in I_{out} values of the memory cell with the time evolution when the gate clock voltage signal was continuously applied to the V_{GO}, which was switched from −5 to 15 V with the duration of 1 s as described in *upper graph*. The programming events for on and off operations were previously performed by applying the program voltages of 20 and −20 V, respectively. **d** Example of NDRO for the on and off states, which were measured with an interval of 1 s by applying the gate clock signal shown in (**b**) [82]

Furthermore, the disturb-free memory operations were also examined by using intentional cross-talk tests.

10.7 Remaining Technical Issues

In previous sections, it was discussed that the proposed OrFOx-MTFTs have great potential as core memory elements for the future large-area electronics. However, some technical challenges still remain for satisfying the suitable specifications required for commercialized applications. In this section, such remaining issues are considered.

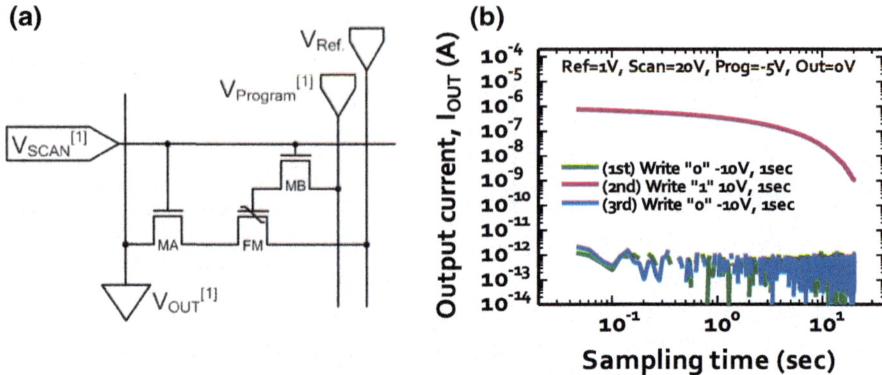

Fig. 10.10 **a** Circuit diagram of the proposed memory array for the disturb-free write and read-out operations. MA and MB correspond to two oxide TFT switches and FM is the OrFOx-MTFT. **b** Microscopic optical image of the fabricated memory cell with two oxide switch TFTs and one OrFOx-MTFT

10.7.1 Low-Voltage Operation

The reported program voltages for the OrFOx-MTFTs are in the range from ± 20 to ± 15 V, assuming that the memory on/off ratio of greater than 100 is required. These values are much lower than those for the organic semiconductor channel memory TFTs. However, further reduction to below ± 10 V is quite demanding. Some strategies to reduce the operating voltage for the OrFOx-MTFT can be available as follows; (1) the device structure of the OrFOx-MTFT should be carefully designed, as discussed in Sect. 10.3.1. The use of thinner film of oxide channel material is desirable; (2) the nano-patterning of the P(VDF-TrFE) was reported to greatly reduce the coercive field to approximately 100 kV/cm for a lower program voltage, which is one-tenth of the conventional value of coercive field; (3) new ferroelectric material with a lower coercive field is also a good approach. Croconic acid exhibited a very low coercive field of 30 kV/cm at 100 Hz. The reduction of program voltage of the OrFOx-MTFT will extend their practical applications.

10.7.2 Turn-on Voltage Control

For the OrFOx-MTFTs, the control of V_{on} is very important, because the device is absolutely desirable to be operated in an enhancement mode. Otherwise, the circuit design is difficult for the memory array configuration due to unwanted current components from unselected memory cell even at the V_G of 0 V. Therefore, the MW of the OrFOx-MTFT is preferred to be located with centering around V_G of

0 V. The problem is that the values of V_{on} for the oxide TFTs are typically determined to be below 0 V. The suitable choice of oxide channel material and/or process optimization will be necessary to shift the V_{on} to the positive direction. The implementation of the OrFOx-MTFT with a double-gate (DG) structure provide a useful solution to control the V_{on} in a more systematic way [83]. The V_{on} of the DG-OrFOx-MTFT was dynamically modulated from −2.1 to 2.0 V when the bias voltage applied to the bottom gate was varied from 6 to −6 V without any changes in the MW.

10.7.3 Program Speed

The program speed of the OrFOx-MTFT is determined by the voltage pulse width required for stable on- and off-program events. According to many previous works, a voltage pulse width as long as 1 s was necessary to drive the ferroelectric switching operations. Relatively short program times for the P(VDF-TrFE)-based memory TFTs were reported to be in the range from 50 µs to 40 ms. However, these values are much longer than those obtained for the P(VDF-TrFE) capacitors. These observed long switching times for the TFTs can originate from following two factors of (1) the formation of a fully-depletion layer in the active channel and (2) the RC time constant caused by the product of S/D channel resistance and gate capacitance. The switching time for the polarization reversal of P(VDF-TrFE) thin film is very sensitively dependent on the electric field applied across the film. Because the switching time is exponentially decayed with 1/E, a slight reduction in the program voltage remarkably impedes the switching operations. Consequently, it is very difficult to simultaneously optimize both issues of lower program voltage and higher speed operations. Although the use of thinner P(VDF-TrFE) gate insulator is desirable in improving the program speed as well as in reducing the program voltage, the drastic increase in the gate leakage current should be considered. Eventually, there is not any best solution to enhance the program speed to be below several µs at a program voltage as low as 10 V. The employment of DG configuration and/or the control of carrier concentration in active channel layer can be effective methods for improving the program speed. Actually, the program time for the DG-structured OrFOx-MTFTs could be reduced to 100 µs to obtain a memory margin of more than 10 with 15-V program voltage [83].

10.7.4 Data Retention

The data retention time is one of the most demanding specifications in employing the OrFOx-MTFTs in practical applications. The reported retention times for the OrFOx-MTFTs are in the range of 2 or 3 h. Important factors influencing the retention characteristics are intrinsic depolarization field, gate leakage current, and the interface quality. The depolarization field is determined by the device structure, as shown in Fig. 10.2a and gives detrimental effects on the retention properties. Therefore, it is important to use a fully saturated ferroelectric hysteresis loop in order to obtain more stable memory operations with time evolution, because minor loops are vulnerable to the application of internal depolarization fields. Consequently, the choice of best program conditions including program pulse amplitude and duration is absolutely desirable to correctly program the on- and off-states so that the polarization can be completely switched at each program operation. However, the use of saturated loop is not so easy for the OrFOx-MTFTs owing to the large coercive field of P(VDF-TrFE) and the formation of the depletion layer within the oxide channel. In other words, the compensation charges at the off-state cannot be sufficiently provided because of the fully depleted semi-conductor channel, and hence it is difficult to guarantee long retention time. Notwithstanding, if new organic ferroelectric material can be secured to have a good interface with an oxide channel layer, the retention characteristics of the OrFOx-MTFTs can be markedly improved. As discussed above, the introduction of the interface controlling layer of ALD-grown Al_2O_3 reduced the gate leakage current and improved interface quality for better retention performance. A low dielectric constant, a low coercive field, and an appropriate remnant polarization are essential requirements for employing full saturated ferroelectric hysteresis loops during the program operations. On the other hand, the OrFOx-MTFTs with DG configuration exhibited larger on/off ratio and longer retention time even with a relatively short signal of 20 ms by controlling the fixed bias of the bottom gate [83]. As discussed in this section, the physical mechanism related to the data retention characteristics for the OrFOx-MTFTs and feasible solutions to enhance the performance should be further investigated.

10.8 Conclusions and Outlooks

In this chapter, the ferroelectric field-effect TFT using a ferroelectric copolymer gate insulator and oxide semiconductor active channel, which was termed OrFOx-MTFT, was overviewed from the viewpoints of device and process technologies. Important issues such as the device design scheme, process optimization, feasible application, and memory array integration were intensively discussed and remaining technological problems were also comprehensively explained. The device performance of the proposed OrFOx-MTFT should be more improved. Transparent and/or flexible memory array implemented on the plastic substrates or the memory-embedded functionalized circuit elements can be final goals of the OrFOx-MTFTs. It was suggested that some technical breakthroughs should be devised for reducing the program voltage and for enhancing the program speed and retention characteristics. However, considering a high degree of difficulties to solve these problems, it would be important that the performance improvements go side by side with exploring suitable killer applications. The OrFOx-MTFT reviewed in this chapter is a promising and interesting approach to realization of an alternative nonvolatile memory for highly functionalized large-area electronic systems.

References

1. H.S. Nalwa, *Ferroelectric Polymers* (Dekker, New York, 1995)
2. T. Furukawa, Phase. Trans. **18**, 143 (1989)
3. R.G. Kepler, R.A. Anderson, Adv. Phys. **41**, 1 (1991)
4. S. Horiuchi, Y. Tokunaga, G. Giovannetti, S. Picozzi, H. Itoh, R. Shimano, R. Kumai, Y. Tokura, Nature **463**, 789 (2010)
5. R.C.G. Naber, P.W.M. Blom, A.W. Marsman, D.M. Leeuw, Appl. Phys. Lett. **85**, 2032 (2004)
6. F. Fang, W. Yang, C. Jia, X. Luo, Appl. Phys. Lett. **92**, 222906 (2008)
7. K. Muller, D. Mandal, K. Henkel, L. Paloumpa, D. Schmeisser, Appl. Phys. Lett. **93**, 112901 (2008)
8. T. Furukawa, T. Nakajima, Y. Takahashi, I.E.E.E. Dielectr, Electr. Insul. **13**, 1120 (2006)
9. H. Xu, X. Liu, X. Fang, H. Xie, G. Li, X. Meng, J. Sun, J. Chu, J. Appl. Phys. **105**, 034107 (2009)
10. H. Xu, J. Zong, X. Liu, J. Chen, D. Shen, Appl. Phys. Lett. **90**, 092903 (2007)
11. R.C.G. Naber, C. Tanase, P.W.M. Blom, G.H. Gelinck, A.W. Marsman, F.J. Touwslager, S. Setayesh, D.M. Leeuw, Nature Mater. **4**, 243 (2005)
12. K.H. Lee, G.B. Lee, K. Lee, M.S. Oh, S. Im, Appl. Phys. Lett. **94**, 093304 (2009)
13. R. Schroeder, L.A. Majewski, M. Grell, Adv. Mater. **16**, 633 (2004)
14. K.N.N. Uni, R. Bettignies, S.D. Seignon, J. Nunzi, Appl. Phys. Lett. **85**, 1823 (2004)
15. R.C.G. Naber, B. Boer, P.W.M. Blom, D.M. Leeuw, Appl. Phys. Lett. **87**, 203509 (2005)
16. K. Muller, K. Henkel, I. Paloumpa, D. Schmeiber, Thin Solid Film **515**, 7683 (2007)
17. S.J. Kang, Y.J. Park, J. Sung, P.S. Jo, C. Park, K.J. Kim, B.O. Cho, Appl. Phys. Lett. **92**, 012921 (2008)
18. C.A. Nguyen, S.G. Mhaisalkar, J. Ma, P.S. Lee, Org. Electron. **9**, 1087 (2008)
19. K. Nomura, H. Ohta, A. Takagi, T. Kamiya, M. Hirano, H. Hosono, Nature **432**, 488 (2004)
20. R.B.M. Cross, M.M. Souza, S.C. Deane, N.D. Young, I.E.E.E. Trans, Electron Dev. **55**, 1109 (2008)
21. A. Suresh, J.F. Muth, Appl. Phys. Lett. **92**, 033502 (2008)
22. J.K. Jeong, H.W. Yang, J.H. Jeong, Y.G. Mo, H.D. Kim, Appl. Phys. Lett. **93**, 123508 (2008)
23. K. Hoshino, D. Hong, H.Q. Chiang, J.F. Wager, I.E.E.E. Trans, Electron Dev. **56**, 1365 (2009)
24. K. Nomura, T. Kamiya, M. Hirano, H. Hosono, Appl. Phys. Lett. **95**, 013502 (2009)
25. J.M. Lee, I.T. Cho, J.H. Lee, H.I. Kwon, Appl. Phys. Lett. **93**, 093504 (2008)
26. S.M. Yoon, S.H. Yang, S.H.K. Park, S.W. Jung, D.H. Cho, C.W. Byun, S.Y. Kang, C.S. Hwang, B.G. Yu, J. Phys. D Appl. Phys. **42**, 245101 (2009)
27. E. Tokumitsu, K. Okamoto, H. Ishiwara, Jpn. J. Appl. Phys. **39**, 2125 (2001)
28. S.M. Yoon, S.H. Yang, C.W. Byun, S.H.K. Park, S.W. Jung, D.H. Cho, S.Y. Kang, C.S. Hwang, B.G. Yu, Jpn. J. Appl. Phys. **49**, 04DJ06 (2010)
29. S.M. Yoon, S.H. Yang, S.W. Jung, C.W. Byun, S.H.K. Park, C.S. Hwang, G.G. Lee, E. Tokumitsu, H. Ishiwara, Appl. Phys. Lett. **96**, 232903 (2010)
30. Z. Hu, M. Tian, B. Nysten, A.M. Jonas, Nat. Mater. **8**, 62 (2008)
31. L. Zhang, S. Ducharme, J. Li, Appl. Phys. Lett. **91**, 172906 (2007)
32. Z. Hu, G. Baralia, V. Bayot, J.F. Gohy, A.M. Jonas, Nano Lett. **5**, 1738 (2005)
33. V.S. Bystrov, I.K. Bdikin, D.A. Kiselev, S. Yudin, V.M. Fridkin, V.M. Kholkin, J. Phys. D Appl. Phys. **40**, 4571 (2007)
34. B.J. Rodriguez, S. Jesse, S.V. Kalinin, J. Kim, S. Ducharme, V.M. Fridkin, Appl. Phys. Lett. **90**, 122904 (2007)

35. C. Rankin, C. Chou, D. Conklin, D.A. Bonnel, ACS Nano **1**, 234 (2007)
36. S.M. Yoon, S.W. Jung, S.H. Yang, S.Y. Kang, C.S. Hwang, B.G. Yu, Jpn. J. Appl. Phys. **48**, 09KA20 (2009)
37. S.H.K. Park, D.H. Cho, C.S. Hwang, S. Yang, M.K. Ryu, C.W. Byun, S.M. Yoon, W.S. Cheong, K.I. Cho, J.H. Jeon, ETRI J. **31**, 653 (2009)
38. S.H.K. Park, C.S. Hwang, M. Ryu, S. Yang, C. Byun, J. Shin, J. Lee, K. Lee, M.S. Oh, S. Im, Adv. Mater. **21**, 678 (2009)
39. S.M. Yoon, S.H.K. Park, C.W. Byun, S.H. Yang, C.S. Hwang, J. Electrochem. Soc. **157**, H727 (2010)
40. S.M. Yoon, S.H.K. Park, S.H. Yang, C.W. Byun, C.S. Hwang, Electrochem. Solid State Lett. **13**, H624 (2010)
41. M.G. Kim, H.S. Kim, Y.G. Ha, J. He, M.G. Kanatzidis, A. Facchetti, T.J. Marks, J. Am. Ceram. Soc. **132**, 10532 (2010)
42. W.H. Jeong, G.H. Kim, H.S. Shin, B.D. Ahn, H.J. Kim, M.K. Ryu, K.B. Park, J.B. Seon, S.Y. Lee, Appl. Phys. Lett. **96**, 093503 (2010)
43. H.S. Shin, G.H. Kim, W.H. Jeong, B.D. Ahn, H.J. Kim, Jpn. J. Appl. Phys. **49**, 03CB01 (2010)
44. J.H. Jeon, Y.H. Hwang, B.S. Bae, K.L. Kown, H.J. Kang, Appl. Phys. Lett. **96**, 212109 (2010)
45. M.K. Ryu, K.B. Park, J.B. Seon, J. Park, I.S. Kee, Y.G. Lee, S.Y. Lee, in *SID International Symposium Digestive Technical Papers* (2009), p. 188
46. S.M. Yoon, S.W. Jung, S.H. Yang, C.W. Byun, C.S. Hwang, H. Ishiwara, J. Electrochem. Soc. **157**, H771 (2010)
47. S.M. Yoon, S.H. Yang, S.W. Jung, C.W. Byun, S.H.K. Park, C.S. Hwang, H. Ishiwara, Electrochem. Solid State Lett. **13**, H141 (2010)
48. S.M. Yoon, S.W. Jung, S.H. Yang, C.W. Byun, C.S. Hwang, S.H.K. Park, H. Ishiwara, Org. Electron. **11**, 1746 (2010)
49. J.Y. Bak, S.M. Yoon, J. Vac. Sci. Technol. B **31**, 040601 (2013)
50. J.Y. Bak, S.W. Jung, S.M. Yoon, Org. Electron. **14**, 2148 (2013)
51. K.K. Banger, Y. Yamashita, K. Mori, R.L. Peterson, T. Leedham, J. Rickard, H. Sirringhaus, Nature Mater. **10**, 45 (2011)
52. M.G. Kim, M.G. Kanatzidis, A. Facchetti, T.J. Marks, Nature Mater. **10**, 382 (2011)
53. Y.H. Kim, J.S. Heo, T.H. Kim, S. Park, M.H. Yoon, J. Kim, M.S. Oh, G.R. Yi, Y.Y. Noh, S.K. Park, Nature **489**, 128 (2012)
54. T. Someya, Y. Kato, S. Iba, Y. Noguchi, T. Sekitani, H. Kawaguchi, T. Sakurai, I.E.E.E. Trans, Electron Dev. **52**, 2502 (2005)
55. K.L. Lin, K. Jain, IEEE Electron Dev. Lett. **30**, 14 (2009)
56. T. Sekitani, T. Yokota, U. Zschieschang, H. Klauk, S. Bauer, K. Takeuchi, M. Takamiya, M. Sakurai, T. Someya, Science **326**, 1516 (2009)
57. G.H. Gelinck, D.M. Leeuw, Nature Mater. **3**, 106 (2004)
58. T. Sekitani, H. Nakajima, H. Maeda, T. Fukushima, T. Aida, K. Hata, T. Someya, Nature Mater. **8**, 494 (2009)
59. S.R. Forrest, Nature **428**, 911 (2004)
60. P.F. Baude, D.A. Ender, M.A. Haase, T.W. Kelley, D.Y. Muyres, S.D. Theiss, Appl. Phys. Lett. **82**, 3964 (2003)
61. M. Jung, J. Kim, J. Noh, N. Lim, C. Lim, G. Lee, J. Kim, H. Kang, K. Jung, A.D. Leonard, J.M. Tour, G. Cho, I.E.E.E. Trans, Electron Dev. **57**, 571 (2010)
62. H. Klauk, M. Halik, U. Zschieschang, F. Eder, D. Rohde, G. Schmid, C. Dehm, I.E.E.E. Trans, Electron Dev. **52**, 618 (2015)
63. I.M. Graz, S.P. Lacour, Appl. Phys. Lett. **95**, 243305 (2009)
64. U. Zschieschang, F. Ante, T. Yamamoto, K. Takamiya, H. Kuwabara, M. Ikeda, T. Sekitani, T. Someya, K. Kern, H. Klauk, Adv. Mater. **22**, 982 (2010)
65. T. Sekitani, K. Zaitsu, Y. Noguchi, K. Ishibe, M. Takamiya, T. Sakurai, T. Someya, I.E.E.E. Trans, Electron Dev. **56**, 1027 (2009)
66. S.T. Han, Y. Zhou, V.A.L. Roy, Adv. Mater. **25**, 5425 (2013)

67. D. Son, J. Lee, S. Qiao, R. Ghaffari, J. Kim, J.E. Lee, C. Song, S.J. Kim, D.J. Lee, S.W. Jun, S. Yang, M. Park, J. Shin, K. Do, M. Lee, K. Kang, C.S. Hwang, N. Lu, T. Hyeon, D.H. Kim, Nature Nanotech. **9**, 397 (2014)
68. S. Kim, J.H. Son, S.H. Lee, B.K. You, K.I. Park, H.K. Lee, M. Byun, K.J. Lee, Adv. Mater. **26**, 7480 (2014)
69. S.M. Yoon, S.W. Jung, S. Yang, S.H.K. Park, B.G. Yu, H. Ishiwara, Curr. Appl. Phys. **11**, S219 (2011)
70. S.M. Yoon, S. Yang, S.H.K. Park, J. Electrochem. Soc. **158**, H892 (2011)
71. H. Yin, S. Kim, C.J. Kim, I. Song, J. Park, S. Kim, Y. Park, Appl. Phys. Lett. **93**, 172109 (2008)
72. A. Suresh, S. Novak, P. Wellenius, V. Misra, J.F. Muth, Appl. Phys. Lett. **94**, 123501 (2009)
73. J.W. Seo, J.W. Park, K.S. Lim, J.H. Yang, S.J. Kang, **93**, 223505 (2008)
74. J.W. Seo, J.W. Park, K.S. Lim, S.J. Kang, Y.H. Hong, J.H. Yang, L. Fang, G.Y. Sung, H.K. Kim, Appl. Phys. Lett. **95**, 133508 (2009)
75. L. Shi, D. Shang, J. Sun, B. Sun, Appl. Phys. Express **2**, 101602 (2009)
76. K.S. Yook, J.Y. Lee, S.H. Kim, J. Jang, Appl. Phys. Lett. **92**, 223305 (2008)
77. K.S. Yook, J.Y. Lee, Thin Solid Film **517**, 5573 (2009)
78. S.M. Yoon, S. Yang, C. Byun, S.H.K. Park, D.H. Cho, S.W. Jung, O.S. Kwon, C.S. Hwang, Adv. Funct. Mater. **20**, 921 (2010)
79. N. Ueda, Y. Ogawa, K. Tanaka, K. Yamamoto, Y. Yamauchi, in *SID International Symposium Digestive Technical Papers* (2010), p. 615
80. L.W. Chu, P.T. Liu, M.D. Ker, G.T. Zheng, Y.H. Li, C.H. Kuo, C.H. Li, Y.J. Hsieh, C.T. Liu, in *SID International Symposium Digestive Technical Papers* (2010), p. 1363
81. S.H. Lee, J. Kim, S.H. Yoon, K.A. Kim, S.M. Yoon, C. Byun, C.S. Hwang, G.H. Kim, K.I. Cho, S.W. Lee, IEEE Electron Dev. Lett. **36**, 585 (2015)
82. S.M. Yoon, C.W. Byun, S. Yang, S.H.K. Park, D.H. Cho, S.W. Jung, S.Y. Kang, C.S. Hwang, IEEE Electron Dev. Lett. **31**, 138 (2010)
83. S.M. Yoon, S. Yang, M.K. Ryu, C.W. Byun, S.W. Jung, S.H.K. Park, C.S. Hwang, K.I. Cho, I.E.E.E. Trans, Electron Dev. **58**, 2135 (2011)

Part VI

Practical Characteristics of Organic Ferroelectric-Gate FETs : Ferroelectric-Gate Field Effect Transistors with Flexible Substrates

Chapter 11
Mechanically Flexible Non-volatile Field Effect Transistor Memories with Ferroelectric Polymers

Richard H. Kim and Cheolmin Park

Abstract Great efforts have been devoted to improve the properties of non-volatile memory with field effect transistor architecture containing ferroelectric polymers (NV-FeFETs) due to the potential advantages of the ferroelectric polymers including their low cost, easy fabrication based on solution processes, and mechanical flexibility. Here, we review the current status of development in particular on mechanically flexible NV-FeFETs. In addition, recent researches that demonstrate the importance of the analysis techniques to characterize the mechanical properties of thin films composing a FeFET are discussed, including nano-indentation and nano-scratch test.

11.1 Introduction

Non-volatile memories are widely used in our daily life such as display, cellular phone, and computer and additional tremendous demand is made on the next generation of smart wearable and patchable on skin electronic applications. To realize future electronic applications, a non-volatile memory mechanically flexible, foldable and stretchable for efficiently adopting non-planar and irregular topological surface of fabrics and skins is required. The development of functional memory elements and appropriate device architectures satisfying the mechanical requirement is, therefore, essential. Currently, the most widely used non-volatile memories are distinguished by various data storage principles such as resistive type [1–3], flash [4–6] and ferroelectric [7–55], and these memories have been extensively

R.H. Kim · C. Park (✉)
Department of Materials Science and Engineering, Yonsei University,
Shinchon-dong, Seodaemun-gu, Seoul 120-749, Korea
e-mail: cmpark@yonsei.ac.kr

© Springer Science+Business Media Dordrecht 2016

B.-E. Park et al. (eds.), *Ferroelectric-Gate Field Effect Transistor Memories*,
Topics in Applied Physics 131, DOI 10.1007/978-94-024-0841-6_11

investigated as candidates for the next generation flexible non-volatile memory market. Ferroelectric-gate field effect transistors (FeFETs) with ferroelectric polymer layers as gate insulators have received specific attention due to their simple device structure of single-transistor, non-destructive read-out, massive memory capacity, compatibility with complementary logic circuits and their potential application as flexible or stretchable charge storage media. Herein, we report the current status of the development on mechanically flexible ferroelectric polymer based memory devices and recent researches that demonstrate the importance of the analysis techniques for evaluating mechanical properties of a memory device containing very thin constituent layers. Particular emphasis will be made on thin film characterization of mechanical properties based on nano-indentation and nano-scratch test.

11.2   FeFETs with Ferroelectric Polymers

11.2.1   Ferroelectric Polymers

The spontaneous polarization of ferroelectric polymers can be switched to the opposite direction with the reversal of the electric field, making it an important element of devices such as non-volatile memories, field-effect transistor sensors, actuators and solar cells. In particular, polyvinylidene fluoride (PVDF) and its copolymers are the most widely used ferroelectric polymers and are commercially manufactured in large quantities for a wide variety of applications. PVDF has its chemical structure in which two fluorine and two hydrogen atoms are alternately bonded to the carbon atoms of the backbone as shown Fig. 11.1a. PVDF exhibits the ferroelectric characteristics because of the strong dipoles which can be switched through molecular rotation by electric field [56, 57]. The carbon (C)-fluorine (F) and carbon- hydrogen (H) bonds compose the dipoles by electronegativity difference between C and F or C and H. PVDF has generally four different types of polymorphic crystalline phases of α [58–60], β [61], γ [62], and δ [63, 64] phases. The conformations of the macromolecules are appointed as the trans (T) and gauche (G) expression. The β phase has the all-consecutive trans (TTTT) conformations, while the α phase consists of alternating trans-gauche (TGTG). The γ phase has an intermediate conformation T^3GT^3G and the parallel version of the α phase is known as δ phase. The β phase shows the best ferroelectric characteristics, however it is not straightforward to obtain the β phase through crystallization process because its curie temperature (T_c) (195–197 °C) [65], at which paraelectric to ferroelectric transition occurs, is higher than the melting temperature (170 °C). In the most of cases, PVDF is mechanically stretched and the resulting aligned polymer chains give rise to the β phase.

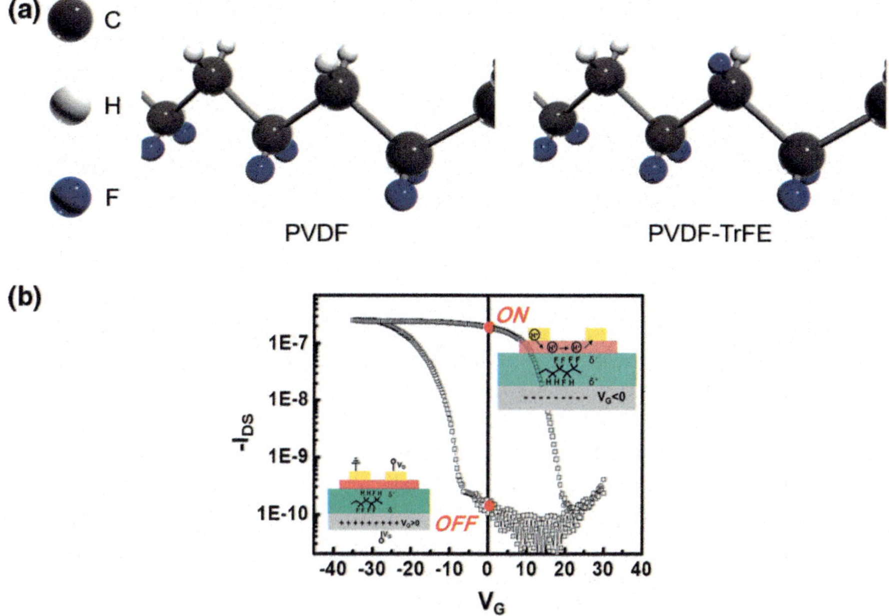

Fig. 11.1 **a** The chemical structure of polyvinylidene fluoride (PVDF) (*left*) and polyvinylidene-trifluoroethylene (PVDF-TrFE) (*right*) with the backbone of carbon atoms and fluoride and hydrogen atoms. **b** Schematic of the operation principle of a FeFET

The development of poly(vinylidene fluoride-co-trifluoroethylene) (PVDF-TrFE) (Fig. 11.1), which is fabricated by copolymerization with TrFE into the PVDF, can lower T_c, making it possible to form the β phase by thermal annealing process below the melting temperature [66]. This is because the interactions between dipole-to-dipole of PVDF are reduced by strong steric hindrance of the fluorine atom in TrFE unit. The remanent polarization (P_r) at zero applied voltage, which is one of the most important parameters for ferroelectric characteristics, of a thermally annealed PVDF-TrFE film is approximately 7–8 $\mu C/cm^2$ large enough for non-volatile data storage.

11.2.2 Operation Principle of FeFETs

In a FeFET, the polarization state of the ferroelectric layer is set by the polarity of the program/erase gate voltage, which controls accumulation or depletion of carriers in the semiconducting channel between the source and drain electrodes. A FeFET with a p-type semiconducting channel layer exhibits p-type transfer characteristics with current hysteresis as a function of V_G arising from the ferroelectric polarization switching of PVDF-TrFE. After the device is turned on, there is a rapid increase of the I_{DS} with a negative bias in V_G due to accumulation of excess holes in the semiconductor layer. When the V_G returns to zero, the I_{DS} still remains saturated due to the non-volatile H-F dipoles, with the fluorine atoms pointing to the semi-conductor layer. The subsequent application of a positive V_G on the device grad-ually switches the H-F dipoles, leading to a decrease in the I_{DS}. The non-volatility of the polarization causes the current to remain constant after the removal of the positive voltage (Fig. 11.1b).

Due to the potential advantages of ferroelectric polymers including their low cost, easy solution processes fabrication, and mechanical pliability (which allows for the production of flexible, foldable, and stretchable devices), great efforts have been devoted to improve the properties of FeFETs using ferroelectric polymers such as PVDF, PVDF-TrFE, poly(vinylidene fluoride-co-chlorotrifluoroethylene) (PVDF-CTFE) as shown in Table 11.1 [7–55].

11.3　FeFET on Flexible Substrates

11.3.1　Flexible FeFETs with Inorganic Semiconductors

Although organic semiconductors are more suitable as semiconductor layers for flexible FeFETs with PVDF-TrFE, FeFETs with inorganic semiconductor channels were demonstrated on flexible substrates such as poly(ethylene naphthalene) (PEN), poly(dimethyl siloxane) (PDMS) due to the relatively small on/off current ratio and low field effect mobility of the organic semiconductors compared with those of inorganic semiconductors. The employment of an oxide semiconductor can be a good approach for realizing excellent memory performance of high mobility as well as relatively low voltage operation. Lee et al. [8], fabricated flexible FeFET with amorphous indium gallium zinc oxide (a-IGZO) channel on the PEN substrate with top gate bottom contact structure using Al as gate and source-drain metal contacts. The flexible FeFET exhibited a memory window of 8.4 V at operation voltages from -20 to 20 V, 7-orders-of-magnitude I_{on}/I_{off} ratio, and showed reliable elec-trical properties up to bending radius ~ 7 mm (Fig. 11.2).

van Breeman et al. [9], demonstrated the fabrication and operation of polymer ferroelectric transistor arrays on the PEN substrate. IGZO was used as semicon-ductor channel layer and PVDF-TrFE as ferroelectric gate insulator. The device structure of FeFET was top gate bottom contact structure with Au top electrode and source-drain metal pads. The memory transistor showed a drain current bistability of 10^5 and data retention time of more than 12 days. The arrays can be bent to a radius below 20 mm in bending radius repeatedly without apparent degradation (Fig. 11.3).

Table 11.1 Characteristics of FeFET memory devices with various semiconducting and ferroelectric layers

Semiconductor	Ferroelectric layer	Operating voltage	On/off ratio	Retention (s)	Endurance (#)	Ref
IGZO	PVDF-TrFE	40	10^2	10^4	–	7
IGZO	PVDF-TrFE	20	10^7	–	–	8
IGZO	PVDF-TrFE	20	10^5	10^6	10^3	9
IGZO	PVDF-TrFE	20	10^4	3×10^3	–	10
ZnO	PVDF-TrFE	20	10^3	10^4	–	11
ZnO	PVDF-TrFE	12	10^8	10^4	–	12
ZnO	PVDF-TrFE	70	10^6	7×10^3	–	13
ZnO NW	PVDF-TrFE	20	10^6	10^4	–	14
MoS_2	PVDF-TrFE	20	10^5	10^3	10	15
SnO	PVDF-TrFE	30	10^2	–	–	16
Si	PVDF-TrFE	15	10^6	10^5	10^5	17
Si NW	PVDF-TrFE	20	10^3	10^4	–	18
Pentacene	PVDF-TrFE	13	10	10^4	–	19
Pentacene	PVDF-TrFE	15	10^3	10^4	–	20
Pentacene	PVDF-TrFE	20	10^3	10^7	–	21
Pentacene	PVDF-TrFE	30	10^4	10^6	2.5×10^3	22
Pentacene	PVDF-TrFE	40	10^3	–	–	23
Pentacene	PVDF-TrFE	60	10^5	10^3	–	24
Pentacene	PVDF-TrFE	60	10^4	10^5	–	25
Pentacene	PVDF-TrFE	20	10^4	–	–	26
Pentacene	PVDF-TrFE	66	10^2	10^6	–	27
Pentacene	PVDF-TeFE	80	10^4	4.5×10^3	–	28
Pentacene	PVDF-TrFE	45	10^2	–	5×10^8	29
Rubrene	PVDF-TeFE	60	10^5	–	–	30

(continued)

Table 11.1 (continued)

Semiconductor	Ferroelectric layer	Operating voltage	On/off ratio	Retention (s)	Endurance (#)	Ref
TIPS-PEN	PVDF-TrFE	30	10^3	5×10^4	–	31
TIPS-PEN	PVDF-TrFE + OS	8	10^2	6×10^3	–	32
TIPS-PEN	PVDF-TrFE + PMMA	40	10^3	–	–	33
TIPS-PEN	PVDF + PMMA	15	10^3	10^4	–	34
TIPS-PEN	PVDF-CTFE	60	10^3	10^3	–	35
TIPS-PEN	PVDF-TrFE + THDA	20	10^3	–	–	36
TIPS-PEN	PVDF-TrFE	70	10^4	10^3	–	37
QQT(CN)4	PVDF-TrFE	50	10^4	6×10^3	10^2	38
QQT(CN)4	PVDF-TrFE	25	10^3	10^3	10^2	39
TIPS-PEN + PαMS	PVDF-TrFE	80	10^4	10^3	10^2	40
MEH-PPV	PVDF-TrFE	150	10^5	6×10^5	10^3	41
MEH-PPV	PVDF-TrFE	150	10^5	–	–	42
P3HT	PVDF-TrFE	18	10^2	10^4	–	43
P3HT	PVDF-TrFE + PVDF-TrFE-CTFE	18	10^4	10^4	1.3×10^2	44
P3HT	PVDF	80	10^2	–	–	45
P3HT	PVDF-TrFE	80	10^3	10^5	1.2×10^2	46
P3HT	PVDF-TrFE + BaTiO$_3$	60	10	2×10^2	–	47
P3HT	PVDF-TrFE	40	10^3	10^4	10^2	48
P3HT	PVDF-TrFE	150	10^4	–	–	49
P3HT	PVDF-TrFE + EMI[TFSA]	7	10^3	10^4	8×10^1	50
P3HT NW	PVDF-TrFE	30	10^2	10^4	10^2	51
PTAA	PVDF-TrFE	110	10^2	10^3	–	52
PTAA	PVDF-TrFE	100	10^3	–	10^4	53
PQT	PVDF-TrFE	50	10^2	10^3	–	54
OC$_1$OC$_{10}$-PPV	PVDF-TrFE	15	10^3	–	–	55

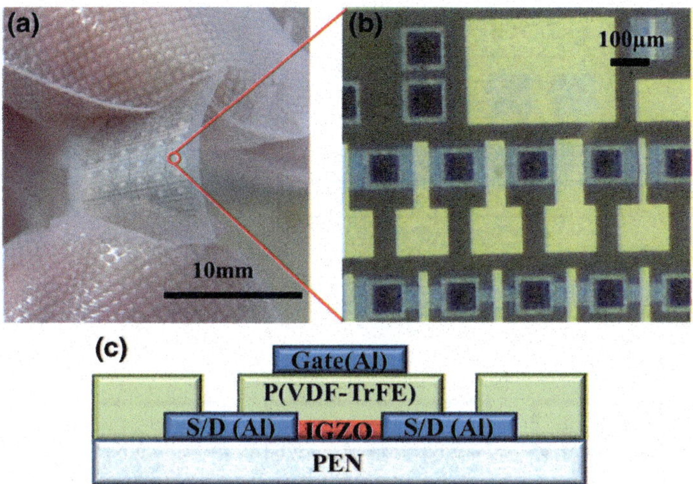

Fig. 11.2 **a** The photograph and **b** photomicrograph of the fabricated Fe-TFTs, respectively. **c** Cross-sectional schematic diagram with Al/PVDF-TrFE/IGZO/Al/PEN structure [8]

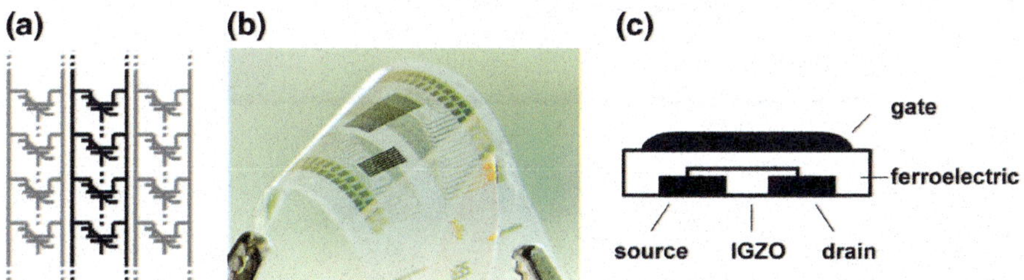

Fig. 11.3 **a** Schematic of the memory arrays. **b** Micrograph of a flexible arrays. **c** Cross section of the ferroelectric transistor, and the materials used in its implementation [9]

Jung et al. [10], demonstrated flexible IGZO semiconductor based FeFET on a PDMS elastomer substrate. Top gate bottom contact device structure was chosen and Mo source-drain contacts and Al gate electrodes were employed. The memory window of 13 V at ±20 V programming was achieved for the device without any interface layer and memory on/off ratio was initially 10^4 and maintained at 10^2 after 3600 s. These memory characteristics did not vary significantly under bending at a radius of 10 mm (Fig. 11.4).

Zinc oxide (ZnO) is another representative inorganic semiconductor material which has high carrier mobility. Yoon et al. [12], proposed an organic/inorganic hybrid-type flexible FeFET on a PEN substrate. The device structure was designed

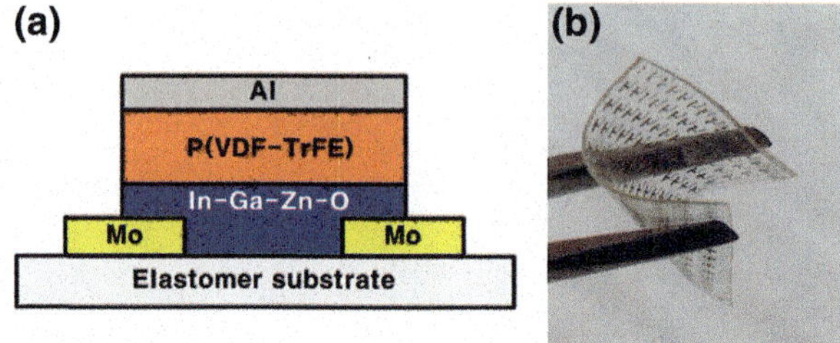

Fig. 11.4 a Schematic cross-sectional diagram and **b** photo image for the Al/PVDF-TrFE/IGZO/Mo structure fabricated on the elastomer substrate [10]

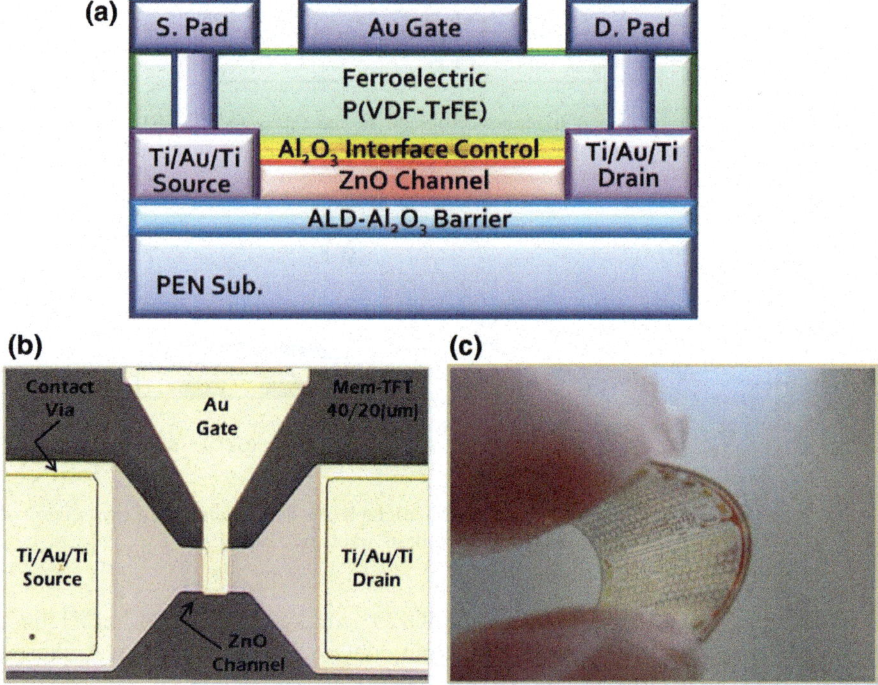

Fig. 11.5 a Schematic cross-sectional diagram of the proposed F-MTFT. **b** Microscopic top-view of the fabricated F-MTFT with the structure of Au/PVDF-TrFE/Al$_2$O$_3$/ZnO/Ti/Au/Ti/Al$_2$O$_3$/PEN. The channel width and length are 40 and 20 mm, respectively. **c** A typical photo image of the PEN substrate under a bending situation. The substrate size is 2×2 cm^2 [12]

to be top gate bottom contact and Al$_2$O$_3$ thin layer was inserted between ZnO channel and PVDF-TrFE insulating layer as an interface controlling layer. The memory window and the on/off ratio of the FeFET were 7.8 V and 10^8. The performance did not show remarkable degradation at bending tests with a bending radius of 9.7 mm and even after repetitive bendings of 20,000 cycles (Fig. 11.5).

11.3.2 Flexible FeFETs with Low Molecule Organic Semiconductors

Pentacene is one of the most representative low molecular organic semiconductors consisting of five linearly-fused aromatic rings and known to show high field effect mobility comparable with inorganic semiconductors. Khan et al. [20], demonstrated arrays of flexible FeFET memories on a banknote paper with pentacene semiconductor channels. A thin PDMS layer was spin-coated onto the banknote to render the surface impermeable upon subsequent solution processes and poly (3,4-ethylenedioxythiophene) polystyrene sulfonate (PEDOT:PSS) layers were used as source-drain contacts. PVDF-TrFE was deposited as a ferroelectric layer, followed by the thermal evaporation of pentacene. The device exhibited performance with memory window of 8 V, I_{on}/I_{off} ratio of 10^3, and retention characteristics up to 10^4 s (Fig. 11.6).

The memory device with thermally deposited pentacene active layer is limited to large area fabrication of memory arrays. Solution-processible organic semiconductors are thus more suitable for large area flexible memory devices. Not only p-type semiconductors but also various small molecular ambipolar and n-type semiconductors have been employed to FeFETs with PVDF-TrFE layers. Recently, Kim et al. [38], reported highly flexible FeFET devices which operated under 0.5 mm of bending radius and even under sharp folded state with an ambipolar dicyanomethylene-substituted quinoidal quaterthiophene derivative (QQT(CN)4) semiconductor channel. The device apparently exhibited both p-type and n-type current hysteresis. Interestingly, p-type to n-type conversion readily occurred in a device by a simple thermal annealing treatment upon which majority hole carrier transport was converted to n-type electron transport in the device (Fig. 11.7).

In spite of the great advance in FeFET performance with QQT(CN)4, there are still technological issues to be developed for the better device performance in particular of lower operation voltage which requires careful design of the interface of ferroelectric gate insulator and metal electrodes to reduce the gate leakage current through a ferroelectric layer. Velusamy et al. [39], introduced a solution-processed interlayer of monolayered, conductive reduced graphene oxides (rGOs) embedded in an insulating conjugated block copolymer, poly(styrene-*block*-paraphenylene) (PS-*b*-PPP) for low voltage operation of a FeFET with QQT(CN)4 channel. An ultra-thin nanocomposite interlayer inserted between ferroelectric PVDF-TrFE and bottom gate electrode allowed not only for significant decrease of operation voltage but also for reduction of gate leakage current. A FeFET with PS-*b*-PPP modified reduced graphene oxide (PMrGO) interlayer exhibited fully saturated ferroelectric p-type hysteresis with on/off current ratio greater than 10^3, memory window of 23 V and leakage current level of -9 nA at the program voltage of ±25 V. The device exhibited excellent data retention over time longer than 2 h and read/write cycle endurance of more than 100 times. The work demonstrated that low voltage operation of the FeFET with interlayer might arise from conductive rGOs which greatly enhanced local electric field upon ferroelectric switching while an insulating

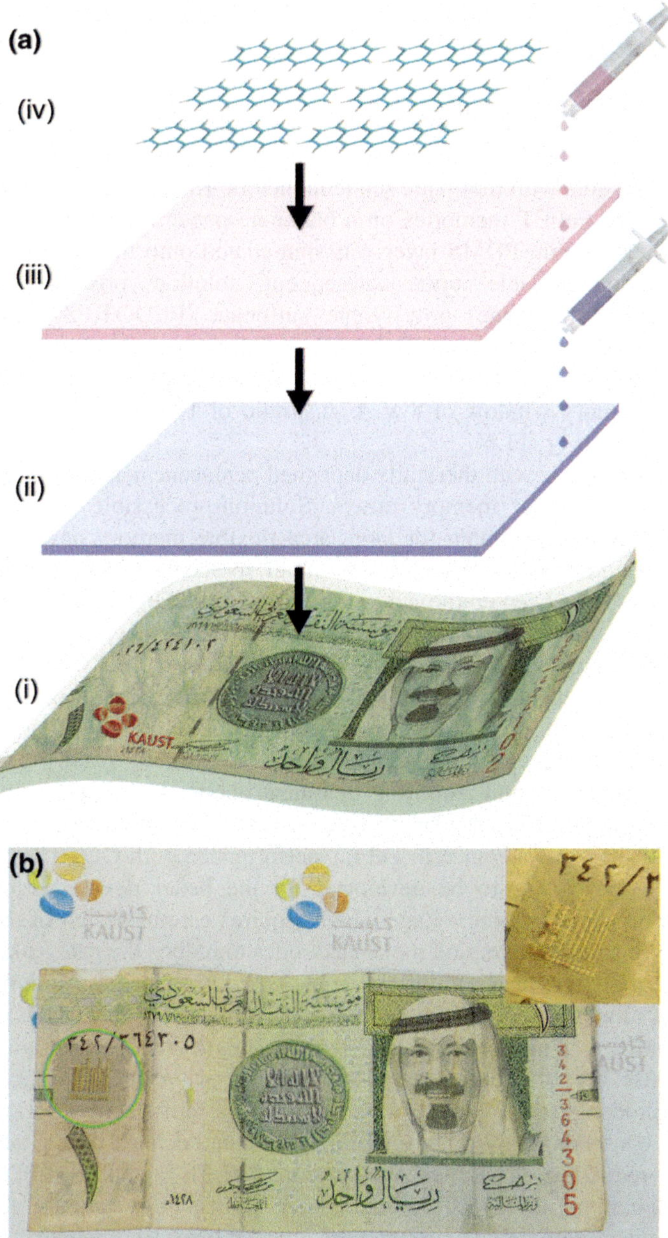

Fig. 11.6 **a** Fabrication of polymer ferroelectric memory devices on banknotes: (*i*) PDMS coated banknote (*ii*) spin-coating PEDOT: PSS bottom electrodes (*iii*) spin-coating PVDF-TrFE ferroelectric layer ∼140 nm (*iv*) thermally evaporated pentacene thin film (∼60 nm) and **b** photograph of a 1-Saudi Riyal note covered with arrays of polymer ferroelectric memory devices [20]

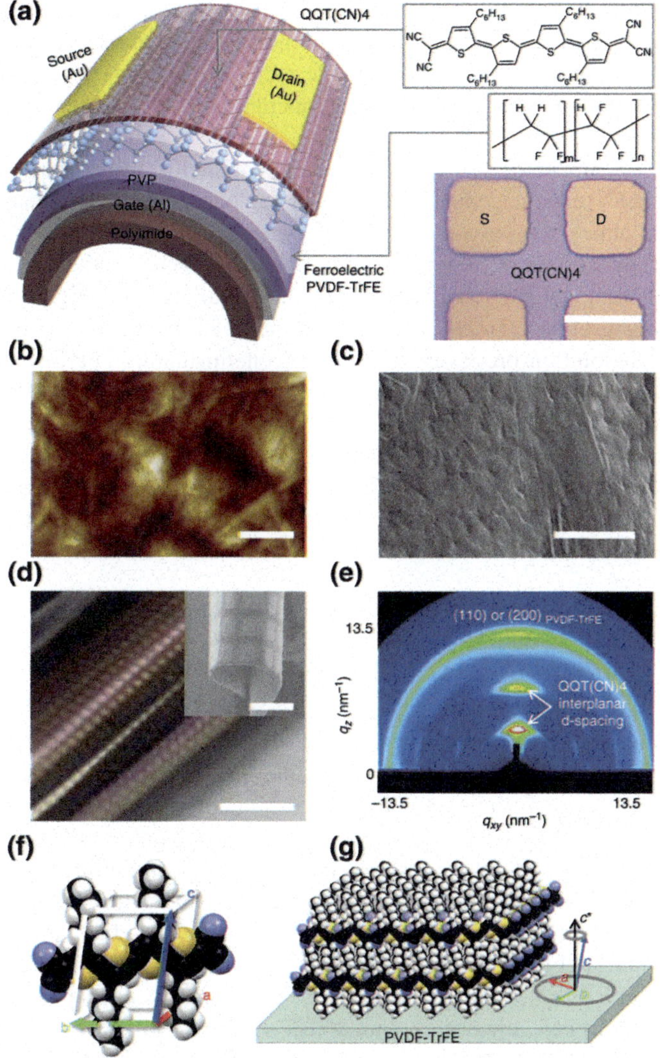

Fig. 11.7 **a** Schematic representation of the devices and molecular structure of the organic semiconductor QQT(CN)4 and the ferroelectric PVDF-TrFE. A photograph of arrays of FeFET devices is also displayed. *Scale bar*, 200μm. **b** Height contrast TM-AFM image of a QQT(CN)4 film spin-coated onto a PVDF-TrFE layer. *Scale bar*, 1μm. **c** SEM image of the surface of a QQT(CN)4 film under extreme bending. The characteristic polycrystalline texture of QQT(CN)4 was also well developed on multilayers of PVDF-TrFE/PVP/Al formed on the PI surface. *Scale bar*, 3μm. **d** Photograph of arrays of flexible FeFET devices prepared on a PI substrate. *Scale bar*, 2mm. An SEM image of extremely flexible arrays fabricated onto a paper substrate is shown in the inset. *Scale bar* in the inset, 500μm. **e** 2D grazing incidence X-ray diffraction image of a PVDF-TrFE/QQT(CN)4 bilayer. **f** Illustration of QQT(CN)4 crystallographic unit cell in as-prepared film. *Thick white lines* indicate the triclinic 3D unit cell and the *coloured arrows* represent the unit cell axes. **g** Schematic representation of QQT(CN)4 molecular packing in as-prepared film on PVDF-TrFE surface and orientation of domains in the film (the *black arrow* points the direction of the c* axis; the colored arrows distributing in a disc or in a cone represent the distribution of direct lattice vectors) [38]

PS-*b*-PPP matrix helped further reducing the gate leakage current. Furthermore, PMrGOs interlayers enabled fabrication of non-volatile transistor memory on a mechanically flexible substrate with reliable memory performance under repetitive bending deformation, making the device operating under the bending radius of 0.5 mm.

Another solution-processed organic semiconductor, triisopropylsilyethynyl acetylene added to pentacene (TIPS-Pentacene) has been widely used as semi-conducting channel of an organic field effect transistor. In addition, a blend film of TIPS-Pentacene with an insulating polymer offers an efficient route for fabricating a uniform and homogeneous semiconducting layer suitable for large area memory arrays. Park's group reported large-area printed arrays of non-volatile FeFET memories with solution-processed TIPS-Pentacene/insulating polymer blends [40]. TIPS-Pentacene/polymer blends were spin-coated onto 4-inch Si wafer and easily micro-patterned by selective contact evaporation printing method they developed. The memory arrays were also fabricated on a polyimide (PI) substrate and the device showed the performance with I_{on}/I_{off} ratio of 10^4, time dependent data retention of 10^3, and write/read endurance cycles of 100 under 5.8 mm of bending radius (Fig. 11.8).

11.3.3 *Flexible FeFETs with Polymer Semiconductors*

Polymer semiconductors have been extensively investigated due to their excellent mechanical ductility when employed in flexible memory devices. One of the most representative polymer semiconductor polymers, poly(3-hexylthiophene-2,5-diyl) (P3HT) has been widely used in organic electronic applications such as solar cells, field effect transistors. In spite of the relatively lower electrical properties of P3HT compared with those of inorganic semiconductors, there are many attempts to develop high performance P3HT based FeFETs. Hwang et al. [46], developed various flexible FeFET devices with P3HT semiconductor channels including multi-level FeFETs in which more than 2 current states were programmed and erased in a single cell, memories with P3HT nanowire channels, memories with vertically defined P3HT channels and 3 dimensionally stacked P3HT FeFETs. Mechanically flexible multilevel FeFET with a thin PVDF-TrFE insulator and P3HT semiconducting channel on PI substrate was fabricated and the device exhibited data retention and endurance of more than 10^5 s and 10^2 cycles. Various remanent polarization states which arose from distributed domain polarization of the ferroelectric layer as a function of the applied gate voltage in turn dictate the interstate levels of the source-drain current, giving rise to a reliable 4-level multi-level FeFET. The multilevel FeFET fabricated on PI substrate exhibited mechanical flexibility with all the 4-levels realized under more than 1000 bending cycles at a bending radius of 5.8 mm (Fig. 11.9).

Operation voltage of a multi-level FeFET was further reduced by employing a solution blended ferroelectric/high k polymer layer as a ferroelectric gate insulator.

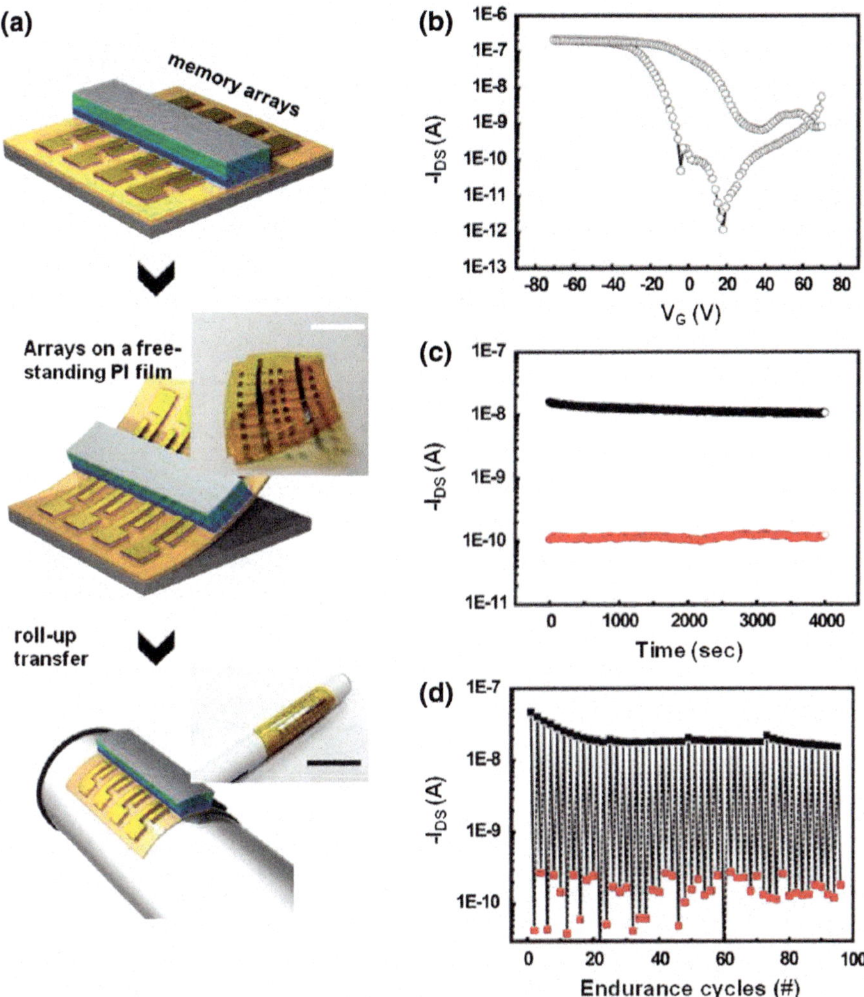

Fig. 11.8 **a** Schematic illustration of roll-up memory arrays fabricated by detaching and transferring FeFET arrays on a flexible substrate with a transient adhesive layer onto the surface of a cylindrical ball-point pen. The *insets* in **a** show the photographs of the FeFET arrays at each fabrication step. The *scale bars* shown in the insets are 1 cm. **b** The I_{DS}–V_G transfer curve of a FeFET mechanically deformed on the curved surface with 50 μm line patterned TIPS-Pentacene: poly(alpha methylstyrene) (PαMS) channel layer. The data retention (**c, d**) write/erase cycle endurance characteristics of the deformed FeFET arrays with 50 μm patterned TIPS-Pentacene channels [40]

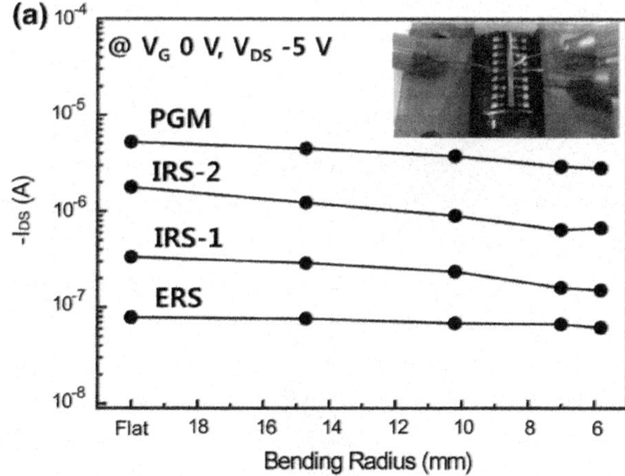

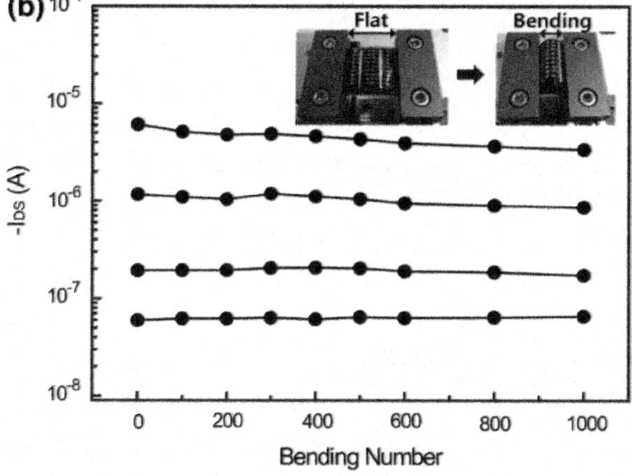

Fig. 11.9 The multilevel I_{DS} values of a flexible FeFET memory: **a** as a function of the bending radius, and **b** as a function of the number of bending cycles under a bending radius of 5.8 mm. The *inset* in **a** exhibits a photograph of a flexible FeFET obtained under bending. The photographs in the *inset* of **b** show arrays of FeFETs in bent and unbent states [46]

The capacitance of the ferroelectric gate insulator layer was enhanced by a simple binary solution-blend of a PVDF-TrFE (k ~ 8) with a relaxer high-k poly (vinylidene-fluoride–trifluoroethylene–chlorotrifluoroethylene) (PVDF-TrFE-CTFE) (k ~ 18). At optimized conditions, a ferroelectric blend insulator of PVDF-TrFE/ PVDF-TrFE-CTFE enabled the discrete six-level multi-state operation of a multi-level FeFET at a gate voltage sweep of ±18 V with excellent data retention and endurance of each state of more than 10^4 s and 120 cycles, respectively [44].

To improve the low cell density of FeFET devices with planar 1T device architecture, Hwang et al. [48], demonstrated 3D-stacked FeFET memory with vertically defined sub-micrometer channels. Simple but robust conformal transfer of a solution processed bilayer of ferroelectric and semiconducting polymer were used to conveniently form a vertical channel on a channel gap area which was controlled by a buffer polymer layer between vertically stacked source and drain electrodes on scales ranging from micrometer to sub-micrometer (Fig. 11.10). The device was

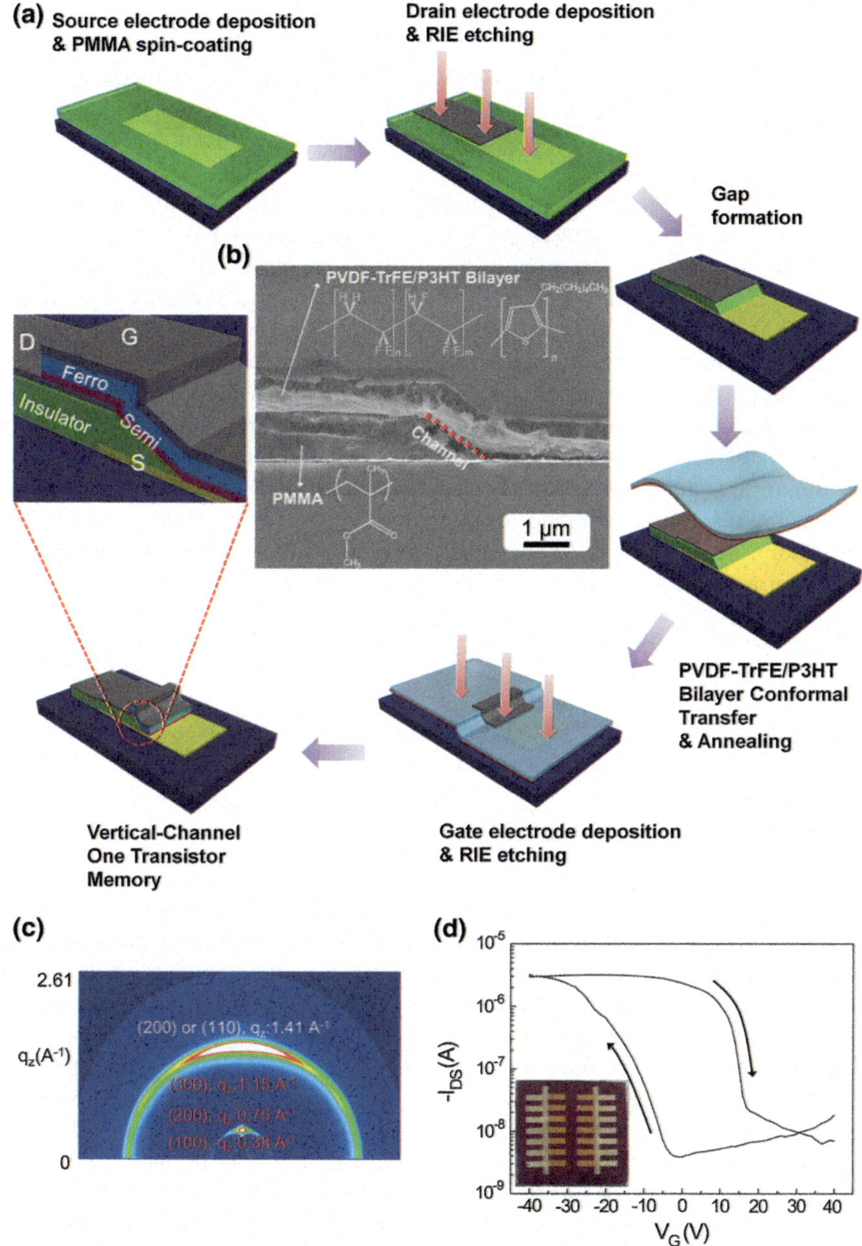

Fig. 11.10 a Schematic illustration of the fabrication of VC 1T polymer memory with a semiconducting P3HT channel and ferroelectric PVDF-TrFE insulator. **b** Cross-sectional SEM image of VC 1T memory, with the chemical structure of P3HT, poly(methyl methacrylate) (PMMA), and PVDF-TrFE. **c** 2D GIWAXS pattern of a PVDF-TrFE/P3HT bilayer film transferred by the floating technique and subsequently annealed at 135°C. **d** I_{DS}–V_G transfer curves of a VC 1T device with 1.9 µm channel length and a PVDF-TrFE layer 630 nm thick. *Inset* Photograph of the device arrays [48]

fabricated on PI substrate and exhibited data retention of more than 10^4 s and endurance of more than 10^2 write/read cycles. To confirm the bending stability of the device, on/off ratio and data retention were measured as a function of the bending radius, when bent in directions both parallel and perpendicular to the channel width; the measured memory performances were no varied significantly down to the bending radius 5.8 mm.

The FeFETs with semiconducting nanowire channels in particular in combination with ferroelectric polymers as gate insulating layers have attracted great attention because of their potential in high density memory integration. However, most of the devices still suffer from low yield of devices mainly due to the ill-control of the location of nanowires on a substrate. Hwang et al. [51] demonstrated the FeFETs with position-addressable P3HT nanowires. P3HT nanowires precisely controlled in both location and number between source and drain electrode were achieved by direct electrohydrodynamic nanowire printing. The polymer nanowire FeFET with a ferroelectric PVDF-TrFE exhibited non-volatile on/off current margin at zero gate voltage of approximately 10^2 with time-dependent data retention and read/write endurance of more than 10^4 s and 10^2 cycles, respectively. Furthermore, device showed characteristic bistable current hysteresis curves when being deformed with various bending radii and multiple bending cycles over 1000 times (Fig. 11.11).

11.4 Characterization of Mechanical Properties of a Flexible FeFET

11.4.1 Bending Characteristics

In the most of cases dealing with flexible electronic devices, device performance is examined as a function of bending radius as well as a function of the number of bending cycles at a fixed bending radius to characterize the flexibility of the device (Table 11.2). One can readily assume that if a device shows a constant the performance, the smaller the bending radius, the better device flexibility. Bending cycle test usually represents the reliability of device flexibility. For instance, Kim et al. investigated the mechanical bending stability of a p-type FeFET on a flexible PI substrate by monitoring on/off current ratios as a function of the bending radius; the results are displayed in Fig. 11.12a [38]. The on/off ratios of the device were measured in situ at various bending states with the apparatus shown in the inset of Fig. 11.12a; no significant variation in the ratios was observed. In particular, in situ current measurements were carefully performed with flexible memory arrays rolled on a commercially available coffee stirrer, as illustrated in Fig. 11.12b. A reasonably

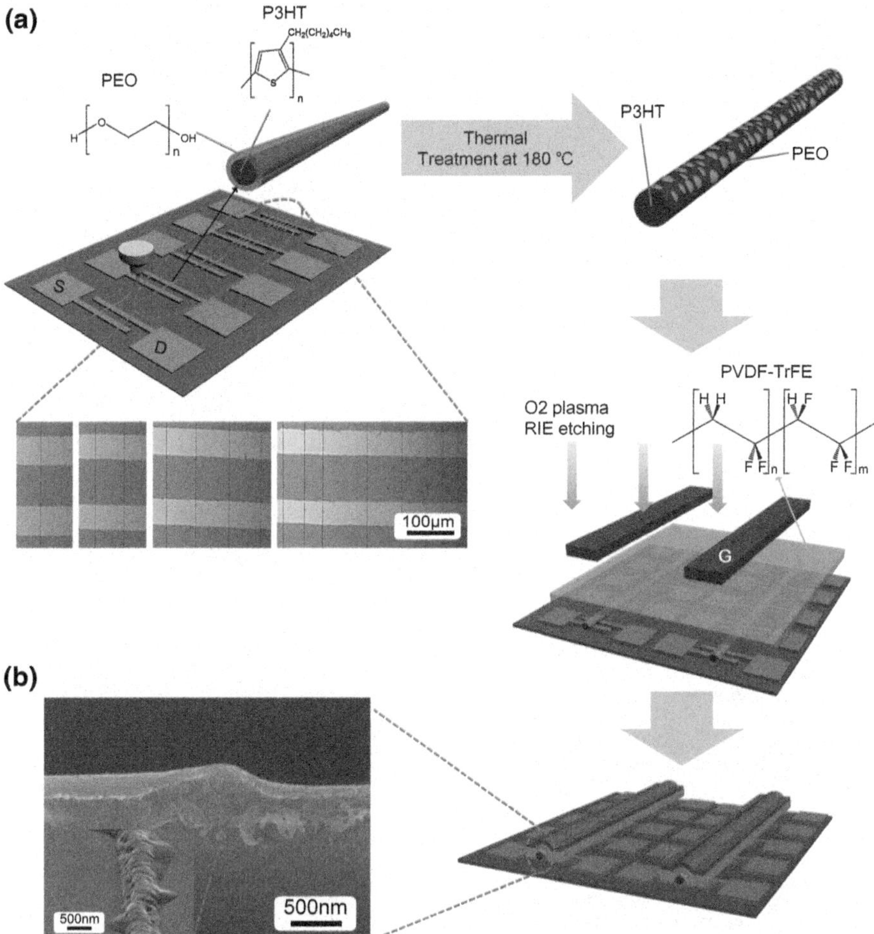

Fig. 11.11 **a** A schematic illustration of the fabrication process for the top-gate bottom contact FeFET memory device with a P3HT:PEO-blend (70:30, w/w) NW channel and ferroelectric insulator. Optical micrograph images of highly aligned polymer NW FeFET memory arrays with precisely controlled NW numbers (*bottom*). Thin PEO shell was dewetted on P3HT NW after thermal treatment. **b** A SEM image showing the cross-sectional of the PVDF-TrFE layer embedded with 300 nm polymer NW of the FeFET. The *inset* shows a SEM image of the thermal treated P3HT NW on the SiO_2 substrate [51]

high on/off ratio of approximately 5×10^3 was obtained even at an extreme bending radius of approximately 500 µm. They examined the memory stability of FeFET as a function of the number of bending cycles. Both outward and inward bending were examined to evaluate the memory reliability under elongation and compression fields at the interface of the organic semiconductor, QQT(CN)4 and PVDF-TrFE layer, respectively. The device exhibited stable on/off characteristics after more than

Table 11.2 Bending characteristics of FeFET memory devices with various semiconducting and ferroelectric layers

Semiconductor	Ferroelectric layer	Substrate	Electrode (G/SD)	Min. bending radius (mm)	Max. bending cycles (no.)	Ref
IGZO	PVDF-TrFE	PEN	Al/Al	7	–	8
IGZO	PVDF-TrFE	PEN	Au/Au	20	–	9
IGZO	PVDF-TrFE	PDMS	Al/Mo	10	–	10
ZnO	PVDF-TrFE	PES	Al/Au	–	–	11
ZnO	PVDF-TrFE	PEN	Au/Au	9.7	20,000	12
Pentacene	PVDF-TrFE	PES	Al/Au	–	–	19
Pentacene	PVDF-TrFE	Banknote/PDMS	PEDOT:PSS/Au	–	–	20
Pentacene	PVDF-TrFE	PEN	Au/Au	20	100	22
Pentacene	PVDF-TrFE	PET	Al/Au	–	–	28
TIPS-PEN	PVDF+PMMA	PI	Al/Au	–	–	34
QQT(CN)4	PVDF-TrFE	PI	Al/Au	0.5 & folding	1,000	38
QQT(CN)4	PVDF-TrFE	PI	Al/Au	5	100	39
TIPS-PEN+PαMS	PVDF-TrFE	PI	Al/Au	5	–	40
P3HT	PVDF-TrFE	PI	Al/Au	5.8	1,000	46
P3HT	PVDF-TrFE	PI	Al/Au, Al	5.8	1,000	48
P3HT	PVDF-TrFE+EMI [TFSA]	PI	Al/Au	5.8	1,000	50
P3HT NW	PVDF-TrFE	PI	Al/Au	5.8	1,000	51
PQT	PVDF-TrFE	PEN	Ag NP	15	–	54

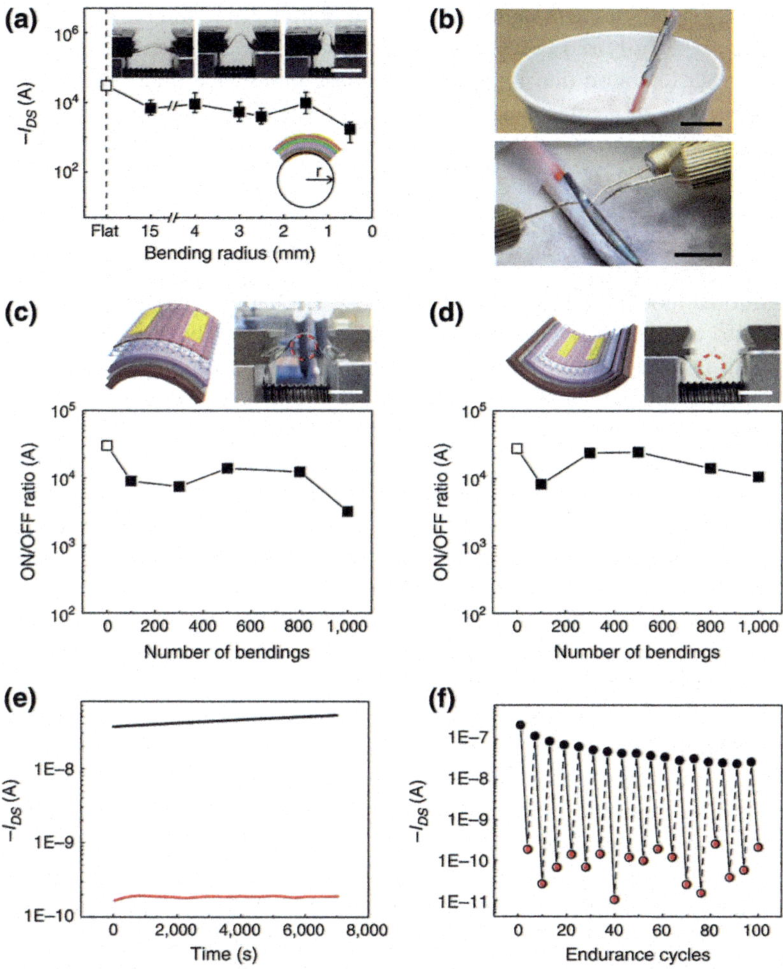

Fig. 11.12 **a** On/off current ratio measured at $V_{DS} = -5$ V as a function of the bending radius in a p-type QQT(CN)4 FeFET device. The data were averaged with 10 devices examined for each bending radius. Photographs of a flexible QQT(CN)4 device at various bending states during these measurements are shown in the inset. Scale bar in the inset, 5mm. **b** Photographs of super-flexible devices rolled on a commercially available coffee stirrer with a bending radius of $\sim 500\mu$m (top image), and in situ measurements of the rolled devices (bottom image). The scale bars of the top and bottom images are 1 cm and 5 mm, respectively. **c, d** On/off current ratio measured in a p-type QQT(CN)4 FeFET as a function of the number of bending cycles in outward and inward directions, respectively. These measurements were performed at $V_{DS} = -5$ V with a bending radius of 4 mm. The scale bars are all 1 cm. **e** Data retention characteristics measured after programming the device with single voltage pulses for the ON and OFF current in a p-type QQT (CN)4 FeFET after 1000 inward bending cycles at a bending radius of 4 mm. **f** Write/erase endurance cycle test as a function of the number of programming cycles in a p-type QQT(CN)4 FeFET after 1000 outward bending cycles at a bending radius of 4mm. For clarity, one cycle out of every four is represented. The programming voltage pulses for switching ON and OFF states were −50 and +50 V, respectively. The read voltages for both ON and OFF states were $V_G = -10$ V and $V_{DS} = -5$ V [38]

1000 bending cycles using a radius of 4 mm in both bending directions, as shown in Fig. 11.12c, d. A n-type FeFET on a flexible substrate also demonstrated reliable long term data retention and multiple write/erase endurance under severe bending cycles at a radius of 4 mm, as shown in Fig. 11.12e, f, respectively.

11.4.2 Nano-Indentation

In a traditional indentation test, a hard tip whose mechanical properties are known is pressed into a sample whose properties are unknown. The load placed on the indenter tip is increased as the tip penetrates further into the specimen and soon reaches a user-defined value. At this point, the load may be held constant for a period or removed. The area of the residual indentation in the sample is measured and the hardness is defined as the maximum load divided by the residual indentation area. Nano-indentation improves on these macro- and micro-indentation tests by indenting on the nanoscale with a very precise tip, high spatial resolutions to place the indents, and by providing real-time load-displacement data while the indentation is in progress. While indenting, various parameters such as load and depth of penetration can be measured. A record of these values can be plotted on a graph to create a load-displacement curve. These curves can be used to extract mechanical properties of the material [67–69].

In order to understand the mechanical behavior of FeFET devices containing thin constituent layers stacked with each other, Kim et al. characterized the mechanical properties of a thin PVDF-TrFE film and a QQT(CN)4 film on PVDF-TrFE by nano-indentation [38]. Samples with ~ 300 nm thick PVDF-TrFE, ~ 100 nm thick QQT(CN)4 film on ~ 300 nm thick PVDF-TrFE, and ~ 100 nm thick other organic semiconductor films were prepared on Si substrates. Representative indentation curves for PVDF-TrFE, QQT(CN)4/PVDF-TrFE and P3HT/PVDF-TrFE are presented in Fig. 11.13a. Instrumented tests were completed and, using both the maximum applied load and the given penetration depth in the indentation curves, the hardness, H_{IT}, was calculated. H_{IT} for PVDF-TrFE, QQT (CN)4/PVDF-TrFE and P3HT/PVDF-TrFE is very similar, yielding values of 126.6 ± 6.4 MPa, 145.1 ± 7.4 MPa and 140.8 ± 6.2 MPa, respectively (Fig. 11.13b). The elastic modulus, noted E_{IT}, was obtained from the slope of indentation curve during instrumented unloading experiments. The E_{IT} values for PVDF-TrFE, QQT(CN)4/PVDF-TrFE and P3HT/PVDF-TrFE were estimated to be 7.8 ± 0.2 GPa, 14.8 ± 4.2 GPa and 11.5 ± 0.6 GPa, respectively (Fig. 11.13b). E_{IT} for QQT(CN)4 is ~ 2 times higher than that for PVDF-TrFE, although their H_{IT} values are similar. Because E_{IT} is estimated from the slope of elastic recovery

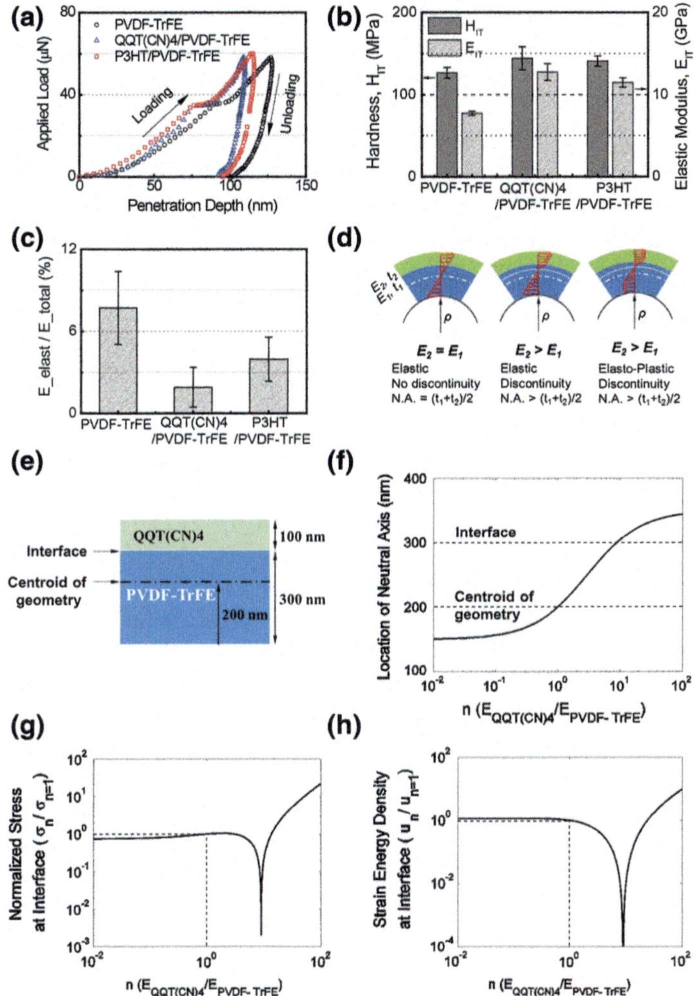

Fig. 11.13 **a** Representative loading and unloading curves in nano-indentation tests. **b** An instrumented testing hardness, H_{IT}, and an instrumented testing elastic modulus, E_{IT} for PVDF-TrFE, QQT(CN)4 coating on PVDF-TrFE and P3HT coating on PVDF-TrFE. *Each bar* is an average value with s.d. for five indentation tests. **c** The ratio of elastic energy (E_{elast}) to the total energy (E_{total}) for PVDF-TrFE, QQT(CN)4 coating on PVDF-TrFE and P3HT coating on PVDF-TrFE. *Each bar* is an average value with s.d. for five indentation tests. **d** Schematic of bending stress distribution of a two-layered film under the condition of elastic deformation with $E_2 = E_1$, elastic deformation with $E_2 > E_1$, and elasto-plastic deformation with $E_2 > E_1$ at the same radius of curvature ρ (centroid of geometry (*dot-dashed line*), neutral axis, N.A. (*dotted line*), Young's modulus (E_1) and thickness (t_1) of substrate, and Young's modulus (E_2) and thickness (t_2) of top layer). **e** Dimension of the composite film consisting of PVDF-TrFE and QQT(CN)4. **f** Location of neutral axis as a function of ratio n ($E_{QQT(CN)4}/E_{PVDF-TrFE}$). **g**, **h**, Normalized stress (**g**) and strain energy density (**h**) at the location of the interface between PVDF-TrFE and QQT (CN)4 (at 300 nm) as a function of ratio n ($E_{QQT(CN)4}/E_{PVDF-TrFE}$). Values were normalized by the value at n = 1 (*dotted line*). When n is equal to 1.6, the normalized stress and the normalized strain energy density at the interface are 1.05 and 0.83, respectively. They are zeroes at n = 9 [38]

during unloading, the difference in E_{IT} might be attributed to a degree of elastic deformation. In this regard, they calculated the ratio of elastic energy to the total energy by dividing the area under the unloading curve by the area under the loading curve (Fig. 11.13c).

Compared to PVDF-TrFE and P3HT, the ratio for QQT(CN)4 is significantly low, which indicates that the material behavior of QQT(CN)4 is more plastic than elastic. To understand how mechanical bending of a composite film depends on material properties and to clarify the role played by the mechanical behavior of QQT(CN)4, Kim et al., developed a model based on a bi-layer film as shown in Fig. 11.13d. They analyzed the normal stress distribution inside the film with parameters obtained from experiments such as film thickness and Young's modulus of PVDF-TrFE and QQT(CN)4, assuming that it undergoes a pure bending. The difference of material properties between QQT(CN)4 and PVDF-TrFE causes the neutral axis to move away from the centroid of geometry and the normal stress to become discontinuous at the interface between the two layers. If QQT(CN)4 is nine times stiffer than PVDF-TrFE, the neutral axis is located at the interface and therefore the stress at the interface becomes zero (Fig. 11.13e–h). According to experimental results, the ratio of Young's modulus for QQT(CN)4 to PVDF-TrFE is 1.6. This suggests that the neutral axis is located at 221 nm from the bottom of the PVDF-TrFE layer. The average stress at the interface is then similar to the case when the stiffness of QQT(CN)4 and PVDF-TrFE is identical.

In addition, plastic characteristic of QQT(CN)4 can form a more complex stress distribution. When the applied moment produces a stress which exceeds a yield stress, stress will remain constant or barely change in the region of plastic deformation in contrary to the linearly varying stress observed in purely elastic materials. The stress analysis implies that the QQT(CN)4 layer helps in reducing the stress and strain at the interface by putting the neutral axis close to the centroid of geometry and by decreasing the maximum stress and maintaining a plastic deformation.

11.4.3 Nano-Scratch

In a scratch test, a well-defined tip is drawn over the surface of the coating while applying a particular normal-load. This normal-load can be constant during the entire scratch, however, it may also be increased during scratching from a low initial value to a maximum value at the end of the scratch. Lateral force trans-ducers can be used to measure the lateral force acting on the scratch tip [70]. If an additional tangential loading component is applied, a partial cone crack can initiate due to the maximum tensile stress at the trailing edge of the spherical indenter [71]. The formation of these partial cone cracks will repeat with a regular interval that depends on the load, the critical flaw size and the fracture tough-ness of the material [72, 73]. The tensile stress at the trailing edge is intensified due to the sliding contact, whereas it is decreased at the front edge of the scratch tip [74].

Kim et al. characterized the interfacial properties of semiconductor/PVDF-TrFE bilayers by nano-scratch technique to understand their excellent memory perfor-mance under severe bending and folding [38]. In these experiments, the adhesion strength of organic semiconductor/PVDF-TrFE bilayers was successfully estimated using a stainless steel ball in contact with a bilayer. With an increasing mechanical force, a line was drawn at constant speed (Fig. 11.14a). Critical load values were identified at failure of the semiconductor layers (Fig. 11.14b). It should be noted that unlike common nano-scratch test where few micro-meter sized tip was used for the measurements, a stainless steel ball with 1.6 mm diameter was used to produce the contact pressure suitable for observing failure at the interface between the semiconductor coatings and PVDF-TrFE films. The ramp loading, from 1 to 70 mN, was applied on the ball while tangentially moving of 2 mm at a speed of 2 mm/min to determine the critical load when the organic semiconductor coatings are mostly delaminated and PVDF-TrFE film started to appear due to the large contact stress. Tearing of the film at failure was confirmed by visualizing the scratched surface using the scanning electron microscopy (Fig. 11.14c). The results of critical load demonstrate the extremely firm interface between a QQT(CN)4 and PVDF-TrFE, compared with those of other bilayers. This convincingly explains the effective mechanical stress release upon severe deformation observed with QQT (CN)4.

To further confirm the crucial role of QQT(CN)4 with fluorinated polymer having semi-crystalline nanostructure, Kim et al., employed another fluorinated polymer, PVDF-TrFE-CTFE, which has the similar crystalline surface structure to PVDF-TrFE. The switching properties of the device were well maintained under repetitive folding deformation (Fig. 11.14d, e) No significant delamination was observed in the nano scratch test, suggesting that QQT(CN)4 and PVDF-TrFE-CTFE form a stable interface (Fig. 11.14f) The results imply again that strong adhesive properties of QQT(CN)4 with PVDF-TrFE were responsible for maintaining the high performance under the deformation in flexible device.

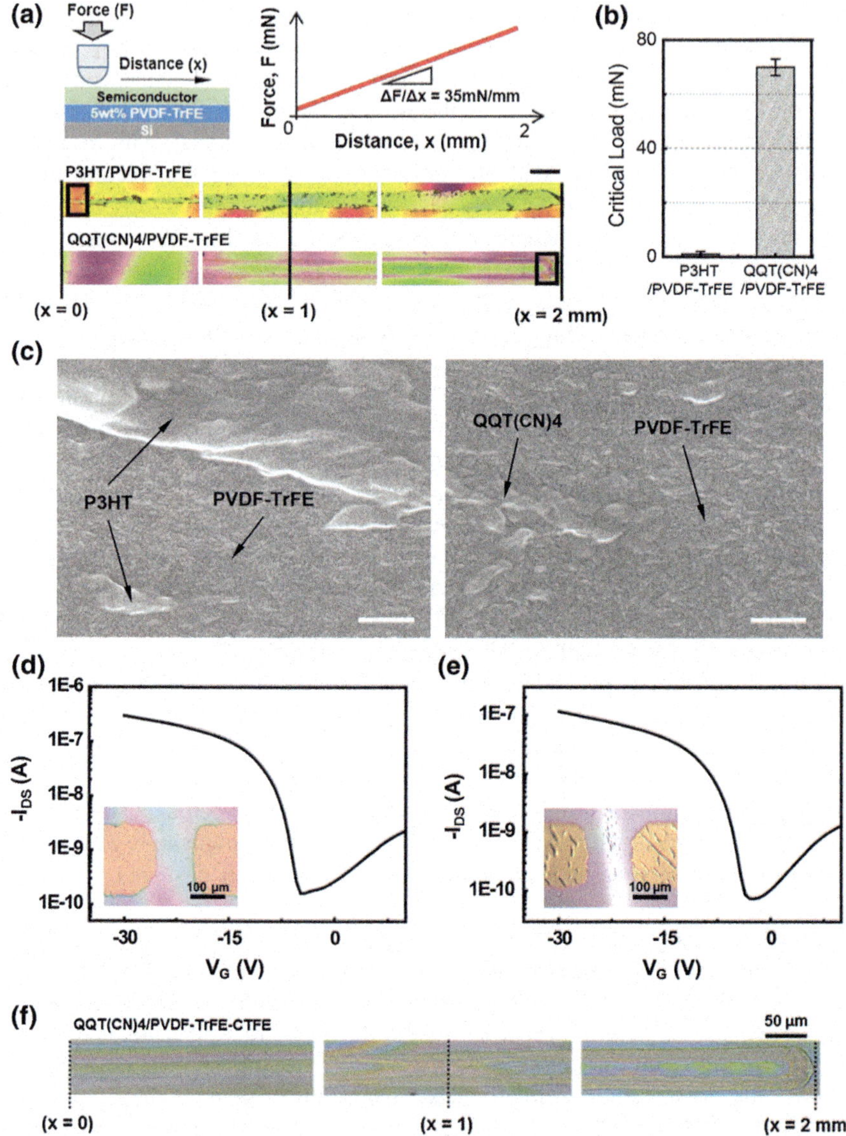

Fig. 11.14 a Schematic of nano-scratch test with a ramp loading and representative optical images along the scratches after tests for P3HT coating on PVDF-TrFE and QQT(CN)4 coating on PVDF-TrFE. A rectangular box in the image indicates a region where a critical load is defined. *Scale bar*, 50 µm. **b** Critical loads for P3HT/PVDF-TrFE and QQT(CN)4/PVDF-TrFE bilayers. Each bar is an average value with s.d. for five nano-scratch tests. **c** The magnified SEM images taken at the region of the *rectangular box* in (**a**). As denoted by *arrows*, semiconductor coatings (P3HT and QQT(CN)4) and the underlying layer of PVDF-TrFE can be readily identified in the images. Scale bar, 300 nm. Transfer characteristics of a PVDF-TrFE-CTFE/QQT(CN)4 field effect transistor **d** before sharp folding and **e** after sharp folding. The inset of each (**d, e**), represents OM image of S/D channel before and after folding test. **f** Image of the scratch on QQT(CN) 4/PVDF-TrFE-CTFE obtained from the nano scratch experiment [38]

11.5   Conclusions

We overviewed the current status of development on mechanically flexible field effect transistor memories with ferroelectric polymer gate insulators. The mechanical flexibility of a device was achieved by appropriate choices of the constituent materials in particular such as semiconducting layers and substrates. Solution-processed low molecular organic semiconductors and polymers are very suitable for high flexibility of a FeFET prepared on a flexible polymer substrate. In addition, device architecture is also one of the important routes for providing the flexibility. A FeFET with vertically defined sub-micron scale channel showed a good mechanical pliability. Position controlled nanowire semiconductors also gave rise to a flexible FeFET. Furthermore, various experimental techniques were reviewed to characterize the mechanical flexibility of a FeFET including bending test, nano-indentation and nano-scratch. There are still a lot of research opportunities for further development of a flexible FeFET. Flexible electrodes such as graphene and conducting polymers can offer more design freedom for the development of a flexible FeFET. Further fascinating research should be done for developing unprecedented non-volatile ferroelectric memories including a mechanically stretchable FeFET, a fiber type FeFET and a mechanically self recovering FeFET. At the same time, detailed analysis techniques of mechanical properties on flexible organic devices should be developed to more properly characterize these future devices.

Acknowledgments This work was supported by the National Research Foundation of Korea (NRF) grant funded by the Korea government (MEST) (No.2014R1A2A1A01005046).

References

1. J. Jang, F. Pan, K. Braam, V. Subramanian, Adv. Mater. **24**, 3573 (2012)
2. S. Lee, H. Kim, D.-J. Yun, S.-W. Rhee, K. Yong, Appl. Phys. Lett. **95**, 262113 (2009)
3. C.-H. Cheng, F.-S. Yeh, A. Chin, Adv. Mater. **23**, 902 (2011)
4. S.-J. Kim, J.-S. Lee, Nano Lett. **10**, 2884–2890 (2010)
5. S.-J. Kim, J.-M. Song, J.-S. Lee, J. Mater. Chem. **21**, 14516 (2011)
6. Y. Zhou, S.-T. Han, Z.-X. Xu, V.A.L. Roy, Nanotechnology **23**, 344014 (2012)
7. A.K. Tripathi, A.J.J.M. van Breeman, J. Shen, Q. Gao, M.G. Ivan, K. Reimann, E.R. Meinders, H. Gelinck, Adv. Mater. **23**, 4146 (2011)
8. G.-G. Lee, E. Tokumitsu, S.-M. Yoon, Y. Fujisaki, J.-W. Yoon, H. Ishiwara, Appl. Phys. Lett. **99**, 012901 (2011)
9. A. van Breemen, B. Kam, B. Cobb, F.G. Rodriguez, G. van Heck, K. Myny, A. Marrani, V. Vinciguerra, G. Gelinck, Org. Electron. **14**, 1966 (2013)
10. S.-W. Jung, J.B. Koo, C.W. Park, B.S. Na, J.-Y. Oh, S.S. Lee, J. Vac. Sci. Technol. B **33**, 051201 (2015)
11. K.H. Lee, G. Lee, K. Lee, M.S. Oh, S. Im, S.-M. Yoon, Adv. Mater. **21**, 4287 (2009)
12. S.-M. Yoon, S. Yang, S.-H.K. Park, J. Electrochem. Soc. **158**, H892 (2011)

13. S.H. Noh, W. Choi, M.S. Oh, D.K. Hwang, K. Lee, S. Jang, E. Kim, S. Im, Appl. Phys. Lett. **90**, 253504 (2007)
14. Y.T. Lee, P.J. Jeon, K.H. Lee, R. Ha, H.-J. Choi, S. Im, Adv. Mater. **24**, 3020 (2012)
15. H.S. Lee, S.-W. Min, M.K. Park, Y.T. Lee, P.J. Jeon, J.H. Kim, S. Ryu, S. Im, Small **20**, 3111 (2012)
16. M.A. Khan, J.A. Caraveo-Frescas, H.N. Alshareef, Org. Electron. **16**, 9 (2015)
17. G.A. Salvatore, D. Bouvet, I. Stolitchnov, N. Setter, A.M. Ionescu, in *Solid-State Device Research Conference, 38th European* (2008), p. 162
18. S. Das, J. Appenzeller, Nano Lett. **11**, 4003 (2011)
19. K.H. Lee, G. Lee, K. Lee, M.S. Oh, S. Im, Appl. Phys. Lett. **94**, 093304 (2009)
20. M.A. Khan, U.S. Bhansali, H.N. Alshareef, Adv. Mater. **24**, 2165 (2012)
21. B. Kam, X. Li, C. Cristoferi, E.C. P. Smits, A. Mityashin, S. Schols, J. Genoe, G. Gelinck, P. Heremans, **101**, 033304 (2012)
22. B. Kam, T.-H. Ke, A. Chasin, M. Tyagi, C. Cristoferi, K. Tempelarrs, A.J.J.M. van Breemen, K. Myny, S. Schols, J. Genoe, G.H. Gelinck, P. Heremans, IEEE Electron Dev. Lett. **35**, 539 (2014)
23. J. Chang, C.H. Shin, Y.J. Park, S.J. Kang, H.J. Jeong, K.J. Kim, C.J. Hawler, T.P. Russell, C. J. Ryu, C. Park, Org. Electron. **10**, 849 (2009)
24. W.-H. Kim, J.-H. Bae, M.-H. Kim, C.-M. Keum, J. Park, S.-D. Lee, J. Appl. Phys. **109**, 024508 (2011)
25. C.W. Choi, A.A. Prabu, Y.M. Kim, S. Yoon, K.J. Kim, C. Park, Appl. Phys. Lett. **93**, 182902 (2008)
26. S.J. Kang, Y.J. Park, J. Sung, P.S. Jo, B.O. Cho, C. Park, Appl. Phys. Lett. **92**, 012921 (2008)
27. T. Kanashima, K. Yabe, M. Okuyama, Jpn. J. Appl. Phys. **51**, 02BK06 (2012)
28. C.A. Nguyen, S.G. Mhaisalkar, J. Ma, P.S. Lee, Org. Electron. **9**, 1087 (2008)
29. Y.J. Park, S.J. Kang, B. Lotz, M. Brinkmann, A. Thierry, K.J. Kim, C. Park, Macromolecules **41**, 8648 (2008)
30. T. Kanashima, Y. Katsura, M. Okuyama, Jpn. J. Appl. Phys. **53**, 04ED11 (2014)
31. S.J. Kang, I. Bae, Y.J. Park, T.H. Park, J. Sung, S.C. Yoon, K.H. Kim, D.H. Choi, C. Park **19**, 1609 (2009)
32. S.J. Kang, I. Bae, Y.J. Shin, Y.J. Park, J. Huh, S.-M. Park, H.-C. Kim, C. Park, Nano Lett. **11**, 138 (2011)
33. I. Bae, S.J. Kang, Y.J. Park, T. Furukawa, C. Park **10**, e54 (2010)
34. S.J. Kang, Y.J. Park, I. Bae, K.J. Kim, H.-C. Kim, S. Bauer, E.L. Thomas, C. Park, Adv. Funct. Mater. **19**, 2812 (2009)
35. R.H. Kim, S.J. Kang, I. Bae, Y.S. Choi, Y.J. Park, C. Park, Org. Electron. **13**, 491 (2012)
36. Y.J. Shin, S.J. Kang, H.J. Jung, Y.J. Park, I. Bae, D.H. Choi, C. Park, A.C.S. Appl. Mater. Interfaces **3**, 582 (2011)
37. I. Bae, R.H. Kim, S.K. Hwang, S.J. Kang, C. Park, A.C.S. Appl. Mater. Interfaces **6**, 15171 (2014)
38. R.H. Kim, H.J. Kim, I. Bae, S.K. Hwang, D.B. Velusamy, S.M. Cho, K. Takaishi, T. Muto, D. Hashizume, M. Uchiyama, P. Andre, F. Mathevet, B. Heinrich, T. Aoyama, D.-E. Kim, H. Lee, J.-C. Ribierre, C. Park, Nat. Commun. **5**, 3583 (2014)
39. D.B. Velusamy, R.H. Kim, K. Takaishi, T. Muto, D. Hashizume, S. Lee, M. Uchiyama, T. Aoyama, J.-C. Ribierre, C. Park, Org. Electron. **15**, 2719 (2014)
40. I. Bae, S.K. Hwang, R.H. Kim, S.J. Kang, C. Park, A.C.S. Appl. Mater. Interfaces **5**, 10696 (2013)
41. R.C.G. Naber, C. Tanase, P.W.M. Blom, G.H. Gelinck, A.W. Marsman, F.J. Touwslager, S. Setayesh, D.M. de Leeuw, Nat. Mater. **4**, 243 (2005)
42. R.C.G. Naber, P.W.M. Blom, G.H. Gelinck, A.W. Marsman, D.M. de Leeuw, Adv. Mater. **17**, 2692 (2005)
43. R.C.G. Naber, B. de Boer, D.M. de Leeuw, P.W.M. Blom, Appl. Phys. Lett. **87**, 203509 (2005)

44. S.K. Hwang, I. Bae, S.M. Cho, R.H. Kim, H.J. Jung, C. Park, Adv. Funct. Mater. **23**, 5484 (2013)
45. S.J. Kang, I. Bae, J.-H. Choi, Y.J. Park, P.J. Jo, Y. Kim, K.J. Kim, J.-M. Myoung, E. Kim, C. Park, J. Mater. Chem. **21**, 3619 (2011)
46. S.K. Hwang, I. Bae, R.H. Kim, C. Park, Adv. Mater. **24**, 5910 (2012)
47. F.A. Yildirim, C. Ucurum, R.R. Schliewe, R.M. Meixner, H. Goebel, W. Krautschneider, W. Bauhofer, Appl. Phys. Lett. **90**, 083501 (2007)
48. S.K. Hwang, S.M. Cho, K.L. Kim, C. Park, Adv. Electron. Mater. **1**, 1400042 (2015)
49. R.C.G. Naber, J. Massolt, M. Spijkman, K. Asadi, D.M. de Leeuw, P.W.M. Blom, Appl. Phys. Lett. **90**, 113509 (2007)
50. S.K. Hwang, T.J. Park, K.L. Kim, S.M. Cho, B.J. Jeong, C. Park, A.C.S. Appl. Mater. Interfaces **6**, 20179 (2014)
51. S.K. Hwang, S.-Y. Min, I. Bae, S.M. Cho, K.L. Kim, T.-W. Lee, C. Park, Small **10**, 1976 (2014)
52. K. Asadi, P.W.M. Blom, D.M. de Leeuw, Appl. Phys. Lett. **99**, 053306 (2011)
53. I. Katsouras, D. Zhao, M.-J. Spijkman, M. Li, P.W.M. Blo, D.M. de Leeuw, K. Asadi, Sci. Rep. **5**, 12094 (2015)
54. T.N. Ng, B. Russo, B. Krusor, R. Kist, A.C. Arias, Org. Electron. **12**, 2012 (2011)
55. G.H. Gelinck, A.W. Marsman, F.J. Touwslager, S. Setayesh, R.C.G. Naber, P.W.M. Blom, D. M. de Leeuw, Appl. Phys. Lett. **87**, 092903 (2005)
56. T. Yagi, M. Tatemoto, J. Sako, Polymer J. **12**, 209 (1980)
57. K. Tashiro, K. Takano, M. Kobayashi, Y. Chatani, H. Tadokoro, Ferroelectrics **57**, 297 (1984)
58. R. Hasegawa, M. Kobayashi, H. Tadokoro, Polym. J. **3**, 591 (1972)
59. R. Hasegawa, Y. Takahashi, Y. Chatani, H. Tadokoro, Polym. J. **3**, 600 (1972)
60. M.A. Bachmann, J.B. Lando, Macromolecules **14**, 40 (1981)
61. H. Ohigashi, Jpn. J. Appl. Phys. **24**, 23 (1985)
62. S. Winhold, M.H. Litt, J.B. Lando, Macromolecules **13**, 1178 (1980)
63. G.T. Davis, J.E. McKinney, M.G. Broadhurst, S.C. Roth, J. Appl. Phys. **49**, 4998 (1978)
64. B. Servet, J. Rault, J. de Phys. **40**, 1145 (1979)
65. T. Furkawa, Phase Transitions **18**, 143 (1989)
66. T. Furukawa, G.E. Johnson, H.E. Bair, Ferroelectrics **32**, 61 (1981)
67. H.J. Kim, D.E. Kim, Tribol. Lett. **49**, 85 (2013)
68. W. Oliver, G. Pharr, J. Mater. Res. **7**, 1564 (1992)
69. M. Kanari, H. Kawamata, T. Wakamatsu, I. Ihara, Appl. Phys. Lett. **90**, 061921 (2007)
70. J. Malzbender, J.M.J. den Toonder, A.R. Balkenende, G. de With, Mater. Sci. Eng. R **36**, 47 (2002)
71. B.R. Lawn, Proc. R. Soc. **299**, 307 (1967)
72. M. Keer, C.H. Kuo, Int. J. Solids Struct. **29**, 1819 (1992)
73. A.F. Bower, N.A. Fleck, J. Mech. Phys. Solids **42**, 1375 (1994)
74. G.M. Hamilton, Proc. Inst. Mech. Eng. **197C**, 53 (1983)

Chapter 12
Non-volatile Paper Transistors with Poly(vinylidene fluoride-trifluoroethylene) Thin Film Using a Solution Processing Method

Byung-Eun Park

Abstract This study demonstrates a new and realizable possibility of 1T-type ferroelectric random access memory devices using an all solution processing method with cellulose paper substrates. A ferroelectric poly(vinylidene fluoride-trifluoroethylene) (P(VDF-TrFE)) thin film was formed on a paper substrate with an Al electrode for the bottom gate structure, and then a semiconducting poly(3-hexylthiophene) (P3HT) thin film was formed on the P(VDF-TrFE)/paper structure using a spin-coating technique. The fabricated ferroelectric gate field effect transistors (FeFETs) on the cellulose paper substrates demonstrated excellent ferroelectric property with a memory window width of 20 V for a bias voltage sweep from -30 to 30 V, and the on/off ratio of the device was approximately 10^2. These results agree well with those of the FeFETs fabricated on a rigid Si substrate. These results will lead to the emergence of printable electron devices on paper. Furthermore, these non-volatile paper memory devices, which are fabricated by a solution processing method, are reliable, very inexpensive, have a high-density, and can be fabricated easily.

12.1 Introduction

Ferroelectric random access memory (FeRAM) consisting of one transistor and one ferroelectric capacitor for storing data (1T1C-type) can be non-volatile, can perform high-speed read/write operation, and requires low power consumption [1–3]. In particular, the ferroelectric gate field effect transistor (FeFET), where the gate insulator film (1T-type) is composed of a ferroelectric material, has attracted considerable attention because of its non-destructive data read-out and the possibility of scaling the devices down in size. The original idea for a ferroelectric gate field effect

B.-E. Park (✉)
School of Electrical and Computer Engineering, University of Seoul,
163 Seoulsiripdae-ro, Dongdaemun-ku, Seoul 130-743, Korea
e-mail: pbe@uos.ac.kr

© Springer Science+Business Media Dordrecht 2016
B.-E. Park et al. (eds.), *Ferroelectric-Gate Field Effect Transistor Memories*,
Topics in Applied Physics 131, DOI 10.1007/978-94-024-0841-6_12

transistor was proposed in 1957 [4]. However, the interface properties of its metal–ferroelectric–semiconductor (MFS) gate structure are generally poor because of the inter-diffusion of the constituent elements, thereby leading to high current leakage, retention, and fatigue problems [5]. Although the metal–ferroelectric–insulator–semiconductor (MFIS) structure that has a buffer layer inserted between the ferroelectric and semiconductor layers seems to have solved the problems associated with the MFS structure, the retention characteristics of the MFIS structure are not reliable for commercialization, despite recent reports indicating considerable advancement in data retention time [6, 7]. Recently, as the idea of organic electronics is extremely attractive for integrated displays, radio-frequency identification (RFID) and other thin film transistors (TFTs), including organic field effect transistors (OFETs), have received considerable interest [8, 9]. An important advantage of organic materials is the low cost of fabrication. Therefore, OFETs are being made to overcome the issue that silicon solutions remain economically out of reach due to the high material costs, high processing costs, and the need for a complex fabrication system. Attempts at ferroelectric memory devices using organic ferroelectric materials and organic thin film transistors (OTFTs) with plastic substrates are expected as key devices for flexible displays and wearable computers [10–12]. The next important step towards practical use would be to realize embedded-type non-volatile memory devices at very low cost and ease of production.

Poly(3-hexylthiophene) (P3HT) is an organic semiconductor used in OTFTs that is soluble in organic solvents. One of the key strategies to achieve high electrical performance is to develop well interconnected crystal structures via annealing temperature because the cross linking of an organic semiconductor is affected by the thermal budget [13]. In particular, the fabrication of organic thin film transistors using paper substrates is very important. Of the various π-conjugated polymers applied to FETs, regioregular P3HT shows efficient field-effect mobility for holes when the active layer is carefully fabricated with respect to the formation of ordered microcrystalline domains by optimizing the physical properties of the polymer and substrate-surface condition [14, 15]. In general, the high mobility in P3HT is related to the degree of order of the polymer chains leading to lamellar structures with 2-D conjugated sheets formed by π–π inter-chain stacking. These π-stacked polymer chains provide an environment for increased inter-chain interactions. The interactions between the π-stacked polymer chains influence the two-dimensional electronic transport and therefore, the mobility of the polymer films. The polymer chain stacking depends on the molecular parameters including the regioregularity, molecular weight, side chain length, and processing conditions, such as film thickness, solvent power, and forming method [16–19].

Organic ferroelectric material, poly(vinylidene fluoride) (PVDF) and its copolymer poly(vinylidene fluoride-trifluoroethylene) (P(VDF-TrFE)) have attracted much attention because they have excellent solution processability, low processing temperature, and excellent ferroelectric properties [20–23]. Various applications have been studied, including in OTFTs, piezoelectric, and pyroelectric devices [24, 25]. Among them, some papers have reported the possibility of non-volatile memory devices using thin polymer films. Möller et al. [26]

demonstrated the polymer/semiconductor write-once–read-many-times memory (WORM) with poly(3,4-ethylenedioxythiopene) (PEDOT). Naber et al. [24] also demonstrated a non-volatile memory device with P(VDF-TrFE) and poly [2-methoxy, 5-(2′-ethyl-hexyloxy)-p-phenylene-vinylene] (MEH-PPV). Although these devices with fabricated organic films showed good ferroelectric properties, most of the reported papers were focused on using plastic and glass substrates for flexible processing [26–30].

This study demonstrates a new and feasible next-generation, organic non-volatile ferroelectric random access memory device on cellulose paper substrates fabricated by an all solution processing method. To date, there has been no report of OTFTs or ferroelectric memory devices fabricated on cellulose paper substrates. Ferroelectric P(VDF-TrFE) thin film was formed on a paper substrate with an Al electrode for the bottom gate structure, and then a semiconducting P3HT thin film was formed on the P(VDF-TrFE)/paper structure using a spin-coating technique [31–34].

12.2 Experimental Procedure

P(VDF-TrFE) polymer film was deposited by spin-coating at 3000 rpm for 25 s on a cellulose paper substrate using a P(VDF-TrFE) (70:30 mol%, Solvay Co.) solution of various wt%, which were diluted in 2-butanol (MEK) to control the film thickness. The spin-coated films were annealed at 140 °C for 1 h in ambient air to enhance the crystallization. In order to measure the polarization-electric field (P-E) characteristics of the fabricated Au/P(VDF-TrFE)/Al/paper structures, top electrodes were fabricated through thermal evaporation using a shadow mask. The P-E characteristics of these structures were measured using a Sawyer–Tower circuit (Toyotech Corporation). Regioregular P3HT (Aldrich Chem. Co.) was used as the active layer, which was solved in chloroform ($CHCl_3$). The thickness of the fabricated films was controlled by changing the weight concentration of P3HT in the solvent. The P3HT solution was spin-coated on the P(VDF-TrFE)/paper substrates at 2500 rpm for 25 s. The P3HT films on the P(VDF-TrFE)/paper structures were annealed at 140 °C for 1 h on a hot-plate to remove the solvent. To enable measurements of the electrical properties of the Au/P3HT/P(VDF-TrFE)/paper structures, Au was deposited as the source and drain electrode by thermal evaporation with a shadow mask. The channel length and width of the fabricated FeFETs were 5 and 100 μm, respectively. The electrical behaviors of the FeFETs were evaluated at room temperature using a semiconductor parameter analyzer (Agilent 4156C). P(VDF-TrFE) films, P3HT films, and electrodes were fabricated on rigid substrates using the same methods and conditions used for fabrication on paper substrates. The cross-section surfaces of the samples were obtained by a Leica Ultracut UC6 ultramicrotome with a diamond knife. Field emission scanning electron microscope (FE-SEM) imaging was performed using a Hitachi S-4300 FE-SEM. Focused ion beam scanning electron microscope (FIB-SEM) imaging was also performed using an FEI Nova Nano Lab 200 FIB-SEM. The surface morphologies of the fabricated devices were observed using a commercial atomic force microscope (AFM) (Seiko Instruments, Inc., SPI-4000 control station).

12.3　Results and Discussion

12.3.1　P(VDF-TrFE)/Al/Paper Structures

The cross-sectional images of P(VDF-TrFE) films were observed using an SEM in order to check whether the P(VDF-TrFE) films with the solution method were formed on the cellulose paper substrate. Figure 12.1 shows a scanning electron microscopy image of a P(VDF-TrFE) film fabricated on a cellulose paper substrate. A 200-µm thick sheet of cellulose paper was used as the substrate. Figure 12.1b also shows a typical field emission scanning electron microscope image of the Au/P(VDF-TrFE)/Al/paper structure. The image was acquired at magnifications of $10,000\times$ in the secondary electron imaging mode using the upper detector, 15.0 mm working distance, and 5.0 kV accelerating voltage. The thicknesses of the P(VDF-TrFE) film and lower Al electrode in these structures were approximately 300 and 270 nm, respectively. Although cellulose paper was used as the substrate, the lower Al electrode had very good adhesive property with paper. Al film was deposited by the vacuum deposition method as the lower electrode. Additionally, the P(VDF-TrFE) film was very uniform on the Al/paper substrate, though it was fabricated by a spin-coating technique. The thickness of the thin P(VDF-TrFE) film was produced by adjusting the solution molar ratio from several ten to several hundreds of nanometers.

The electrical properties were investigated by forming the capacitors on the cellulose paper substrates to confirm the basic ferroelectric properties of the fabricated P(VDF-TrFE) film. Typical polarization-versus-electric field (P-E) characteristics for an Au/P(VDF-TrFE)/Al/paper structure are shown in Fig. 12.2a. Even though a paper substrate was used, a very good hysteresis loop due to the ferroelectricity of the P(VDF-TrFE) film with a 180-nm thickness was observed in the P-E curve of the metal-ferroelectric-metal (MFM) structure. Typical remanent polarization and coercive field in these capacitors were approximately 8 $\mu C/cm^2$ and 800 kV/cm at a measurement frequency of 100 Hz. The remanent polarization was slightly smaller than those of the film on the $Pt/Ti/SiO_2/Si$ substrate, which is likely due to the difference in the surface morphologies of the substrates. The frequency dependency of the remanent polarization and coercive electric field are shown in Fig. 12.2b. The remnant polarization decreases slightly with an increase in frequency, whereas the coercive field increases more rapidly with increasing frequency. This result means that the polarization reversal speed in the P(VDF-TrFE) film is slower because of the slower molecular ordering than that in the inorganic ferroelectric materials.

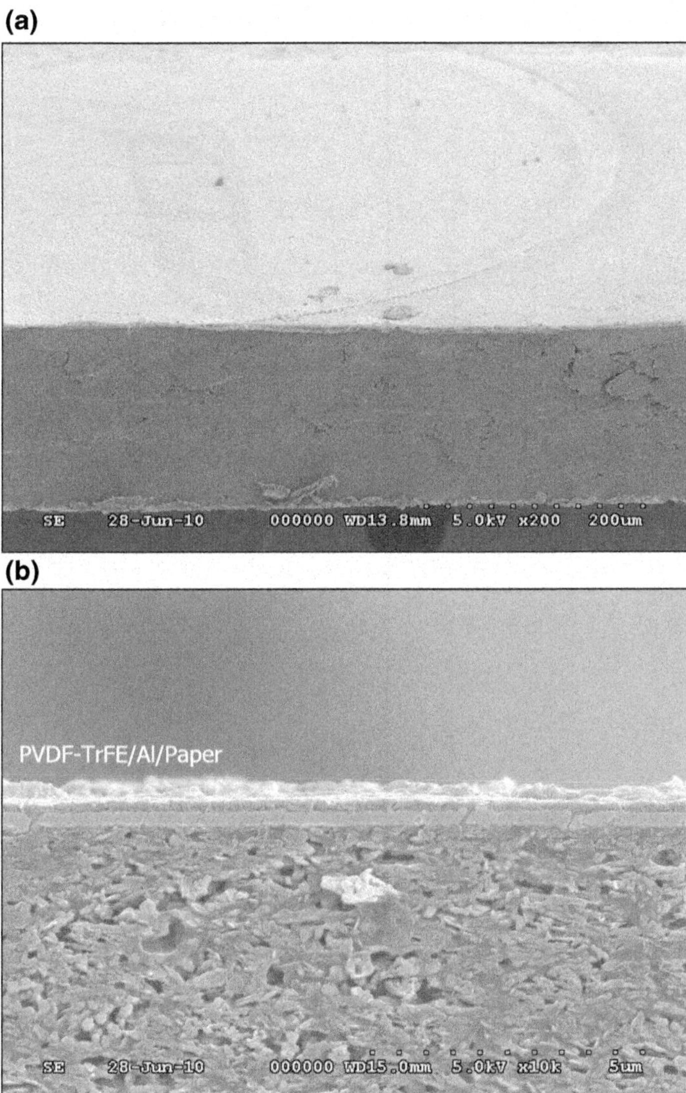

Fig. 12.1 **a** A typical scanning electron microscopy image of P(VDF-TrFE) film fabricated on a cellulose paper substrate. A 200-μm thick sheet of cellulose paper was used as the substrate. **b** A typical field emission scanning electron microscope image of the Au/P(VDF-TrFE)/Al/paper structure. The image was acquired at magnifications of 10,000× in the secondary electron imaging mode using the upper detector, a 15.0 mm working distance, and a 5.0 kV accelerating voltage

The surface morphologies of the P(VDF-TrFE) films were observed using atomic force microscopy (AFM). Figure 12.3 shows the surface AFM images of the P(VDF-TrFE) films. The measured area was $1 \times 1 \ \mu m^2$. The root mean-squared

Fig. 12.2 a Typical polarization-versus-electric field (P-E) characteristics for both Au/P(VDF-TrFE)/Al/paper and Pt/Ti/SiO$_2$/Si substrates. The measurement was performed at 100 Hz. b The remanent polarization (P_r) and coercive field (E_c) values as a function of the supplied frequency

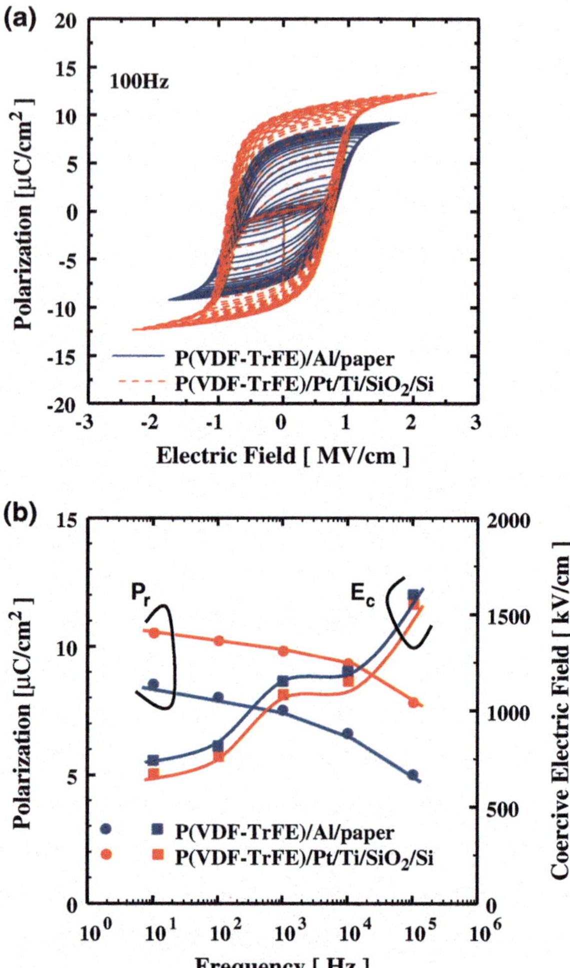

(RMS) surface roughness values of the P(VDF-TrFE) films were approximately 5.8 and 4.1 nm for the Al/Paper and Al/SiO$_2$/heavily doped Si substrates, respectively. These values indicate that both P(VDF-TrFE) films had a good surface morphology. Even though the P(VDF-TrFE) thin film was formed on a paper substrate, it had a good surface morphology comparable to the thin film with a rigid Si substrate. These AFM images agree with the P-E measurements. The polarization value from the P-E curves for the P(VDF-TrFE)/Al/paper structure was less than that of the P(VDF-TrFE)/Pt/Ti/SiO$_2$/Si structure. For the P(VDF-TrFE) films using paper as the substrate, lower polarization and wider coercive field values were observed in this experiment.

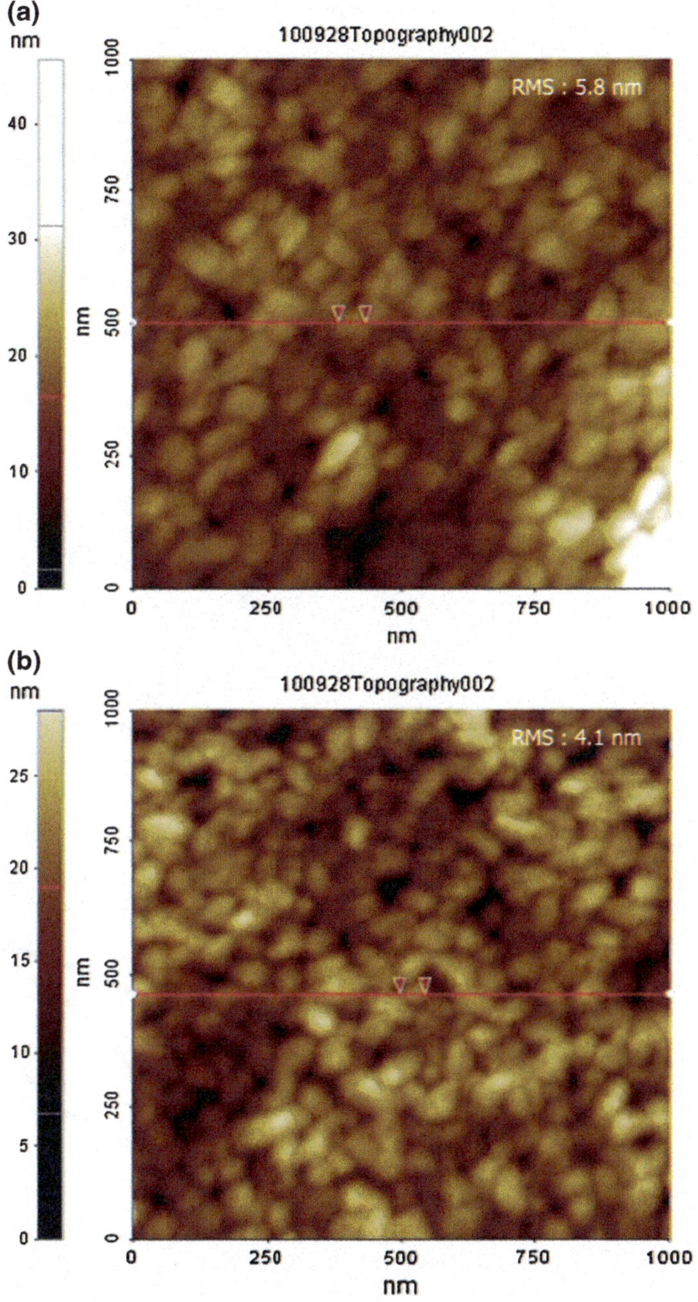

Fig. 12.3 **a** A typical surface atomic force microscopy (AFM) image of P(VDF-TrFE) film on Al/Paper substrates. The measured area was $1 \times 1\ \mu m^2$. **b** A typical AFM image of the P(VDF-TrFE)/Al/SiO$_2$/heavily-doped Si substrate

12.3.2 *P3HT/PVDF-TrFE/Al/Paper Structures*

Based on the electrical properties of the capacitors with the P(VDF-TrFE)/Al/paper structures, FeFETs were fabricated on the paper substrate with a semiconducting P3HT film as the active layer on the SiO_2/heavily doped Si substrate. The typical thickness of the SiO_2 film in this structure was approximately 100 nm. First, a series of OTFTs using P3HT films with different concentrations was fabricated. The schematic structure of the OTFTs with a bottom-gate/top-contact device is shown in Fig. 12.4a.

Figure 12.4b shows the typical drain current-gate voltage characteristic $(I_D–V_G)$ curve of OTFTs using P3HT films. The off current values were approximately -2.7×10^{-10}, -7.7×10^{-10} and -4.3×10^{-9} A for the P3HT molar ratio at 0.1, 0.3, and 0.7 wt%, respectively. The drain current on/off ratio values, which are simply divided as on current and off current, were approximately 144, 303, and 65

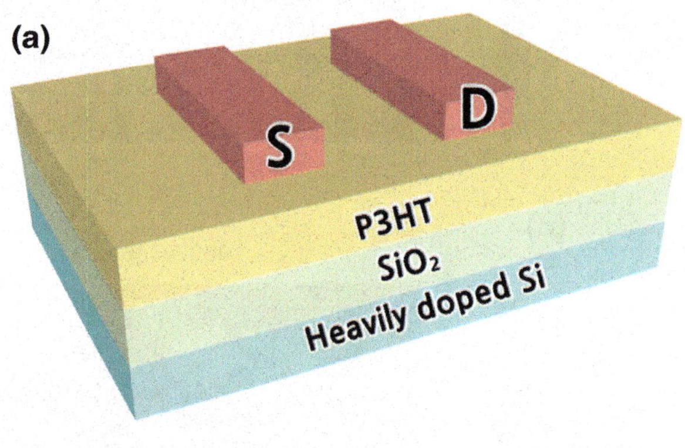

Fig. 12.4 **a** Schematic structure of the fabricated organic thin film transistors (OTFTs) with a bottom-gate/top-contact structure. **b** Typical drain current-gate voltage characteristic $(I_D–V_G)$ curves of the fabricated OTFTs using P3HT films with a molar ratio of 0.1, 0.3, and 0.7 wt%

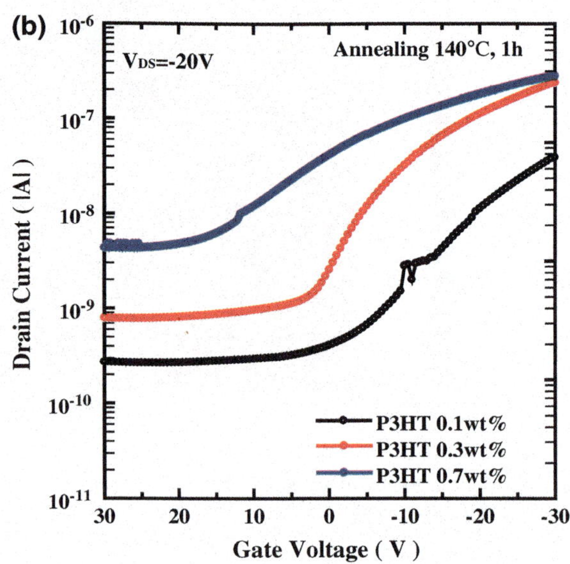

for the P3HT molar ratios at 0.1, 0.3 and 0.7 wt%, respectively. In the case of OTFTs with the 0.7 wt% P3HT mole ratio film, a relatively low drain current on/off ratio value with an approximate 10^1 order was shown in comparison to the on/off ratio of 0.3 and 0.1 wt% P3HT films with an approximate 10^2 order. Additionally, the threshold voltage gradually shifted the off current side depending on the increasing P3HT molar ratio. As the thickness of the P3HT film channel layer was increased, the electrical property of the OTFTs showed an undesirable tendency such as increasing off current and shift of the threshold voltage in this experiment. It is expected that there is a leakage current that is directly connected to the source and drain electrodes. The off currents might have two current paths; a channel path at the interface between the SiO_2 film and P3HT film, and another path directly connected to the source and drain electrode with P3HT film. The off current is strongly dependent on the increasing P3HT molar ratio in this experiment, indicating that a relatively thinner P3HT film shows enhanced electrical property. Alternatively, in the OTFTs with a thinner P3HT film, even though it exhibits better electrical properties compared to the thick film, there is some difficulty in the reproducibility of the device fabrication.

Based on the reviews of the electrical properties of OTFTs using P3HT films as the channel layer on the SiO_2/heavily doped Si substrate, FeFETs were fabricated on the paper substrate. For comparison with the paper substrate, FeFETs were also fabricated on the rigid Si substrate. In order to investigate the differences in the electrical properties between the cellulose paper and rigid Si substrates, FeFETs with the P3HT/P(VDF-TrFE)/Al structures on the SiO_2/heavily doped Si substrates were also prepared. The schematic structure of the fabricated FeFETs with a bottom-gate/top-contact device is shown in Fig. 12.5a.

Figure 12.5b shows the typical drain current-gate voltage characteristic (I_D–V_G) curves of the OTFTs and FeFETs on the rigid Si and paper substrates. The off current values of the OTFTs on the rigid Si substrate were approximately -6.4×10^{-10} and -2.8×10^{-9} A for a P3HT molar ratio of 0.2 and 0.7 wt%, respectively. The drain current on/off ratio values were approximately 344 and 54 for the P3HT molar ratios of 0.2 and 0.7 wt%, respectively. These values are very consistent with the results shown in Fig. 12.4b. The drain current I_D of the fabricated FeFETs on the rigid Si and paper substrates were also plotted against the sweeping gate voltage, as shown by the red and black circles, respectively. Both drain currents I_D of the FeFETs on the rigid Si and paper substrates exhibited clear hysteresis curves, as shown in Fig. 12.5b. This hysteresis curve illustrates that the turn-on voltage of the FeFETs was shifted by the ferroelectric nature of the P (VDF-TrFE) gate insulator. Both drain current on/off ratios on the rigid Si and paper substrates were approximately 10^2 and both memory window widths were approximately 20 V for the gate voltage sweep from -30 to 30 V. Both FeFETs on the rigid Si and paper substrates showed the same tendency in ferroelectric properties such as memory window width and drain current on/off ratio. In the case of the fabricated FeFETs on the rigid Si substrate, a P3HT film with 0.2 wt% was used because of the high leakage current of the FeFETs fabricated with the 0.7 wt% P3HT film. However, as shown in the figure, the off current values on the rigid Si

Fig. 12.5 a Schematic structure of the fabricated ferroelectric gate field effect transistors (FeFETs) with a bottom-gate/top-contact structure. **b** Typical drain current-gate voltage characteristic (I_D–V_G) curves of the fabricated OTFTs and FeFETs on rigid Si and paper substrates

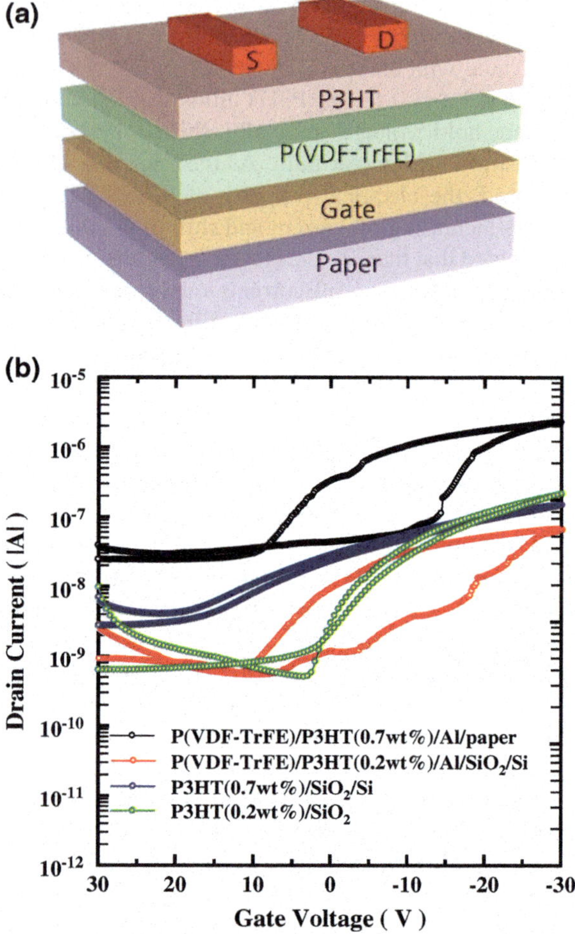

and paper substrates were approximately -9.1×10^{-10} and -2.4×10^{-8} A, respectively. The lower off drain current in the P3HT/P(VDF-TrFE)/Al/SiO$_2$/ heavily doped Si structures might be due to the thinner P3HT film thickness, as the same tendency was also observed in the OTFTs, in which the P3HT film was directly deposited on the SiO$_2$/heavily doped Si substrate, as shown in Fig. 12.4b. Based on these results, the fabricated FeFETs showed good ferroelectric properties because of the P(VDF-TrFE) film, regardless of the rigid Si or cellulose paper substrate. Additionally, the electrical properties of the fabricated devices are determined by the nature of the channel layer such as the drain current, on/off ratio, and leakage currents. These results illustrate that if the channel layer improves, better FeFETs can be fabricated, regardless of the substrate flexibility.

Figure 12.6 shows the focused ion beam scanning electron microscopy (FIB-SEM) images for the same samples shown in Fig. 12.5b. As shown in

Fig. 12.6 a Focused ion beam scanning electron microscope (FIB-SEM) image of fabricated FeFETs with an Au/P3HT/P(VDF-TrFE)/Al/paper structure. The image was acquired at a magnification of 50,000× in the secondary electron imaging mode using the upper detector, a 4.9 mm working distance, and a 5.0 kV accelerating voltage. **b** FIB-SEM image of fabricated FeFETs with an Au/P3HT/P(VDF-TrFE)/Al/SiO$_2$/heavily-doped Si structure. The image was acquired at a magnification of 120,000× in the secondary electron imaging mode. **c** FIB-SEM image of OTFTs with an Au/P3HT/SiO$_2$/heavily-doped Si structure. The image was acquired at magnifications of 100,000× in the secondary electron imaging mode

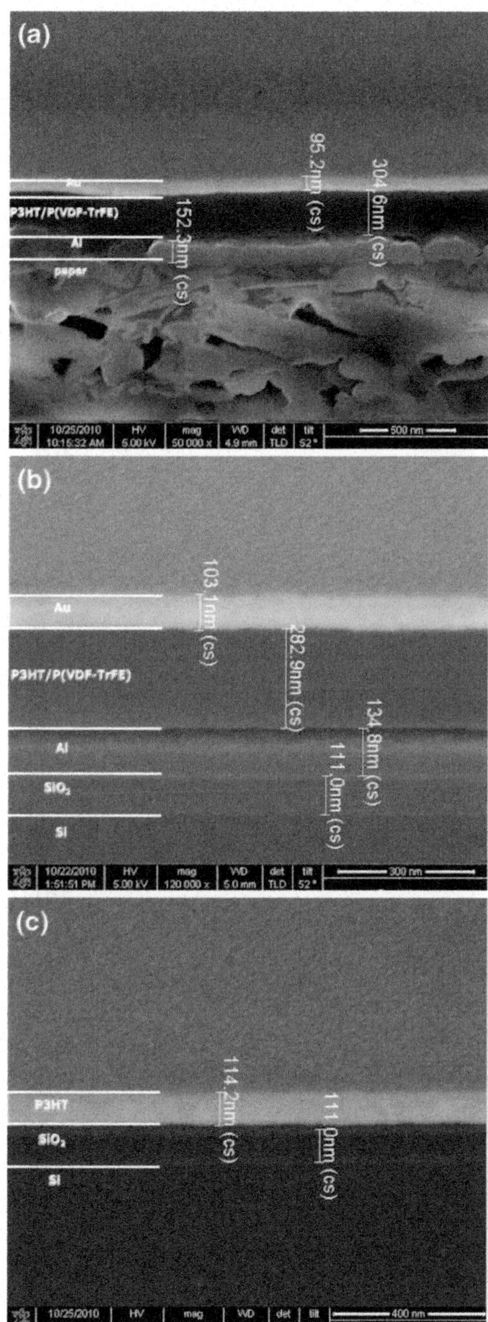

Fig. 12.7 **a** A typical surface atomic force microscopy (AFM) image of fabricated FeFETs with P3HT/PVDF-TrFE/Al/Paper structures. The measured area was 1×1 μm^2. **b** A typical AFM image of an FeFET with P3HT/PVDF-TrFE/Al/SiO$_2$/ heavily-doped Si structures. **c** A typical AFM image of an OTFT with P3HT/SiO$_2$/ heavily-doped Si structures

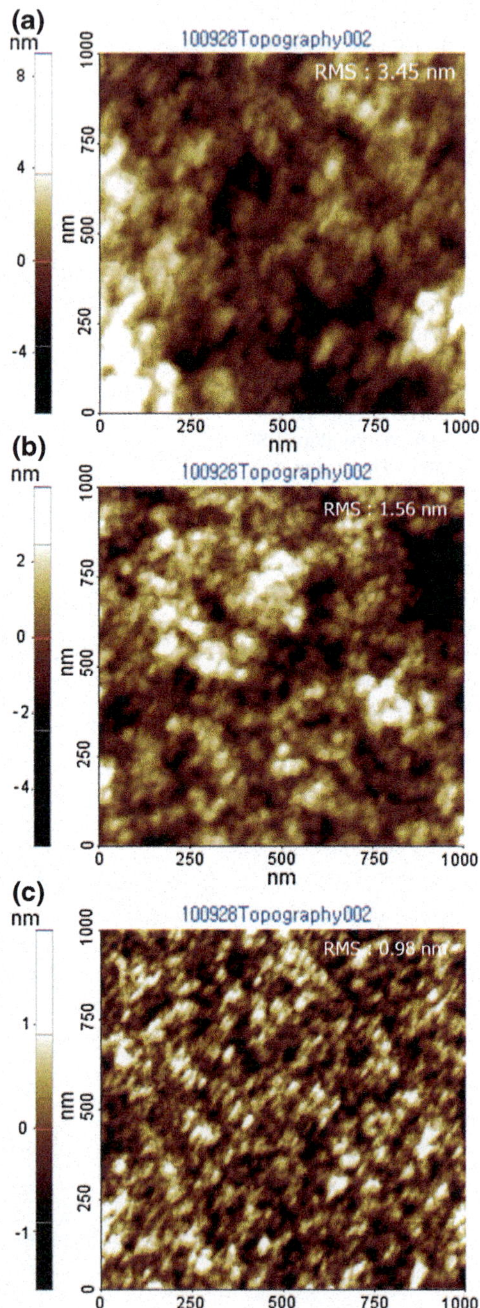

Fig. 12.6a, the FIB-SEM image of the structures shows that the P(VDF-TrFE) and P3HT films were formed very uniformly on a cellulose paper substrate, exhibiting excellent adhesion characteristics. The thickness of the P3HT/P(VDF-TrFE) films in this stack structure was approximately 305 nm. An FIB-SEM image of the FeFETs with the Au/P3HT/P(VDF-TrFE)/Al/ SiO_2/heavily doped Si structures was shown in Fig. 12.6b. This image was taken at 120,000× magnification in the secondary electron imaging mode. The thickness of the P3HT/P(VDF-TrFE) films in this stack structures was approximately 283 nm. The FIB-SEM image of the OTFT with the Au/P3HT/SiO_2/heavily doped Si substrate, which was taken at 100,000× magnification in the secondary electron imaging mode, is shown in Fig. 12.6c. The thickness of the P3HT film with 0.7 wt% was approximately 114 nm. Based on these images, the thicknesses of the films were approximately 180, 130, and 100 nm for the P(VDF-TrFE) film, P3HT film with 0.7 and 0.2 wt%, respectively. There is the likelihood of an observational error. Depending on the fabrication conditions, the formed films have a slight thickness difference. Nevertheless, those FE-SEM images clearly show the interface properties of each film. Additionally, the organic films with stack structures on the paper substrate can be formed with very good adhesive properties by a spin-coating technique.

The surface morphologies of the fabricated FeFETs and OTFT were observed using atomic force microscopy (AFM). Figure 12.7 shows the atomic force microscopy images of the same samples as that in Fig. 12.6. The measured area was 1×1 μm^2. The root mean-squared (RMS) surface roughness values of the fabricated FeFETs were approximately 3.45 and 1.56 nm for P3HT/PVDF-TrFE/Al/Paper and P3HT/PVDF-TrFE/Al/SiO_2/heavily doped Si structures, respectively. These values indicate that both fabricated FeFETs have a good surface morphology. Even though the FeFETs were fabricated on a paper substrate, it has a good surface morphology comparable to the FeFETs with a rigid Si substrate. The surface morphology of the fabricated OTFT is also shown in Fig. 12.7c. The RMS surface roughness of the fabricated OTFT was approximately 0.98 nm, indicating that the fabricated OTFT has a very good surface morphology.

12.4 Conclusion

In this work, non-volatile memory transistors were fabricated on paper substrates for the first time. As compared with the fabricated devices on rigid Si substrates, those of the fabricated devices on paper substrates showed nearly the same electrical properties and tendencies. These results suggest that the electrical properties of the fabricated FeFETs are mostly determined by the nature of the ferroelectric and semiconductor materials, irrespective of the substrates. These results indicate that a better channel layer can be used to fabricate a better FeFET, regardless of the substrate flexibility. Additionally, non-volatile memory devices can be easily fabricated on cellulose paper substrates by a solution processing method. Through the solution processing method, a non-volatile memory device can be fabricated at low temperatures, at a low cost and over a large area. The fabricated FeFETs on the paper substrates using a solution processing method can be used in many electronic systems, such as paper tags and paper displays, and will provide new potential for electronic devices.

Acknoledgement I would like to thank Dr. Seung Pil Han, Dr. Jeong-Hwan Kim, Dr. Gwang-Geun Lee, Dr. Dae-Hee Han and Mr. Min Gee Kim for valuable discussions and excellent experimental assistance.

References

1. C.A. Araujo et al., Nature **374**, 627–629 (1995)
2. J.F. Scott, C.A. Araujo, Science **246**, 1400–1405 (1989)
3. O. Auciello, J.F. Scott, R. Ramesh, Phys. Today **51**, 22–27 (1998)
4. D.H. Looney, U.S. Patent 2791758 (1957)
5. Nalwa HS: Handbook of Thin Film Materials (2002) (Academic Press, New York)
6. M. Takahashi, S. Sakai, Jpn. J. Appl. Phys. **44**, L800–L802 (2005)
7. K. Takahashi, K. Aizawa, B.E. Park, H. Ishiwara, Jpn. J. Appl. Phys. **44**, 6218–6220 (2005)
8. C.D. Dimitrakopoulos, P.R.L. Malenfant, Adv. Mater. **14**, 99–117 (2002)
9. F. Antonio, M.H. Yoon, T.J. Marks, Adv. Mater. **17**, 1705–1725 (2005)
10. G. Horowitz, Adv. Mater. **10**, 365–377 (1998)
11. G. Malliaras, R.H. Friend, Phys. Today **58**, 53–58 (2005)
12. H. Klauk, *Organic Electronics: Materials, Manufacturing and Applications* (Wiley-VCH, 2006)
13. H. Yang, S.W. LeFevre, C.Y. Ryu, Z. Bao, Appl. Phys. Lett. **90**, 172116 (2007)
14. H. Sirringhaus, T. Kawase, R.H. Friend, T. Shimoda, M. Inbasekaran, W. Wu, E.P. Woo, Science **290**, 2123–2126 (2000)
15. K. Marumoto, Y. Muramatsu, Y. Nagano, T. Iwata, S. Ukai, H. Ito, K. Shin-ichi, Y. Shimoi, S. Abe, J. Phys. Soc. Jpn. **74**, 3066–3076 (2005)
16. F. Xue, Z. Liu, Y. Su, K. Varahramyan, Microelectron. Eng. **83**, 298–302 (2006)
17. M. Surin, Ph Leclere, R. Lazzaroni, J.D. Yuen, G. Wang, D. Moses, A.J. Heeger, S. Cho, K. Lee, J. Appl. Phys. **100**, 033712 (2006)
18. S.W. Jeong, D.H. Han, B.E. Park, J. Ceramic Soc. Jpn. **118** [11], 1094–1097 (2001)
19. D.H. Kim, Y.D. Park, Y. Jang, H. Yang, Y.H. Kim, J.I. Han, D.G. Moon, S. Park, T. Chang, C. Chang, M. Joo, C.Y. Ryu, K. Cho, Adv. Funct. Mater. **15**, 77 (2005)
20. K. Asadi et al., Nat. Mater. **7**, 547–550 (2008)
21. R.C.G. Naber, Adv. Mater. **17**, 2692–2695 (2005)
22. T. Furukawa, Phase Transit. **18**, 143–211 (1989)
23. F.J. Balta Calleja, et al., Adv. Polym. Sci. **108**, 1–48 (1993)
24. R.C.G. Naber et al., Nat. Mater. **4**, 243–248 (2005)
25. A.J. Lovinger, Science **220**, 1115–1121 (1983)
26. S. Moller et al., Nature **426**, 166–169 (2003)
27. R.C.G. Naber et al., Appl. Phys. Lett. **85**, 2032–2034 (2004)
28. T.J. Reece, S. Ducharme, A.V. Sorokin, M. Poulsen, Appl. Phys. Lett. **82**, 142–144 (2003)
29. A. Bune et al., Appl. Phys. Lett. **67**, 3975–3977 (1995)
30. M. Zirkl et al., Adv. Mater. **23**, 2069–2074 (2011)
31. L.L. Chua et al., Nature **434**, 194–199 (2005)
32. M. Wohlgenannt et al., Nature **409**, 494–497 (2001)
33. H. Sirringhaus et al., Nature **401**, 685–688 (1999)
34. Z. Bao, A. Dodabalapur, A.J. Lovinger, Appl. Phys. Lett. **69**, 4108–4110 (1996)

Part VII
Applications and Future Prospects

Chapter 13
Novel Application of FeFETs to NAND Flash Memory Circuits

Shigeki Sakai and Mitsue Takahashi

Abstract A 64 kbit (kb) one-transistor-type ferroelectric memory array and peripheral logic circuits were integrated and characterized. N-channel Pt/SrBi$_2$Ta$_2$O$_9$/Hf–Al–O/Si ferroelectric-gate field-effect transistors (FeFETs) were used as the memory cells. The array was designed as a NAND flash memory, which had 32 blocks and 8 word lines × 256 bit lines per block. The bit-line- and block-selector logic circuits were represented by logic NOT and NAND units that were constructed by the complimentary structure of an n-channel FeFET and a p-channel FeFET. The erase, program, and nondestructive read operations were demonstrated for all blocks. The reading of the memory cells showed a clear separation of their erased and all "1"-programmed states. Threshold-voltage retention of one block showed no significant degradation after two days. To program arbitrary "1" and "0" patterns a single-cell self-boost program scheme was introduced, that can reduce program-disturb and power dissipation, and in fact very low program-inhibit-bit-line voltage (≤ 1.0 V) was achieved. Using this scheme the memory cells were programmed in "1" and "0" checkered pattern, and two distinguishable threshold-voltage distributions of 1 block (2 k cells) could be read out.

13.1 Introduction

Ferroelectric-gate field effect transistors (FeFETs) have attracted attention as nonvolatile memory cells [1–10]. Owing to their one-transistor-type (1T) structure, the FeFETs are potentially scalable and the read operation is nondestructive. After long data retention and high endurance were realized using a Pt/SrBi$_2$Ta$_2$O$_9$ (SBT)/Hf–Al–O (HAO)/Si metal/ferroelectric/insulator/semiconductor FeFET, [11–14] various characteristics of the FeFETs have been intensively investigated. Normal operations and long retention at elevated temperatures 85 and 120 °C were

S. Sakai (✉) · M. Takahashi
National Institute of Advanced Industrial Science and Technology,
Central 2, 1-1-1, Umezono, Tsukuba, Ibaraki 305-8568, Japan
e-mail: shigeki.sakai@aist.go.jp

© Springer Science+Business Media Dordrecht 2016
B.-E. Park et al. (eds.), *Ferroelectric-Gate Field Effect Transistor Memories*,
Topics in Applied Physics 131, DOI 10.1007/978-94-024-0841-6_13

exhibited [15, 16]. The control and distribution of threshold voltages (V_{th}s) were shown [17, 18]. The downsized FeFETs with up to 100-nm-long metal gate were demonstrated [19–22]. Complementary-FeFETs and their application to nonvolatile-logic circuits have been developed [23, 24].

The FeFET performance was estimated experimentally as a memory cell of NAND flash memories, and the voltage conditions for erase-, program- and read-operations of ferroelectric-NAND flash memory (Fe-NAND) were proposed in 2008 [25]. As the conventional floating-gate NAND flash memory (FG-NAND) has it, the Fe-NAND has a good scalable property based on the $4F^2$ downsizing rule, where F is the feature size. The Fe-NAND has program voltages of 6–7.5 V, being about 1/3 smaller than that of the FG-NAND, and the allowed endurance cycles of 10^8–10^9 are also at least 10^3 times larger than those of the FG-NAND [25, 26]. The Fe-NAND is expected to have much lower power dissipation than the FG-NAND, as indicated by a simulation using a single-cell self-boost (SCSB) program scheme [27]. Therefore, the Fe-NAND is, in principle, suitable and profitable for applications of high reliability use such as solid-state-drive application in data centers because of their low power dissipation, high endurance [26, 27].

In this article the experimental details of the 64 kb FeNAND array that we researched and developed are reviewed [28–30]. As the first trial of the Fe-NAND, the chosen memory array size was 64 kb. This size enabled us to evaluate statistically read operation results after erase- and program-operations. To avoid process problems such as etching damage at gate side walls, the most reliable fabrication technique at that time was adopted to Pt/SBT/HAO/Si type FeFETs. Sections 13.2, 13.3, and 13.4 describes the fabrication process, the test element group characteristics, and the 64 kb Fe-NAND flash memory itself, respectively. The article is concluded in Sect. 13.5. The contents in the single cell performance (Sect. 13.3.1) and 64 kb FeNAND flash memory array (Sect. 13.4) are based on [30]. Regarding the NOT-gates part in the logic element performance (Sect. 13.3.2) and the test circuits for block selector and bit-line selector, [29] should be cited. The ring oscillator part in it is not published before. The content of the single cell self-boost operations using a 4×2 miniature array (Sect. 13.3.4) is based on [28]. The fabrication (Sect. 13.2) is described commonly in [29, 30].

13.2 Fabrication Process

A photograph of a fabricated Fe-NAND-flash-memory array was shown in Fig. 13.1a. Active fields and LOCOS (local oxidation of silicon) areas were formed in a p-type Si substrate. P- and n-type wells were formed by ion implantation into the Si substrate, as shown in Fig. 13.1b. Layers of 7-nm-thick high-k HAO and 500-nm-thick ferroelectric SBT were successively deposited on the patterned Si substrate by a large-area-type pulsed-laser deposition technique. The 220-nm-thick Pt gate metal layer was deposited by electron-beam evaporation. The ion-milling technique was used to form gates and contact holes. Annealing for SBT

Fig. 13.1 **a** Photomicrograph of the 64 kb Fe-NAND memory array with a magnified schematic layout, and **b** the schematic cross section showing the p- and n-wells in the substrate [30]. Copyright (2012) The Japan Society of Applied Physics

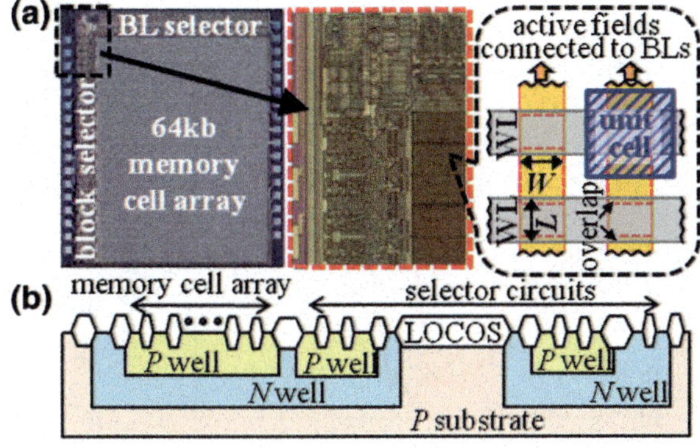

crystallization was performed at 800 °C for 1 h in O_2. Circuit wiring above the gate layers in the 64 kb Fe-NAND-flash-memory array was carried out using two metal-wire layers of Ti prepared by a lift-off process. The FeFETs in the memory array had a gate length L and a width W of $L = W = 5$ μm. The gate patterns overlapped with the sources and drains by 0.5 μm.

The circuits of the bit-line (BL) selector and block selector were designed in the periphery of the memory array. The devices and circuits of the test element group were also formed in the chip of the 64 kb Fe-NAND memory array integration. All transistors in the chip were made of FeFETs.

13.3 Test Element Group Characteristics

13.3.1 Single Cell Performance

The single FeFETs were placed in the test element group (TEG) area in the chip of the 64 kb Fe-NAND flash memory array. Figure 13.2a shows a drain current (I_d)— gate voltage (V_g) curve of an n-channel FeFET for the memory cell. The curve showed a memory window of 1.18 V for a V_g scanning between −6 and 6 V with $V_d = 0.1$ V. As shown in Fig. 13.2b, the p- and n-channel FeFETs for the logic circuits that were designed to show symmetrical I_d-V_g curves with respect to the axis of $V_g = 0$ V had very small memory windows because V_g was scanned in a narrow range between 0 and 6 V with $V_d = 6$ V for the n-channel FeFET and between 0 and −6 V with $V_d = -6$ V for the p-channel. The logic-circuit FeFETs showed little change in their I_d-V_g curves after 30 times of the I_d-V_g curve drawing tests.

Figure 13.3 shows a result of an erase-and-program (E/P) endurance test where bipolar voltage pulses of ±7.5 V and 10 μs were given to a memory-cell-type n-channel FeFET. After 10^8 cycles, no significant V_{th} shifts appeared. The 10 μs pulse width were suitable for the endurance test because 10 μs is a time scale used

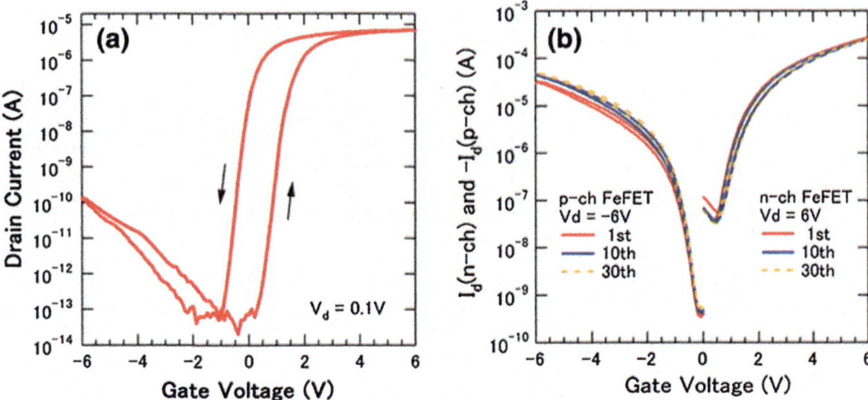

Fig. 13.2 a Drain current (I_{d})—gate voltage (V_{g}) curve of an n-channel FeFET for the memory cell, and **b** p- and n-channel FeFETs for the logic circuits. Modified from [30]

Fig. 13.3 V_{th} versus erase-and-program (E/P) endurance cycles. *Inset* **a** I_{d}-V_{g} curves after one cycle and after 10–10^5- and 10^8-cycles. *Inset* **b** one cycle of ± 7.5 V and 10 μs bipolar gate pulse. Modified from [30]

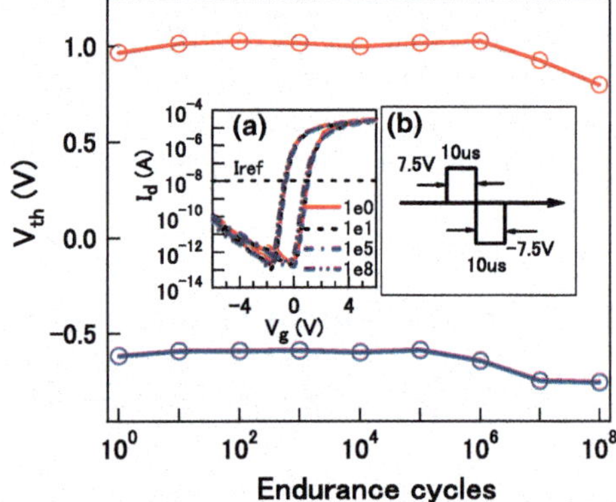

for program operations of the conventional FG-NAND flash memory [31]. The allowed maximum endurance cycles of the FG-NAND are at most 10^5 [32]. The obtained 10^8 endurance cycles of the Fe-NAND were much better than those of the FG-NAND.

13.3.2 *Logic Element Performance: NOT-Gates and a Ring Oscillator*

Figure 13.4a, b are the photomicrograph, and circuit representation of the NOT logic element. As shown in Fig. 13.4b, the NOT is constructed by a p-channel FeFET and an n-channel FeFET which is very similar to NOT by the conventional

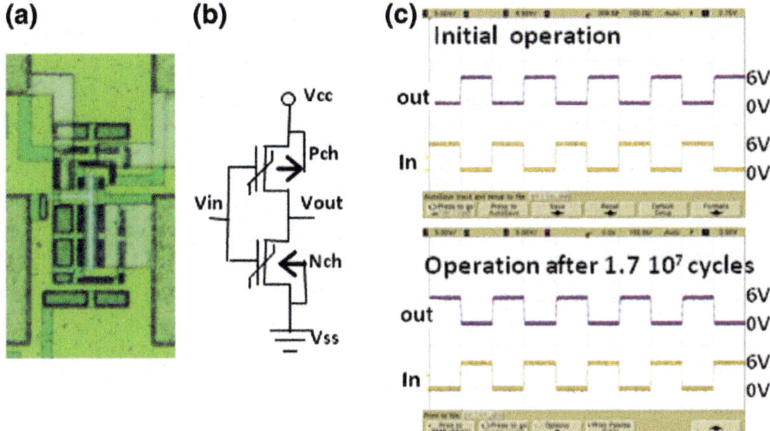

Fig. 13.4 **a** Photomicrograph of the fabricated NOT logic element and **b** its equivalent circuit that is constructed by a p-channel FeFET and an n-channel FeFET. **c** Pulse operation results, where the upper is the result at the initial and the lower is that after 1.7×10^7 pulse cycles are experienced. Modified from [29]

CMOS. Figure 13.4c shows the correct pulse operation results. The upper- and lower-graphs are the results at the initial state and the state after 1.7×10^7 pulse cycles were given to V_{in}, respectively. The correct inverted output V_{out} (6 V or 0 V) corresponding to the input V_{in} (0 V or 6 V) can be seen for both graphs.

A 29-stage ring oscillator ($N_{stage} = 29$) was designed, fabricated, and tested [33]. The oscillator was constructed by 28 NOT elements and one NAND logic element as shown in Fig. 13.5a. When control signal input of the NAND was in a high level, the oscillation was observed. Figure 13.5b show an oscillation waveform at $V_{cc} = 4$ V, and Fig. 13.5c is a graph of the oscillation frequency (f) and the propagation delay ($1/(2fN_{stage})$) as a function of V_{cc}. Since the delay is expected to be proportional to $1/L^2$ and $L = 5$ μm, the obtained results are reasonable and a faster delay less than 1 ns is promised by downsizing the FeFETs.

13.3.3 Test Circuits for Block Selector and Bit-Line Selector

Figure 13.6a, b show a photomicrograph and schematic drawing of the test circuit for the block selector. The test circuit, that is constituted by NOT and NAND elements, has 5 inputs of E_0, E_1, E_2, E_3, and E_4 and one output. Among 32 ($=2^5$) cases of the logic high (H) and low (L), one combination of E_0–E_4 selects one block whose output is logic H, while the other 31 combinations make the logic L output. The output is connected to all the gates of ten transfer-gate (TG) transistors which are simply n-channel FeFETs. When the output is logic H, the input voltages can be transferred from the I/O-pads: *sgd*, wl$_0$–wl$_7$ and *sgs*, to the internal signal lines:

Fig. 13.5 **a** Equivalent circuit of a 29-stage ring oscillator, consisting of 28 NOT elements and one NAND logic element formed by FeFETs. **b** Oscillation waveform at $V_{cc} = 4$ V, and **c** oscillation frequency (f) and the propagation delay ($1/(2fN_{\text{stage}})$) as a function of V_{cc} [33]

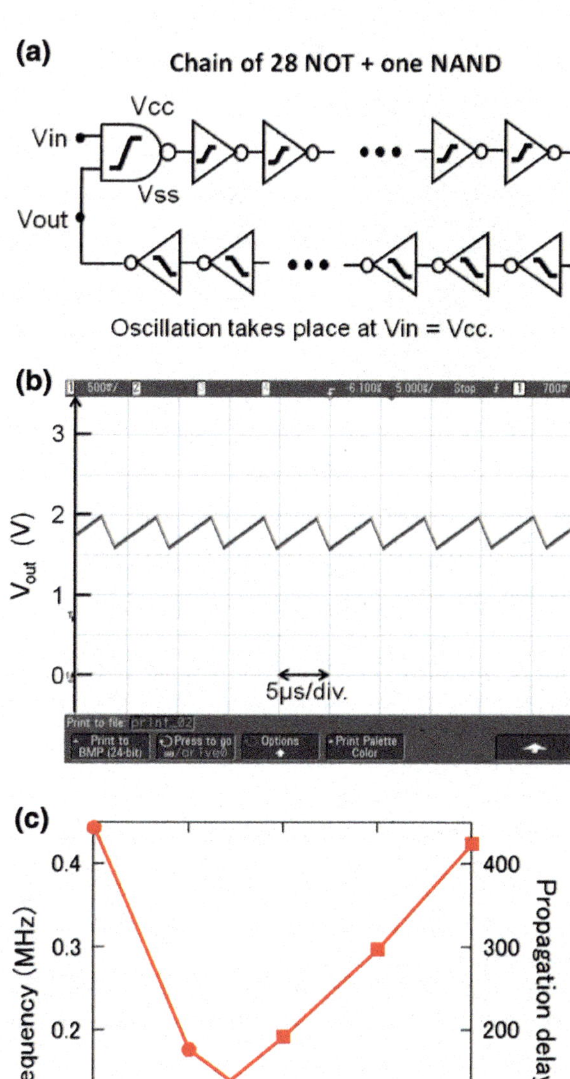

SGD(j), WL_0(0,j)-WL_7(7,j), and SGS(j). Figure 13.6c shows the correct operation results. Only when the combination $(E_0,\ldots, E_4) = (6$ V, 6 V, 6 V, 6 V, 6 V) was selected, the output V_{SGD} followed the input V_{sgd} from $V_{\text{sgd}} = 0$–5.4 V. In the range of 5.4 V $< V_{\text{sgd}} \leq 6$ V, V_{SGD} was clipped at 5.4 V. The voltage drop of 0.6 V

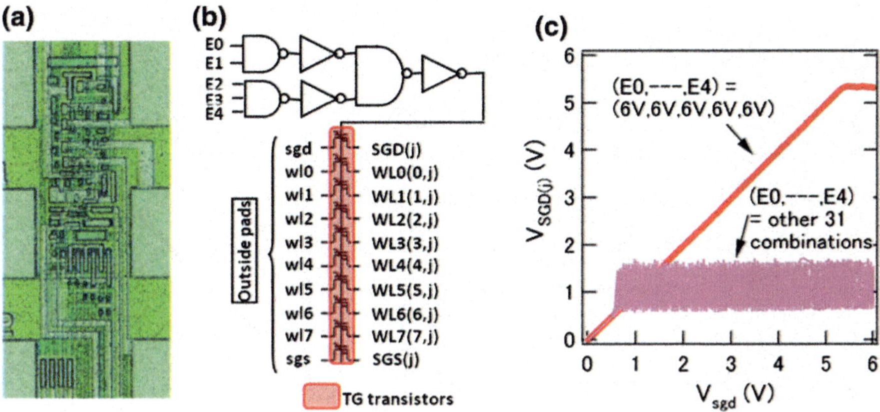

Fig. 13.6 **a** Photomicrograph of the fabricated block-selector test circuit in TEG, **b** its equivalent circuit and **c** operation results. Only when the combination E0–E4 = (6, 6, 6, 6, 6) V was selected, the input voltage was transferred to the output in the range of 0 V < V_{sgd} ≤ 5.4 V. For 5.4 V < V_{sgd} < 6.0 V, V_{SGD} was clipped at 5.4 V. This voltage drop of 0.6 V is corresponding to the V_{th} of the TG transistors. When the other combinations were selected, the output V_{SGD} did not follow the input V_{sgd} since the corresponding TG transistors were in off-states. Modified from [29]

corresponded to the V_{th} of the TG transistors. Other TG transistors labeled wl_0, …, wl_7, and *sgs* showed the same results as those of the *sgd* TG transistor. When the other combinations were selected, the output V_{SGD} did not follow the input V_{sgd} since the gate voltages of the TG transistors were logically low.

In the 64 kb Fe-NAND memory array with 32 block selector circuits that is described in Sect. 13.4, each selector circuit is adjacent to each block of the memory array. Input pad signals of e_0, e_1, e_2, e_3 and e_4 and their inverted signals are connected by wires via wire layers properly to the 5 inputs of E_0, E_1, E_2, E_3, and E_4 of all 32 block selectors. One combination of H or L of the input pad signals forms all H of E_0, E_1, E_2, E_3, and E_4 only for one block selector circuit, meaning that only one block is selected among all the blocks.

Figures 13.7a, b show a photomicrograph and schematic drawing of the test circuit for the BL selector. As found in Fig. 13.7b, the test circuit, that is constituted by NOT and NAND elements, has 5 inputs of C_0, C_1, C_2, C_3, and C_4 and one output. Among the 32 cases of the logic H and L combinations of C_0–C_4, one combination of C_0–C_4 selects one BL-set whose the output is logic H, while the other 31 combinations make the logic L output. When the output logic is H, the voltages coming from the I/O-pads, *ldat* and *rdat*, reach the source sides of the TG transistors. Four select lines: GBL01, GBL23, GBL45, and GBL67, are prepared to select a BL pair to be connected to the *ldat* and *rdat*. One BL-selector circuit controls 8 bit lines. For an example, the voltage can be transferred from *ldat* and *rdat* to a pair of odd- and even-BLs, BL(0,i) and BL(1,i) in case of BLSET = GBL01 = logic   H   and   GBL23 = GBL45 = GBL67 = logic   L. Figure 13.9c shows the BL voltage V_{BL0} versus *ldat* voltage (V_{ldat}) curve for 32

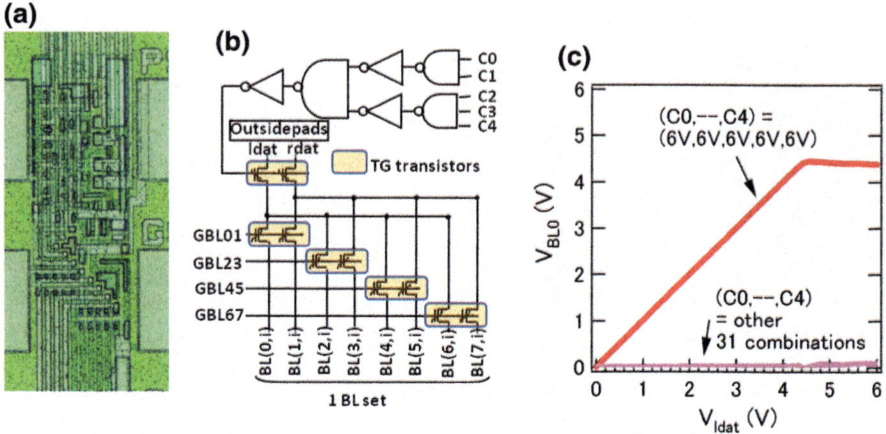

Fig. 13.7 **a** Photomicrograph of the fabricated BL-selector test circuit in TEG and **b** its equivalent circuit. As found in (**b**), the BL has a logic circuit that has 5 inputs and one output. Among the 32 cases of the logic H and L combinations of C0–C4, only one combination of C0–C4 selects the output H, while the other 31 combinations make the output L. When the output is H, the voltage transferred from ldat and rdat, to the sources of the TG transistors. Four select lines: GBL01, GBL23, GBL45, and GBL67, are prepared to select a BL pair to be connected to the ldat and rdat. **c** Operation results. The BL voltage V_{BL0} versus ldat voltage (V_{ldat}) curves for 32 combinations of C0–C4 are shown. When BL01 was logic H, only combination C0–C4 = (6, 6, 6, 6, 6) V made output V_{BL0} followed V_{ldat}, while the other 31 combinations made output V_{BL0} of logic L. Modified from [29]

combinations of C_0–C_4. When BL01 was logic H, only combination (C_0, ..., C_4) = (6, 6, 6, 6, 6) V made an output V_{BL0} followed V_{ldat}, while the other 31 combinations made outputs V_{BL0} of logic L. A V_{th} drop 1.5 V was found in the voltage transfer curve. For the actual measurement of the memory array that is described later, BL voltages were about 3.2 V for the conventional bit-line program scheme and ≤ 1 V for the single-cell-self-boost program scheme. Therefore, the maximum transfer voltage 4.5 V was sufficient. For the voltage transfer from *ldat* and *rdat* to the other BLs, the same transfer results were obtained.

In the 64 kb Fe-NAND memory array that has 32 BL-selector circuits, each selector circuit controls a set of 8 bit lines. Input pad signals of c_0, c_1, c_2, c_3 and c_4 and their inverted signals are connected by wires via wire layers properly to the 5 inputs of C_0, C_1, C_2, C_3, and C_4 of all the 32 BL-selector circuit. One combination of H or L of the input pad signals forms all H of C_0, C_1, C_2, C_3, and C_4 only for one BL-selector circuit, meaning that one group with 8 bit lines is selected among all the 32 groups.

13.3.4 Single Cell Self-boost Operations Using a 4 × 2 Miniature Array

A miniature 4 × 2 cell array of the Fe-NAND flash memory was prepared in the TEG (Fig. 13.8a). As shown in Fig. 13.8b, the equivalent circuit consists of 12 n-channel FeFETs on the cross points of 2 bit lines and 6 control lines which were named an SGD, 4 word lines WL_i (i = 0, 1, 2, 3) and an SGS.

A series of successive operations of Erase, Read, Program and Read was performed. The 4 × 2 memory cell array has $2^4 \cdot 2^4 = 256$ program patterns by the combination of "1" and "0". The series operations of Erase, Read, Program and Read were repeated 256 times by changing the program pattern entirely. At Erase, as shown in Fig. 13.9a, a voltage pulse of V_{erase} in height and t_{erase} in width was given to the p-well and also to the n-well surrounding the p-well, while 0 V was given to the control lines SGD and WL_i (i = 0, 1, 2, 3). At Read, as shown in Fig. 13.9b, V_{read} was sequentially applied on WL_is from WL_0 to WL_3 in this order. Bit-line currents were measured by scanning V_{read} typically between 2 and 0 V with constant bit-line voltages $V_{bl0} = V_{bl1}$.

The 4 × 2 memory cell array was programmed from WL_3 to WL_0 in this order as shown in Fig. 13.10a, b. The voltage profiles with time were in Fig. 13.10c. $V_{sgs} = 0$ V was set at the SGS control line, and as the bit-line voltages of V_{bl0} and V_{bl1}, 0 V or V_{pibl} (the program-inhibit-bit-line voltage) was given, depending on the selected program patterns. At the first stage of Program (Fig. 13.10a), we applied a voltage pulse with V_{sgd} in height and t_{sgd} in width to the control line SGD, pulses with V_{pass} and t_{pass} to WL_0 and WL_1, a pulse with V_{cut} and t_{cut} to WL_2. The conditions of $V_{sgd} = V_{pass}$ and $t_{sgd} = t_{pass} = t_{cut}$ were taken. Then, a program pulse with V_{pgm} and t_{pgm} was supplied to WL_3. On the bit line on which 0 V was given, the bit-line voltage was transferred and at the V_{pgm}-application timing "1" was programmed at the selected cell on WL_3. On the other hand, on the bit line on which V_{pibl} was given, the bit-line voltage transfer was inhibited by the transistor on

Fig. 13.8 **a** Photomicrograph of a fabricated 4 × 2 miniature FeNAND array and **b** schematic equivalent circuit. Modified from [28]

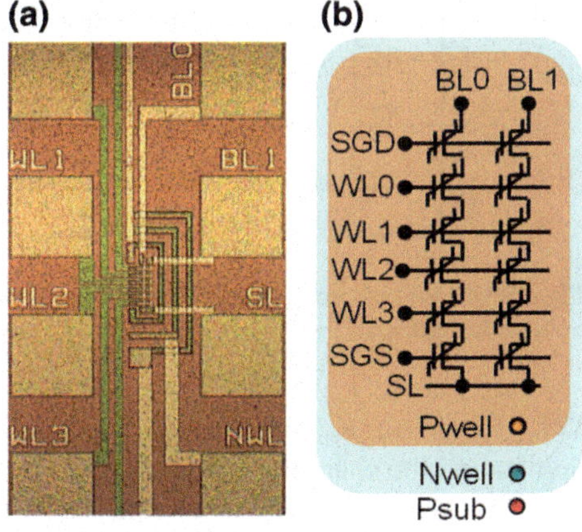

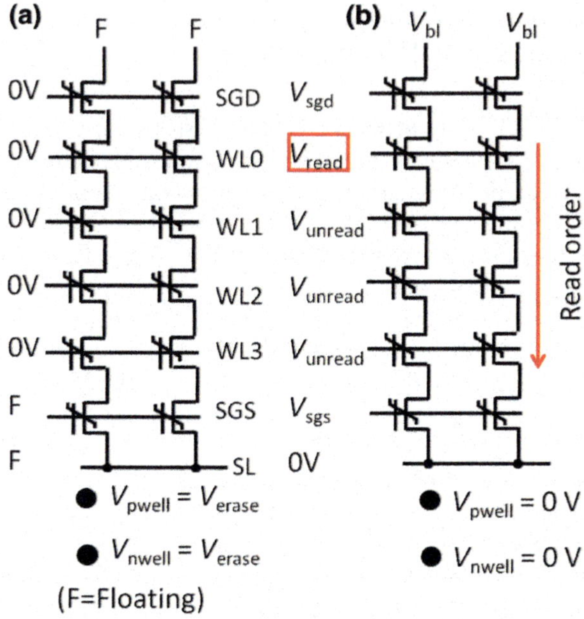

Fig. 13.9 Voltage conditions for **a** Erase and **b** read of the 4 × 2 miniature FeNAND array. Modified from [28]

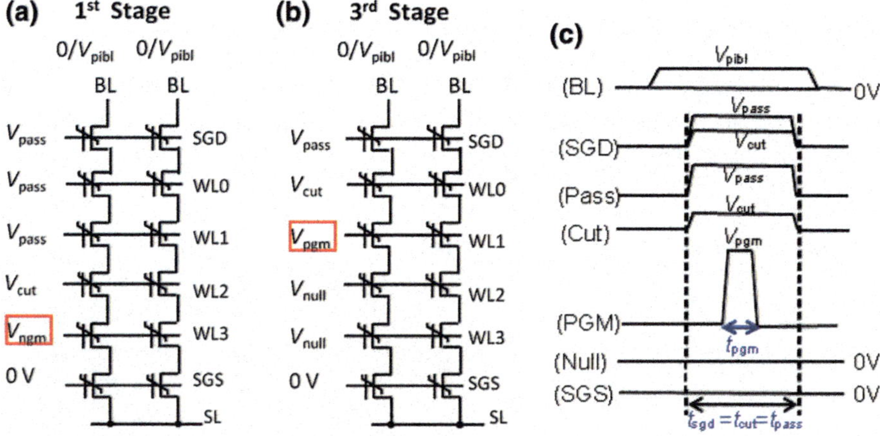

Fig. 13.10 Voltage conditions for program to the cells belonging to **a** WL3 and to **b** WL1. Schematic time chart of voltage pulses for program. Modified from [28]

WL_2 biased to V_{cut} because $V_{cut} - V_{pibl} < V_{th}$, and thus at the timing of V_{pgm} application the channel voltage was boosted at the unselected cell on WL_3, meaning that "0" was programmed. At the second, third, and fourth stages, the program

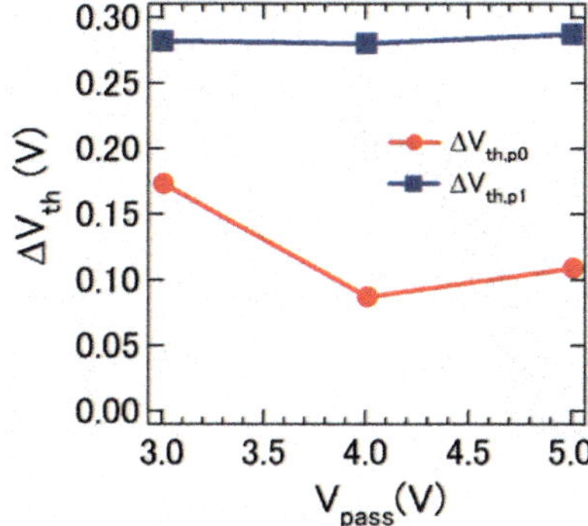

Fig. 13.11 V_{th} changes from erase to P0 and to P1 by a conventional bit-line program scheme while keeping a condition of $V_{pass} - V_{pibl} = 2$ V. Modified from [28]

pulse to be applied was shifted to WL_2, WL_1 and WL_0. In the figure, only the third stage is shown in Fig. 13.10b. The V_{cut} was also shifted in sequence, and at the 4th step V_{cut} was given to the SGD control line. Voltages $V_{nul} = 0$ V were given to the WLs to which V_{pgm} was already applied at the prior steps.

First, operation results by a conventional bit-line program scheme ($V_{cut} = V_{pass}$) were summarized. Figure 13.11 shows the V_{th} changes ($\Delta V_{th,p1}$ and $\Delta V_{th,p0}$). Namely $\Delta V_{th,p1} = V_{th,e} - V_{th,p1}$ and $\Delta V_{th,p0} = V_{th,e} - V_{th,p0}$, where $V_{th,e}$, $V_{th,p1}$, and $V_{th,p0}$ were the V_{th}s of the erase, program "1" (P1), and program "0" (P0) states, respectively. As shown in Fig. 13.11, the P0 state V_{th} change from the erase state was very sensitive to the pass voltage V_{pass} where $V_{pass} - V_{pibl} = 2$ V was kept. At $V_{pass} = 5$ V, $\Delta V_{th,p0}$ increased due to the pass voltage disturb, and at $V_{pass} = 3$ V, $\Delta V_{th,p0}$ increased due to the program voltage disturb. To get the largest V_{th} difference (0.21 V) between P1 and P0 states, a high V_{pass} (4 V) and a high V_{pibl} (2 V) were necessary.

Operation results (Fig. 13.12) using the SCSB program scheme were obtained on the conditions of $V_{pgm} = 7$ V, $t_{pgm} = 10$ μs, $V_{sgd} = V_{pass} = 3$ V, $V_{cut} = 2.1$ V, $t_{sgd} = t_{pass} = t_{cut} = 100$ μs, $V_{pibl} = 0.6$ V at Program, and $V_{pwell} = 7$ V and $t_{pwell} = 1$ ms at Erase [34]. Figure 13.12a shows accumulated Read curves of the erase, P0, and P1 states for all 256 patterns. Figure 13.12b shows the V_{th} distributions. The average V_{th}s of the erase—P0, and P1 states are 1.388, 1.273, and 1.027 V, and the standard deviations σ_e, σ_{p0} and σ_{p1} of them are 0.050, 0.033, and 0.029 V, respectively. The P1-state V_{th} distribution was separated clearly from the P0-state V_{th} distribution. More quantitatively, $(V_{th,p0} - V_{th,p1})/(\sigma_{p0} + \sigma_{p1}) = 4.0$. The P0-state V_{th} distribution did not change much from the erase state distribution, indicating that the program disturbs could be suppressed. Figure 13.12 clearly showed that the program-inhibit-bit-line voltage can be reduced much less than 1 V. V_{pibl} was varied while V_{cut} was kept at 1.8, 2.1 and 3 V, and the same operations as those to get the results in Fig. 13.12 were performed. Figure 13.13a shows $V_{th,e}$, $V_{th,p0}$ and $V_{th,p1}$, where the average values of the distributions obtained

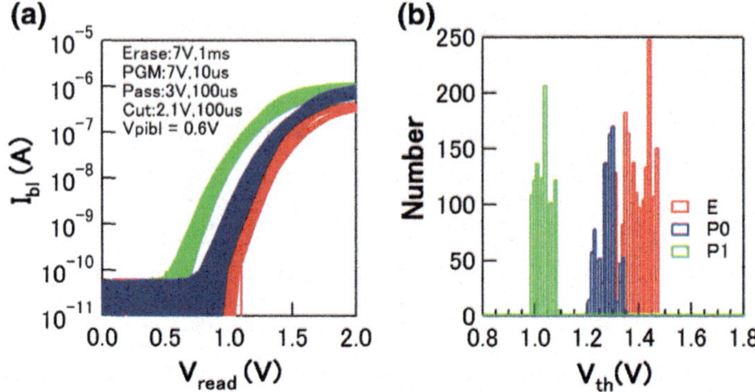

Fig. 13.12 **a** All read curve (I_{bl} vs. V_{read}) and **b** V_{th} histogram after erase and after program for the entire 256 patterns of 4 × 2 memory cell using the SCSB scheme and **b** V_{th} histogram after Erase (E), P0 and P1 states. At Erase, $V_{erase} = V_{pwell} = V_{nwell} = 7$ V. At Program, $V_{pgm} = 7$ V, $t_{pgm} = 10$ μs, $V_{pass} = 3.0$ V, $V_{cut} = 2.1$ V, and $V_{pibl} = 0.6$ V [34]

by all 256 patterns were taken. Figure 13.13b shows $\Delta V_{th,p1}$ and $\Delta V_{th,p0}$ as a function of V_{pibl}. Wide V_{pibl} ranges to ensure clear V_{th} differences (>0.3 V) between P1 and P0 states for the cases at $V_{cut} = 1.8$, 2.1 and 3 V can be found in Fig. 13.13a, b. With decrease of V_{pibl}, the decreases of $V_{th,p0}$ started around $V_{pibl} = 0.4$, 0.6 and 1.2 V for $V_{cut} = 1.8$, 2.1 and 3 V, respectively, below which the SCSB effect was weakened.

Similar experiments to those for Fig. 13.12a, b were done at $t_{sgd} = t_{pass} = t_{cut} = 10$ μs. The obtained V_{th}s as a function of V_{pibl} were almost same as those shown in Fig. 13.13a, meaning that t_{sgd}, t_{pass} and t_{cut} can be reduced to the same pulse width of the program, $t_{pgm} = 10$ μs.

The difference between the P0- and P1-state V_{th}s of the SCSB programming was compared with those of the conventional bit-line programming in Fig. 13.14. $V_{th,p0} - V_{th,p1}$ in the SCSB case was 32 % larger than that in the conventional case. Figure 13.14 shows that the SCSB case results were robust against the fluctuation of setting voltage (V_{cut}). Namely, the variation of $V_{th,p0} - V_{th,p1}$ in the SCSB case was 11 % for the V_{cut} variation from 1.8 to 3 V, while that in the conventional bit-line programming was 26 % for the V_{pass} variation from 3.4 to 4.6 V.

13.4 64 Kb Fe-NAND Flash Memory

13.4.1 Architecture

The circuits of the Fe-NAND-flash-memory cell array with the block- and BL-selectors are schematically shown in Fig. 13.15a–c. The 64 kb memory cell array has a structure of 8 WLs × 32 blocks × 256 BLs. The NAND string consists

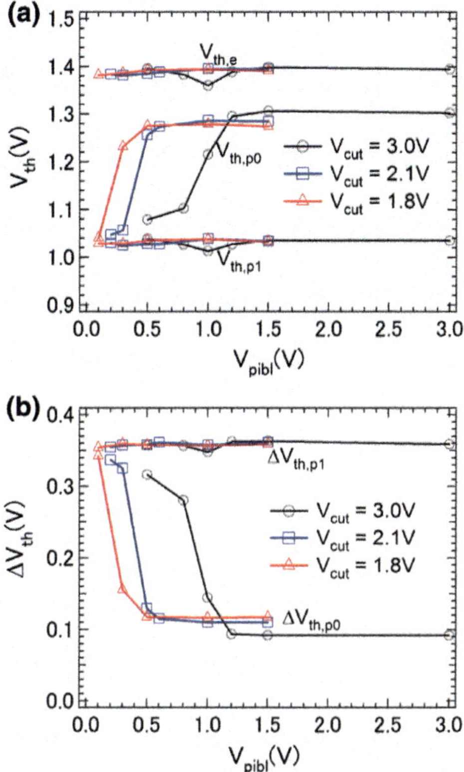

Fig. 13.13 **a** Average values of V_{th} distributions for E-, P0-, and P1-states as a function of V_{pibl}, while V_{cut} = 1.8, 2.1, and 3.0 V with V_{pgm} = 7 V, t_{pgm} = 10 μs and $t_{sgd} = t_{pass} = t_{cut}$ = 100 μs. At Erase, $V_{erase} = V_{pwell} = V_{nwell}$ = 7 V. **b** Extracted $\Delta V_{th,p1}$ and $\Delta V_{th,p0}$, (differences between $V_{th,p1}$, $V_{th,p0}$ and $V_{th,e}$) versus V_{pibl}. Modified from [28]

Fig. 13.14 V_{th} differences between P0- and P1-states of the SCSB program scheme. Those by conventional bit-line program were also added. Modified from [28]

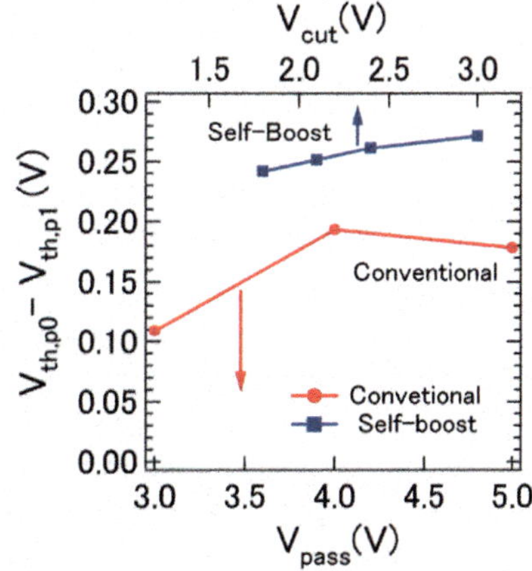

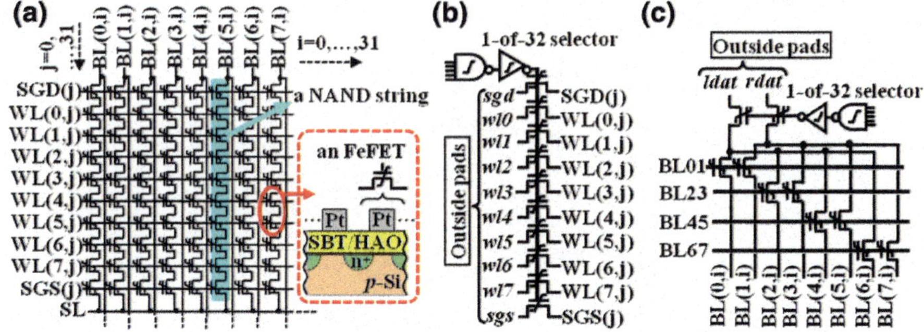

Fig. 13.15 Partial equivalent circuits of the 64 kb Fe-NAND flash memory array. **a** Memory cell array, **b** block selector, and **c** BL selector [30]. Copyright (2012) The Japan Society of Applied Physics

of a select-gate transistor (SGD) of the drain- or BL-side, eight memory cells, and a select-gate transistor (SGS) of the source-side. A block (j) includes 2 kb memory cells on the cross points of 8 WLs from WL(0,j) to WL(7,j) and all 256 BLs. A source line (SL) is connected to the sources of all the SGS. A block selector (Fig. 13.15b) and a BL selector (Fig. 13.15c) were placed on the two sides of the array, that is shown in Fig. 13.1a. Five-input logic NAND-function circuits, the details of which were described in Sect. 13.3.3, can make $2^5 = 32$ cases. The integer j from 0 to 31 in Fig. 13.15b is the block index, and the integer i from 0 to 31 in Fig. 13.15c is the index for 32 bit-line group, one of which has 8 BLs. We also have one important choice of selecting all blocks and all BL sets. When block j is selected, input voltages on the I/O pads, *sgd*, *wl0*, …, *wl7* and *sgs*, are transferred to the lines, SGD(j), WL(0,j), …, WL(7,j), and SGS(j). The BL selector has 4 control lines BL01, BL23, BL45, and BL67. When bit-line group i is selected and the BL01 control line is high, input voltages on the I/O pads *ldat* and *rdat* are transferred to a pair of BL(i,0) and BL(i,1).

The SGS and SGD transistors at both ends of the NAND string were also formed by FeFETs. As we previously reported in a study of nonvolatile logic, p- and n-channel-type FeFETs can work as logic transistors [23, 24].

13.4.2 Basic Operations

This sub-section describes Erase, Read and Program operations of the 64 kb Fe-NAND-flash-memory cell array.

Erase: An erase pulse of V_{erase} in height and t_{erase} in width was given to the p-well for the memory cell array and the surrounding n-well in Fig. 13.1b. All cells in a selected block j were erased at once. During Erase, the I/O pad voltages from *wl0* to *wl7* were set to 0 V. The voltages on the SL and I/O pads, *sgd*, *sgs*, *ldat*, and *rdat* were floating.

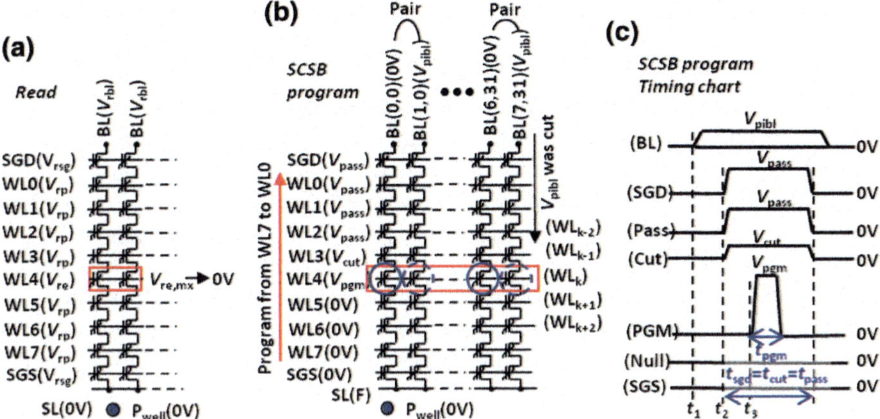

Fig. 13.16 Diagrams of explaining Fe-NAND operations for **a** read, **b** program, **c** timing chart of SCSB program scheme [30]. Copyright (2012) The Japan Society of Applied Physics

Read: A pair of BLs was selected (Fig. 13.16a) and a bit-line voltage for Read, V_{rbl}, was given to the selected pair of BLs via the *ldat* and *rdat* pads. A select-gate voltage for Read, V_{rsg}, was given to the SGD and SGS lines to let the SGD and SGS transistors ON, while voltages of the SL, p-well, and n-well were 0 V. A WL was selected and a word-line voltage for Read, V_{re}, was applied to the selected WL. To the other unselected WLs the read-pass voltage V_{rp} was given. The proper V_{rp} value was chosen to let all the unselected memory-cell-transistor channels be conductive. Bit-line currents, I_{rbl}s, were measured on both the *ldat*- and *rdat*-pads by scanning V_{re} between 0 V and $V_{re,max}$. Normally, the setting voltages have the relationships of $V_{re,max} \leq V_{rp} \leq V_{rsg}$. Such the read operation was performed successively from WL_0 to WL_7 so that all memory cells in the block could be measured. Using the measured I_{rbl} versus V_{re} curves the threshold voltages (V_{th}s) were determined as the voltages which gave a defined reference current (I_{ref}).

Program: All 256 BLs in a block were selected, and the SCSB program scheme was used. As shown in Fig. 13.16b, let us assume that select memory cells to be programmed exist on the cross point of WL_k and all left BLs of all the BL pairs, and that unselect cells to be program-inhibited exist on the cross point of the WL_k and all right BLs of all the BL pairs. Program-inhibit-bit-line voltage, V_{pibl}, is given to the BLs of the program-inhibit memory cells at the time t_1 while 0 V is given to the BLs of the selected memory cells for the program. At t_2 in Fig. 13.16c, a pulse of the height V_{cut} and width t_{cut} is given to WL_{k-1}, and a pulse of the height V_{pass} and width t_{pass} is given to all WL_ms where m $\leq$ k $-$ 2. At t_3, a program pulse of the height V_{pgm} and width t_{pgm} is given to WL_k. Throughout the programming, $V_{null} = 0$ V is applied to all WL_ns, where n $\geq$ i + 1. The p-well for the memory cell array and surrounding n-well are 0 V. SGS voltage V_{sgs} is 0 V. SL is floating. The program operations to be done to WL_k are repeated sequentially to all WLs from WL_7 to WL_0 of this order. During the WL_0 programming, the SGD is set to V_{cut}. In this experiment, we set $V_{\text{sgd}} = V_{\text{pass}}$ and $t_{\text{sgd}} = t_{\text{pass}} = t_{\text{cut}}$. On the program inhibit BL, the V_{pibl} transfer is cut off by the V_{cut} on WL_{k-1}. The channel voltage of the program-inhibit memory cell is self-boosted during the voltage pulse application on the WL_k by the coupling of gate-channel capacitance and channel-body capacitance.

As a special program operation, we have an all-program option where all memory cells in a block are programmed on at once. All the 256 BLs and 8 WLs in a block are selected. A program pulse with a height V_{pgm} and a width t_{pgm} is applied to the *wls*, *wl0*, …, *wl7*. During the program, the p-well for the memory cell array and the surrounding n-well were set to 0 V. A pass voltage on *sgd* was applied to the *sgd* line, and the voltages on *sgs*, *lda*, *t* and *rdat* are 0 V. The SL is floating.

In this article, the "0" states are defined as the states after Erase or the states of program-inhibit (or unselect) cells after Program. The "1" states are defined as the states of select cells after Program.

13.4.3 Results and Discussion

Using the integration and fabrication processes described in Sect. 13.2, a Fe-NAND-flash-memory array chip was made. In the chip, five identical memory arrays were prepared. The obtained arrays included defects due to immature technology. Nevertheless, two of them named array *A* and array *B* could be measured. In array *A*, a set of 8 BLs was defective, that was 3 % of the memory cells in every block. In array *B*, a WL in a block was defective, that was 12.5 % of the memory cells in a block. In this article, the contributions from such defective cells were eliminated when counting V_{th} distributions.

64 memory cells located at the cross points of 8 WLs and 8 BLs in a block in array *A* were operated and 64 I_{rbl} versus V_{re} read curves were shown on the same graph (Fig. 13.17). First Erase and Read were operated, and next all-"1"-Program

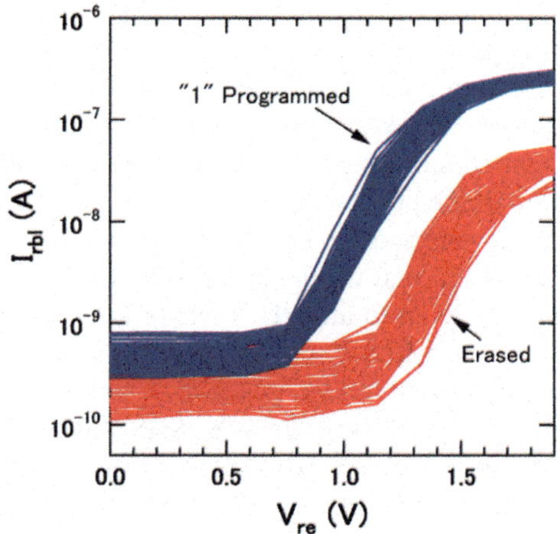

Fig. 13.17 I_{rbl} versus V_{re} curves of 64 cells of 1 BL set at 1 block in array *A* after erase and all-"1"-program operations [30]. Copyright (2012) The Japan Society of Applied Physics

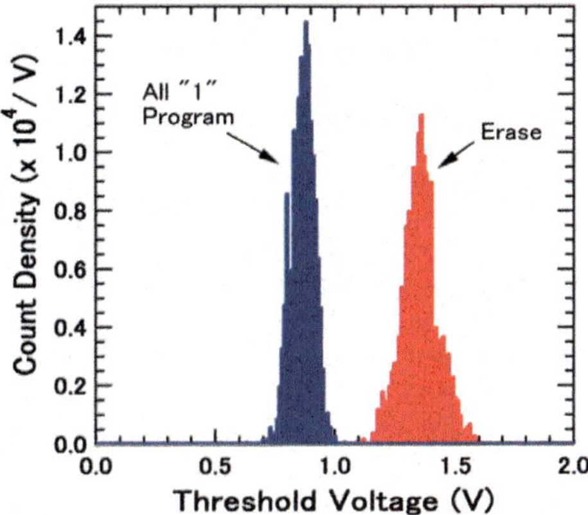

Fig. 13.18 V_{th} histograms of 1 block (2 kb) Fe-NAND in array B after erase and all-"1"-program operations [30]. Copyright (2012) The Japan Society of Applied Physics

and Read were operated. The pulse conditions for Erase were $V_{erase} = 7$ V and $t_{erase} = 100$ ms. The conditions for all-"1"-program were $V_{pgm} = 7.5$ V and $t_{pgm} = 1$ ms. The read conditions were $V_{rbl} = 0.2$ V, $V_{rsg} = 2.5$ V, and $V_{rp} = 1.9$ V. V_{re} was swept from $V_{re,max}$ to 0 V where $V_{re,max} = 1.9$ V. As indicated in Fig. 13.17, one finds a clear separation between the curves after Erase and after all "1"-Program.

Figure 13.18 shows the V_{th} distributions of 1 block (2 kb cells) in array B, which were read after Erase and after the all-"1"-program. The erase pulse conditions were $V_{erase} = 7$ V and $t_{erase} = 100$ ms. All-"1"-program conditions were $V_{pgm} = 7.5$ V and $t_{pgm} = 1$ ms. The read operations were done on the condition of $V_{rbl} = 0.2$ V, $V_{rsg} = 2.5$ V, and $V_{rp} = 2.0$ V. V_{re} was varied from $V_{re,max}$ (2.0 V) to 0 V. The V_{th}s were decided at $I_{ref} = 3 \times 10^{-8}$ A. The V_{th} distributions in Fig. 13.18 showed a good separation between the erase- and all-"1"-program-states. The contributions from 19 % defective cells were eliminated. Defective cells increased from the initial 12.5 to 19 % as we repeated the measurements. For the V_{th} distributions of the erase states, the V_{th} average ($V_{the,av}$) and the standard deviation (σ_e) were 1.349 and 0.076 V, respectively. In the V_{th} distributions of the all-"1"-programm states, the V_{th} average ($V_{thA1P,av}$) and the standard deviation (σ_{A1P}) were 0.860 and 0.048 V, respectively. The difference between the two average V_{th}s was $V_{the,av} - V_{thA1P,av} = 0.489$ V. We also obtained a value for statistical estimation: $(V_{the,av} - V_{thA1P,av})/(\sigma_e + \sigma_{A1P}) = 3.9$. This estimated value is practically convenient for guessing how large scale integration is possible.

Figure 13.19a shows the program-pulse-height V_{pgm} dependence of the average V_{th}s. The V_{th}s were changeable with V_{pgm} because unsaturated polarization switching of the ferroelectric SBT was used in the FeFETs [3, 12]. The V_{th}s after the program decreased as V_{pgm} increased. The decrease rate was $\Delta V_{th}/\Delta V_{pgm} = -0.1$ for all program-pulse width of $t_{pgm} = 10$, 100 μs, and 1 ms. Figure 13.19b shows the program-pulse-width t_{pgm} dependence of average V_{th}s. The V_{th} after the program decreased with t_{pgm} increasing. The decrease rate was $\Delta V_{th}/\Delta \log(t_{pgm}) = -0.1$ V/decade with $V_{pgm} = 7.5$ V. For the measurement shown

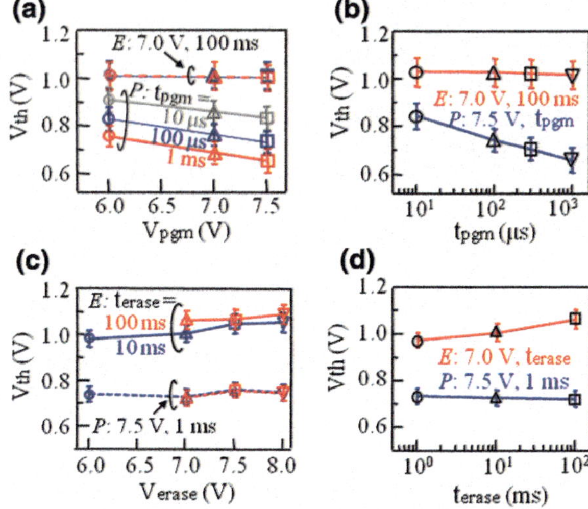

Fig. 13.19 V_{th}-distribution transitions depending on **a** program-pulse height, **b** program-pulse width, **c** erase-pulse height, and **d** erase-pulse width. 56 memory cells were measured in 7 WLs × 8-BL set in a block. A pair of identical symbols in every figure indicated a series of operations that were read after erase and read after program. The *symbols* indicate average V_{th}s and *error bars* are doubled standard deviations of the V_{th} distributions. Modified from [30]

in Fig. 13.19a, b, the erase conditions were fixed at V_{erase} = 7.0 V and t_{erase} = 100 ms. These indicate that the read margins can be increased by increasing V_{pgm} and/or t_{pgm}. On the other hand, the read margins may not be affected very much by erase-pulse conditions. As indicated in Fig. 13.19c, d, the V_{th} varied only slightly while increasing V_{erase} and t_{erase} at the common program conditions of V_{pgm} = 7.5 V and t_{pgm} = 1 ms.

The entire 64 kb (exactly 65,536) memory cells were investigated in both A and B arrays. By performing the erase, read, all-"1"-program, and read operations in all the 32 blocks, I_{re} versus V_{rbl} read curves of the memory cells were gathered, and V_{th}s were obtained from the curves. The V_{th} distribution of array A is shown in Fig. 13.20a and that of array B is in Fig. 13.20b, respectively. The erase operations were done on a pulse condition of V_{erase} = 7 V and t_{erase} = 100 ms. The all-"1"-program operations were done by V_{pgm} = 7.5 V and t_{pgm} = 1 ms. The read operations were performed by V_{rbl} = 0.3 V, V_{rsg} = 2.5 V, V_{rp} = 2.0 V, and $V_{re,mx}$ = 2.0 V with I_{ref} = 10^{-7} A for array A. Those for array B were performed by V_{rbl} = 0.2 V, V_{rsg} = 2.5 V, V_{rp} = 2.0 V, and $V_{re,mx}$ = 2.0 V with I_{ref} = 10^{-8} A. The eliminated-defective-cell rates were 3 % in array A (Fig. 13.20a) and 29.6 % in array B (Fig. 13.20b). The high defective-cell rate in the array B was due to several defective blocks in addition to one defective WL and several defective BL sets. The broad V_{th} distributions in array A (Fig. 13.20a) were characterized as $V_{the,av}$ = 1.344 V, σ_e = 0.165 V, $V_{thA1P,av}$ = 0.875 V, and σ_{A1P} = 0.124 V. The relatively narrow V_{th} distributions in array B (Fig. 13.20b) were characterized as $V_{the,av}$ = 1.315 V, σ_e = 0.098 V, $V_{thA1P,av}$ = 0.859 V, and σ_{A1P} = 0.068 V. The difference between the two averaged V_{th}s in array B was evaluated as $V_{the,av} - V_{thA1P,av}$ = 0.45 V and the value for statistical estimation was $(V_{the,av} - V_{thA1P,av})/(\sigma_e + \sigma_{A1P})$ = 2.7.

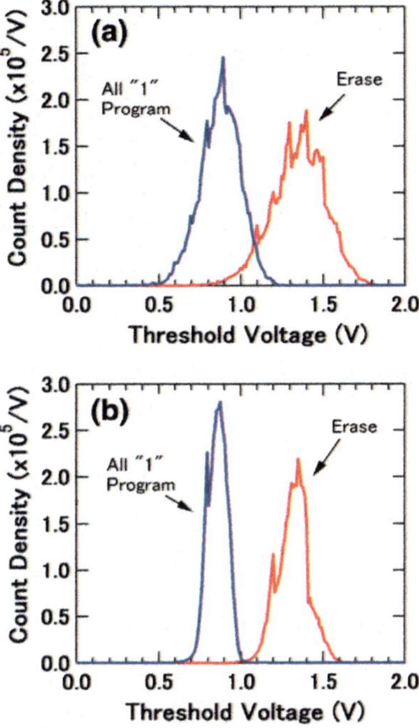

Fig. 13.20 V_{th} histograms of 64 kb Fe-NAND in **a** array A and **b** array B after erase and all-"1"-program operations [30]. Copyright (2012) The Japan Society of Applied Physics

Data retentions of a block of 2 kb in Fe-NAND-flash-memory array B were investigated by nondestructive read operations. First after Erase, the retentions (i.e., V_{th} variation with time) were measured for 2 days. Then, all-"1"-program operation was done and the retention measurement was started and continued for 2 days. No WL hold voltages were given to the memory cells during the retention-measurement time except during the reading periods. The contributions from 19 % defective cells were eliminated from counting. The erase pulse conditions were $V_{erase} = 7$ V and $t_{erase} = 100$ ms. The all-"1"-program conditions were $V_{pgm} = 7.5$ V and $t_{pgm} = 1$ ms. Figure 13.21a shows the V_{th} distributions measured after 1 min, 1 day, and 2 days. One finds good separations between the two distributions for erase- and all-"1"-programm cells. Figure 13.21b shows the time dependences of the average V_{th}s with error bars of doubled standard deviations. The distributions did not change significantly after 2 days. A potentially very long retention (10 years) could be suggested by the extrapolation lines shown in Fig. 13.21b. Note that the retention characteristics of 2 kb memory cells shown in Fig. 13.21a, b included read-disturb effects, indicating that the read operation did not significantly degrade the retention.

Finally, let us make programs with a "1" and "0" checkerboard pattern (Fig. 13.22a). The checkerboard pattern includes program disturbs fairly, that are phenomena in the Fe-NAND case such that V_{th} of the "0" states is decreased [25]. If the disturb effect is strong, the separation of V_{th} distributions between the "1" and "0" states gets small. Using the SCSB program scheme described in Sects. 13.3.4

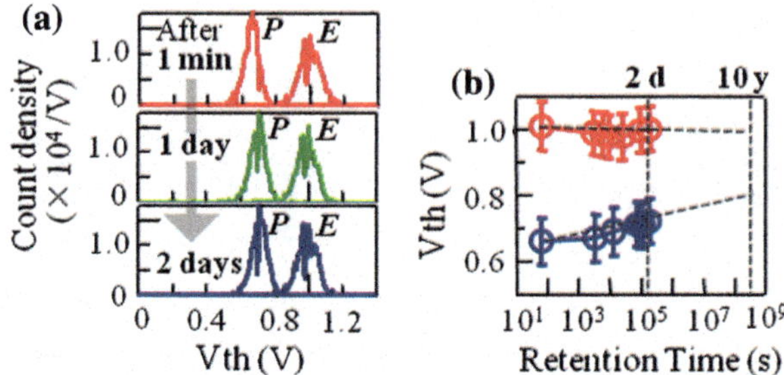

Fig. 13.21 One block V_{th} retention up to 2 days in array *B*. **a** V_{th} distributions are shown for read operations performed 1 min (*top*), 1 day (*middle*), and 2 days (*bottom*) after erase and all-"1"-program operations. **b** V_{th} variation with time using the averages (*symbols*) and doubled standard deviations (*bars*). Modified from [30]

and 13.4.2, the checkerboard pattern was programmed in a block of 2 kb in array *B*. We used the conditions of $V_{\text{erase}} = 7.0$ V and $t_{\text{erase}} = 100$ ms at Erase, and those of $V_{\text{rbl}} = 0.2$ V, $V_{\text{rsg}} = 2.5$ V, $V_{\text{rp}} = 2.0$ V, and $V_{\text{re,mx}} = 2.0$ with $I_{\text{ref}} = 3 \times 10^{-8}$ A at Read. At Program the conditions used were $V_{\text{pgm}} = 7.5$ V, $t_{\text{pgm}} = 10$ μs, $V_{\text{pass}} = V_{\text{sgd}} = 3$ V, $V_{\text{cut}} = 2.1$ V, $t_{\text{sgd}} = t_{\text{pass}} = t_{\text{cut}} = 100$ μs, and $V_{\text{pibl}} = 1.0$ V. A very small voltage of $V_{\text{pibl}} = 1.0$ V could achieved using the SCSB program scheme. As a result of applying $V_{\text{pibl}} = 1.0$ V, a small supply voltage V_{cc} of 1 V can be used, which realizes small power consumption of Fe-NAND. Reference [27] indicated that charge-pump circuits in the Fe-NAND using $V_{\text{cc}} = 1$ V will consume much a lower power than those in the conventional FG-NAND. Therefore, the Fe-NAND has a clear advantage in lower-power dissipation. The chosen program-pulse width $t_{\text{pgm}} = 10$ μs was a value that was used in the FG-NAND operations [31]. Three distinguishable V_{th}-distributions of erased (E), half-"1"-programmed (P1) and the-other-half-"0"-programmed (P0) cells were obtained, as shown in Fig. 13.22b. The excluded-defective-cell rate was 19 %. The V_{th} average and standard deviation after Erase were $V_{\text{the,av}} = 1.211$ V and $\sigma_{\text{e}} = 0.066$ V. After Program of the checkerboard pattern, the V_{th} average and standard deviation of P1 were $V_{\text{thP1,av}} = 0.947$ V and $\sigma_{\text{P1}} = 0.042$ V, and those of P0 were $V_{\text{thP0,av}} = 1.098$ V and $\sigma_{\text{P0}} \approx 0.062$ V. Hence, the V_{th} difference in average between P1- and P0-states was $V_{\text{thP0,av}} - V_{\text{thP1,av}} = 0.15$ V. The read margin (i.e., the difference between the two average V_{th}s) was smaller than the value of 0.33 V achieved using the 4 × 2 miniature array in Sect. 13.3.4. The small P0–P1 separation in this case was probably because insufficient gate voltages were applied to the selected cells for the "1" program due to voltage drops by V_{th} after passing through transfer-gate FeFETs in the block-selector circuits. In the 64 kb Fe-NAND-flash-memory array n-channel FeFETs with positive V_{th}s were used as the transfer gates, while the 4 × 2 miniature Fe-NAND array in TEG did not include any transfer gates. The V_{th}-drop problem will be improved by using another type of a transfer gate that is composed of a pair of n- and p-channel FeFETs. Using the transfer gates without the V_{th} drops will result in the widening of the read margin of the Fe-NAND.

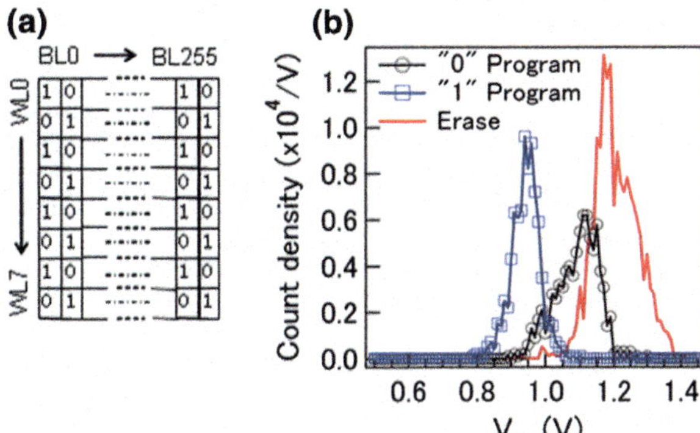

Fig. 13.22 **a** Schematic checkerboard program pattern. **b** V_{th} distribution histograms after erase operation and after "1" and "0" checkerboard program operation. The *left* and *middle* histograms are those of the "1" and "0" states after the program, respectively. The *right* histogram is that after the erase. Modified from [30]

Another method for widening the read margin is to use a verify-read technique during Program which can make narrow V_{th} distributions. As we discussed on the basis of Fig. 13.19a–d, the read margins will be controlled by increasing V_{pgm} and/or t_{pgm}. The results in Fig. 13.19 indicate that the read-verify technique which is commonly used in FG-NAND [35, 36] can also be used in Fe-NAND.

13.5 Conclusion

The 64 kb Fe-NAND-flash-memory array was integrated. The memory cells consisted of n-channel FeFETs. The BL- and block-selector logic circuits were constructed by n- and p-channel FeFETs. All the FeFETs were of Pt/SrBi$_2$Ta$_2$O$_9$/Hf–Al–O/Si metal-ferroelectric-insulator-semiconductor type. The states after Erase and Program could be nondestructively read. The accesses to all memory cells in all the 32 blocks were done. The reading of the memory cells showed a clear separation between the V_{th} distributions of the erased- and all "1"-programmed-states. V_{th} data retentions of all-programmed and all-erased cells in a block of 2 kb showed a clear separation for at least 2 days with no hold voltages. A single-cell self-boost program (SCSB) scheme was introduced and very low program-inhibit-bit-line voltage ($V_{pibl} \leq 1.0$ V) was achieved. Using the SCSB scheme, a "1" and "0" checkerboard pattern was programmed in a block of 2 kb, which showed three V_{th} distributions of all-erase, half-"0", and the-other-half-"1". The achieved small bit-line voltages (V_{pibl}) with small program voltages of 7.5 V indicated that the Fe-NAND has a great potential as a semiconductor silicon-based storage with a feature of very lower power dissipation.

Acknowledgments This work was partially supported by New Energy and Industrial Technology Development Organization.

References

1. Y. Tarui, T. Hirai, K. Teramoto, H. Koike, K. Nagashima, Appl. Surf. Sci. **113**, 656 (1997)
2. J.F. Scott, *Ferroelectric Memories* (Springer, Berlin, 2000) Chap. 12, p. 175
3. S. Sakai, R. Ilangovan, IEEE Electron Device Lett. **25**, 369 (2004)
4. M. Okuyama, Y. Ishibashi (eds.), *Ferroelectric Thin Films—Basic Properties and Device Physics for Memory Applications* (Springer, 2005) Part 4, p. 219
5. S. Sakai, Adv. Sci, Technol. **45**, 2382 (2006)
6. H. Ishiwara, Curr. Appl. Phys. **9**, S2 (2009)
7. S. Sakai, M. Takahashi, Materials **3**, 4950 (2010)
8. T. Hatanaka, R. Yajima, T. Horiuchi, S. Wang, X. Zhang, M. Takahashi, S. Sakai, K. Takeuchi, IEEE J. Solid-State Circ. **45**, 2156 (2010)
9. T.S. Böscke, J. Müller, D. Bräuhaus, U. Schröder, U. Böttger, IEDM Tech. Dig., 547 (2011)
10. S. Sakai, X. Zhang, L.V. Hai, W. Zhang, M. Takahashi, in *Proceedings of the 12th IEEE Annual Non-Volatile Memory Technology Symposium*, 2012, p. 55
11. S. Sakai, US Patent 7,226,795 (2005)
12. S. Sakai, R. Ilangovan, M. Takahashi, Jpn. J. Appl. Phys. **43**, 7876 (2004)
13. S. Sakai, M. Takahashi, R. Ilangovan, IEDM Tech. Dig., 915 (2004)
14. M. Takahashi, S. Sakai, Jpn. J. Appl. Phys. **44**, L800 (2005)
15. Q.-H. Li, S. Sakai, Appl. Phys. Lett. **89**, 222910 (2006)
16. T. Horiuchi, M. Takahashi, Q.-H. Li, S. Wang, S. Sakai, Semicond. Sci. Technol. **25**, 055005 (2010)
17. Q.-H. Li, M. Takahashi, T. Horiuchi, S. Wang, S. Sakai, Semicond. Sci. Technol. **23**, 045011 (2008)
18. Q.-H. Li, T. Horiuchi, S. Wang, M. Takahashi, S. Sakai, Semicond. Sci. Technol. **24**, 025012 (2009)
19. L.V. Hai, M. Takahashi, S. Sakai, Semicond. Sci. Technol. **25**, 115013 (2010)
20. L. V. Hai, M. Takahashi, S. Sakai, in *Proceedings of 3rd IEEE International Memory Workshop*, 2011, p.175
21. L. V. Hai, M. Takahashi, W. Zhang, S. Sakai: Semicond. Sci. Technol. **30**, 015024 (2015)
22. L. V. Hai, M. Takahashi, W. Zhang, S. Sakai: Jpn. J. Appl. Phys. **54**, 088004 (2015)
23. M. Takahashi, T. Horiuchi, Q.-H. Li, S. Wang, K.-Y. Yun, S. Sakai: Electron. Lett. **44**, 467 (2008)
24. M. Takahashi, S. Wang, T. Horiuchi, S. Sakai: IEICE Electron. Express **6**, 831 (2009)
25. S. Sakai, M. Takahashi, K. Takeuchi, Q.-H. Li, T. Horiuchi, S. Wang, K.-Y. Yun, M. Takamiya, T. Sakurai, in *Proceedings of 23rd IEEE Non-Volatile Semiconductor Memory Workshop: 3rd International Conference on Memory Technology and Design*, 2008, p. 103
26. S. Wang, M. Takahashi, Q.-H. Li, K. Takeuchi, S. Sakai, Semicond. Sci. Technol. **24**, 105029 (2009)
27. K. Miyaji, S. Noda, T. Hatanaka, M. Takahashi, S. Sakai, K. Takeuchi, Solid-State Electron. **58**, 34 (2011)
28. X.-Z. Zhang, K. Miyaji, M. Takahashi, K. Takeuchi, S. Sakai, in *Proceedings of 3rd IEEE International Memory Workshop*, 2011, p. 155
29. X. Zhang, M. Takahashi, S. Sakai, Integr. Ferroelectr. **132**, 114 (2012)
30. X. Zhang, M. Takahashi, K. Takeuchi, S. Sakai, Jpn. J. Appl. Phys. **51**, 04DD01 (2012)
31. K. Imamiya, H. Nakamura, T. Himeno, T. Yamamura, T. Ikehashi, K. Takeuchi, K. Kanda, K. Hosono, T. Futatsuyama, K. Kawai, R. Shirota, N. Arai, F. Arai, K. Hatakeyama, H. Hazama, M. Saito, H. Meguro, K. Conley, K. Quader, J.J. Chen, IEEE J. Solid-State Circuits **37**, 1493 (2002)
32. *International Technology Roadmap for Semiconductors 2007 Edition*. Process integration, devices, and structures, 2007, p. 36 (Table PIDS5a)
33. S. Wang, M. Takahashi, S. Sakai, unpublished
34. X. Zhang, M. Takahashi, S. Sakai, unpublished

35. T. Tanaka, Y. Tanaka, H. Nakamura, K. Sakui, H. Oodaira, R. Shirota, K. Ohuchi, F. Masuoka, H. Hara, IEEE J. Solid-State Circuits **29**, 1366 (1994)
36. M. Momodomi, T. Tanaka, Y. Iwata, Y. Tanaka, H. Oodaira, Y. Itoh, R. Shirota, K. Ohuchi, F. Masuoka, IEEE J. Solid-State Circuits **26**, 492 (1991)

Chapter 14
Novel Applications of Antiferroelectrics and Relaxor Ferroelectrics: A Material's Point of View

Min Hyuk Park and Cheol Seong Hwang

Abstract In this chapter, various applications based on pyroelectricity of antiferroelectrics and relaxor ferroelectrics are presented from the material's point of view. While the pyroelectricity of the conventional antiferroelectric materials, such as $PbZrO_3$, has been reported from the 1960s, the pyroelectricity of antiferroelectric $Hf_xZr_{1-x}O_2$ (x = 0.1–0.4) thin films was first reported in 2014. The antiferroelectric $Hf_xZr_{1-x}O_2$ (x = 0.1–0.4) thin films are believed to be highly promising for various applications, including electrostatic energy storage, electrocaloric cooling, pyroelectric energy harvesting, and infrared sensing. The theoretical background and basic operation principles of these energy-related applications will be briefly presented in the introduction. In this chapter, the material properties of only these new antiferroelectric thin films are dealt with because those of the conventional materials have been reported extensively elsewhere. In Sects. 14.2–14.5, the performance of $Hf_xZr_{1-x}O_2$ thin films in the aforementioned applications will be presented, and the performance of $Hf_xZr_{1-x}O_2$ films will also be compared with that of the conventional antiferroelectric and relaxor ferroelectric materials. In Sect. 14.6, the perspectives of antiferroelectric HfO_2-based films in the aforementioned applications will be presented, with focus on the recently suggested future multifunctional monolithic device.

14.1 Introduction

In this book, most of the chapters focus on the operation principles and applications of ferroelectric (FE) field effect transistors (FeFETs), which utilizes two remanent polarization states ($\pm P_r$) of FE materials. There is another type of polarization-ordered state, which is stable under certain bias conditions, whereas the

M.H. Park · C.S. Hwang (✉)
Department of Materials Science and Engineering and Inter-University Semiconductor
Research Center, College of Engineering, Seoul National University, Seoul 151-744,
Republic of Korea
e-mail: cheolsh@snu.ac.kr

© Springer Science+Business Media Dordrecht 2016

B.-E. Park et al. (eds.), *Ferroelectric-Gate Field Effect Transistor Memories,*
Topics in Applied Physics 131, DOI 10.1007/978-94-024-0841-6_14

alignment of the polarization vector inside the material becomes anti-parallel under the zero-bias condition. The materials that have these properties are named anti-ferroelectric (AFE) materials. The phenomenological treatment for AFE materials was first proposed by Kittel in the early 1950s [1]. Unlike FE materials, AFE materials are not appropriate for non-volatile memory applications due to the absence of macroscopic remanent bistable polarizations owing to the absence of a bias field. In the polarization-electric field (P-E) curves of AFE materials, the characteristic double hysteretic loops resulting from the field-induced reversible transitions between the FE and AFE phases are generally observed [1]. The other polar material class is the relaxor ferroelectrics (in short, relaxors), which was more recently found compared to the aforementioned FE and AFE materials. Owing to the relatively short history of the research on them, the fundamental physics of relaxors have not yet been clearly elucidated. Burns et al. suggested that the electrical properties of relaxors might come from the polar nano regions (PNRs) with randomly oriented polarizations when the electric field is absent [2]. When a sufficiently high electric field is applied, however, the NPRs whose polarization is oriented parallel with the applying electric field selectively grow [2]. When the electric field is reduced, on the other hand, the original configuration of PNRs with randomly oriented polarization states is recovered [2]. Xu et al. [3] first experimentally probed how the PNRs are present in the FE phase in $Pb(Zn_{1/3}Nb_{2/3})O_3$. These PNRs are regarded as embryos of the FE phase and are widely believed to be essential in the relaxor properties [3]. Various arguments on the fundamental physics of relaxors can be found in other review papers [4–7] and thus will not be the main focus of this chapter.

AFE materials and relaxors can be used for various applications, such as for electrostatic energy storage (EES) [8–13], pyroelectric energy harvesting (PEH) [8, 13–20], solid-state cooling based on the electrocaloric effect (ECE) [13, 20–29], and infrared (IR) sensing for thermal imaging [30, 31]. These applications are based on the coupling of the thermal and electric properties of polar materials, and the coupling property is called pyroelectricity. Pyroelectricity was first observed by the Greek philosopher Theophrastus about 24 centuries ago [32]. Pyroelectricity should be distinguished from the thermoelectric effect, which is another form of coupling of the thermal and electric properties. The pyroelectricity originates from the change in the electrical polarization states with time-dependent variations in temperature whereas the thermoelectric effect refers to the conduction of electric current with the spatial gradient of temperature. For the aformentioned applications based on pyroelectricity, AFE materials and relaxors with a large transition electric field, a large breakdown field, and a large pyroelectric constant are preferred. The Zr-rich $Pb(Zr,Ti)O_3$(PZT)-based inorganic materials and PVDF-based polymers have been the most promising candidates [12, 13]. The use of PZT films is limited in many countries, however, due to the environmental concerns associated with them, and PVDF-based polymers are generally not reliable at high temperatures due to their low melting point ($\sim$443 K) [12, 13, 20, 22, 23]. Therefore, developing new lead-free AFE materials with a large transition field, a large breakdown field, and robust thermal stability is an urgent task in this field. Park et al. [12, 13, 20] first examined the various energy-related applications of the AFE Zr-rich $Hf_xZr_{1-x}O_2$(HZO) films and showed that they can be another promising candidate for these applications. They also reviewed the researches on the FE and AFE properties of doped HfO_2-based films in their recent review paper based on their collaboration with German researchers [13]. In this chapter, the various applications of AFE and relaxor materials are presented from the material's point of view. Especially, the interesting material properties of AFE doped HfO_2-based thin films are focused on, and they will be compared with the conventional AFE or relaxor materials. The ferroelectricity in HfO_2-based films, and their non-volatile memory application, were studied by Schroeder et al. and will be discussed in another chapter.

Figure 14.1a shows a schematic thermodynamic diagram of the reversible interactions that may occur among the thermal and electrical variables of polar materials [20]. First, the AFE films can work as dielectric materials, which can store a large amount of charges and electrostatic energy [12]. In fact, pure (non-hysteretic) HfO_2 or ZrO_2 films have been studied and used as a high-charge capacitive layer in the metal oxide semiconductor field effect transistor (MOSFET) and dynamic random access memory (DRAM), but the AFE HfO_2-based films are believed to be inappropriate for these applications due to the large hysteretic loss involved [33]. It should be noted that pure ZrO_2 films can also be AFE-based, but this was observed only very recently probably owing to the large electric field needed for the AFE-FE phase transition [11, 34]. In contrast, the large magnitude of the electric field for FE-to-AFE phase transition and its inverse, large maximum polarization, and the small P_r of the AFE HfO_2-based films are appropriate for EES applications [12]. Figure 14.1b shows the polarization-electric field (P-E) hysteresis curve, which depicts the energy storage behavior of AFE materials [12]. The blue area refers to the recoverable part of the stored energy, which is called energy storage density (ESD), and the green area refers to the unrecoverable loss due to the polarization switching. The ESD can be calculated using (14.1).

$$ESD = \int_0^{P_{max}} EdP, \tag{14.1}$$

where P_{max} refers to the maximum polarization value in the P-E hysteresis curve. As shown in Fig. 14.1b, the efficiency of energy storage can be calculated from the relative ratio of ESD to the summation of ESD and hysteretic loss, where hysteretic loss refers to the internal area of P-E hysteresis.

The pyroelectricity and ECE of the AFE HfO_2-based films that use the coupling of the thermal and electrical properties can be used for various applications, such as ECE coolers, PEH devices, and IR sensors for thermal imaging [13, 20]. It should be noted that ECE is the physical inverse of pyroelectricity. ECE generally uses the isothermal entropy change (ΔS) and/or adiabatic temperature change (ΔT) from the poling/depoling of polar materials by applying/withdrawing an electric field to them [21, 22]. In fact, the direct measurement of configurational entropy is a difficult task

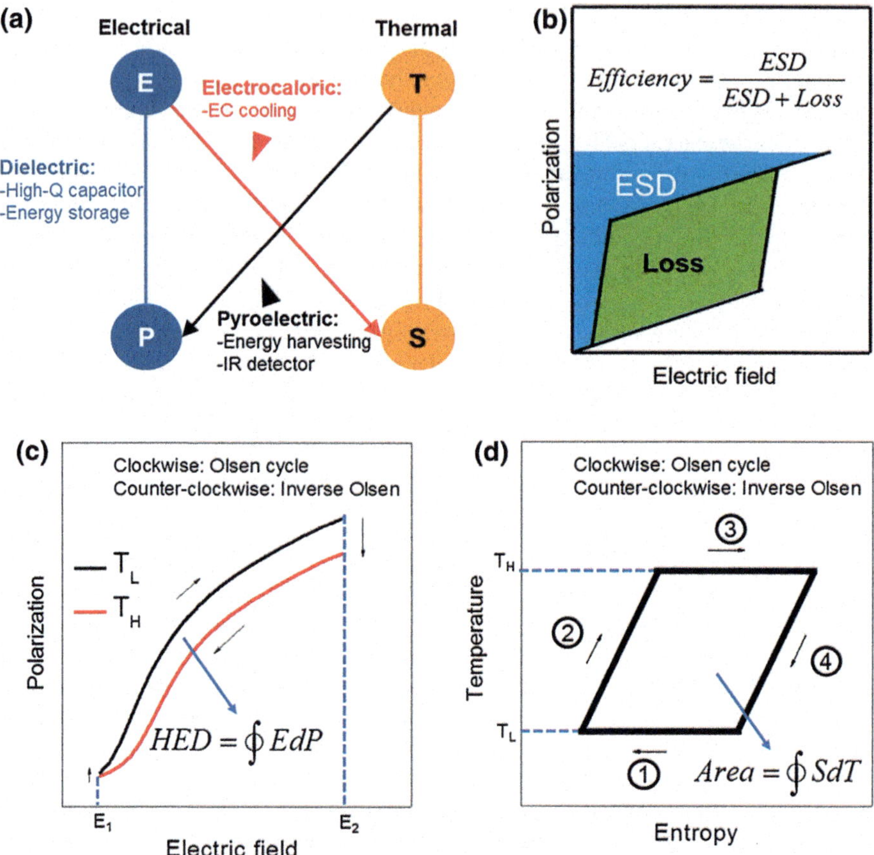

Fig. 14.1 **a** Schematic diagram showing the coupling of the electrical and thermal properties. Reproduced with permission [20]. Copyright 2015, Elsevier. **b** Schematic diagram for the calculation of the energy storage density and loss from the polarization-electric field hysteresis. Reproduced with permission [12]. Copyright 2015, John Wiley & Sons, Inc. **c** Schematic diagrams showing the Olsen cycle for energy harvesting and electrocaloric cooling in temperature-dependent polarization-electric field curves. **d** Schematic diagram showing the Olsen cycle in the temperature-entropy plane. Panels (**c**, **d**) were reproduced with permission [20]. Copyright 2015, Elsevier

to accomplish; thus, another indirect methodology is used to evaluate the entropy change [21]. The change in the polarization state of polar materials is strongly correlated with configurational entropy by the well-known Maxwell's relation, $(\partial P/\partial T)_E = (\partial S/\partial E)_T$. Based on this relation, the field-induced change in entropy can be calculated based on the temperature-dependent change in polarization. This indirect methodology is the most frequently used methodology in studies on ECE. The direct measurement of ΔT was also used in some studies, but it could be applied only to bulk or thick films [35–37].

For actual cooling, various thermodynamic cycles, such as the Carnot, Ericsson, and Olsen cycles, can be used, and among these, the Olsen cycle is discussed in detail in this chapter for the direct comparison of ECE to the PEH [14, 18]. As can be seen in Fig. 14.1c, the Olsen cycle for ECE consists of the following four steps (anti-clockwise): (1) poling of the pyroelectric material by increasing the electric field ($E_1 \rightarrow E_2$) at a low T (T_L); (2) partly depoling it by increasing T ($T_L \rightarrow T_H$); (3) depoling it by decreasing the electric field ($E_2 \rightarrow E_1$) at T_H; and (4) partly poling it by cooling it to the initial T_L. As previously mentioned, the degree of polarization of the polar materials is directly related to their entropy (S), and as such, the poling/depoling of a pyroelectric material means the extraction/absorption of the heat or entropy [14, 20, 38]. Figure 14.1d shows the schematic Olsen cycle for ECE in the T-S diagram [20]. In this case, the pyroelectric material releases the entropy at a high temperature (T_H) and absorbs it at a low temperature (T_L) [20]. As the magnitude of heat in an isothermal reversible process can be calculated from the product of the ΔS and T, the released heat ($T_H \Delta S$) is larger than the absorbed heat ($T_L \Delta S$) in the case of the single Olsen cycle for ECE [20]. Meanwhile, the Olsen cycle can also be used for the PEH (clockwise in Fig. 14.1c, d) [20]. In the Olsen cycle for PEH, the pyroelectric materials absorb heat at T_H and release it at T_L, and as such, they can transform the absorbed heat into electrical energy.

14.2 Electrostatic Supercapacitors

With the development and modification of new technologies for generating and handling energy, high-power electrical-energy-storing devices, including electro-chemical and electrostatic capacitors, are becoming more important for advanced energy storage systems [39–41]. It should be noted that the term "supercapacitors" generally refers to electrochemical supercapacitors, and, thus, the term "electrostatic supercapacitors" will be used in this chapter to distinguish their pure electronic charging/discharging mechanism from that of electrochemical supercapacitors. The energy storage per unit mass of the electrochemical capacitors is generally higher than that of the electrostatic capacitors by $\sim$2–3 orders of magnitude. On the contrary, the achievable power density of the electrochemical capacitors is smaller than that of the electrostatic capacitors by $\sim$3–5 orders of magnitude due to the slow charging and discharging mechanism based on the electrochemical reactions [39–41]. Thus, the electrostatic supercapacitors are considered appropriate for applications that require high power delivery and uptake owing to their high power density (up to 10^8 W kg^{-1}), which results from the extremely high charging and discharging speeds [41]. Haspert et al. [40] recently demonstrated that the hybrid energy storage circuits of heterogeneous devices can exploit the extremely high power of electrostatic devices, which is comparable to that of electrochemical capacitors. Moreover, it was reported that the capacitance per projected area and the resulting energy storage of electrostatic capacitors can be increased by the use of patterned nanostructures such as Si nanotrenches, anodized aluminum oxide (AAO), and self-rolling metallic structures [42–44].

The concurrent high-power capacitors are made mainly of linear dielectric polymers with an ESD of ~ 1–2 J cm^{-3} [41]. Even though the ε_r values of linear dielectric polymers are generally small (~ 2–5), their electric breakdown fields are quite high, allowing the application of high electric fields, which result in a relatively high ESD [41]. Nonetheless, the ESD values of dielectric polymers are still insufficient, and as such, many other materials have been sought to replace them. Most of the previous researches on this topic focused on AFE PZT-based materials because their ESD values are as large as ~ 1 and ~ 10–20 J cm^{-3} for bulk and thin films, respectively [41]. In a number of studies, ESD values as high as ~ 50 J cm^{-3} were reported for thin PZT-based films when a high electric field of ~ 3.5 MV cm^{-1} was applied [45, 46]. An electric field as high as that, however, might induce a significant reliability concern for PZT films, which usually have a relatively low breakdown field [47]. Moreover, the commercial use of lead-containing materials is restricted in many countries due to their environmental impact. Another serious problem of electrostatic PZT capacitors is the ~ 20–40 % decrease in their ESD when their operating temperature increases to 423 K [48, 49]. Other promising candidates are AFE PVDF-based materials. They can have a large breakdown field and maximum polarization value and are relatively easy to fabricate. In the case of AFE PVDF-based films, a maximum ESD of ~ 14 J cm^{-3} has been reported, with a higher efficiency of ~ 70 % [50]. Nevertheless, the remaining most serious problem of PVDF-based materials is thermal stability. Their relatively low melting point (~ 443 K) seriously restricts their practical application for energy storage [12, 13]. Therefore, new materials for energy storage with high ESD, robust thermal stability, and three-dimensional structure compatibility are highly required.

Park et al. [12] first reported the energy storage behavior of AFE Zr-rich HZO films in 2014. Figure 14.2a shows the variations in the ESD values of ~ 9.2-nm-thick HZO films as a function of the height of the triangular double pulses used for the P-E analysis [12]. The ESD values of the Zr-rich HZO films were all higher than 30 J cm^{-3}, and the ESDs were even higher compared to those of the intensively studied Pb- and Ba-based perovskite AFE materials or PVDF-based AFE polymers [12, 51]. The low thickness of AFE Zr-rich HZO films should be a crucial advantage for mass production. The ESD of the 9.2-nm-thick $Hf_{0.3}Zr_{0.7}O_2$ capacitor was the largest among the HZO films with various Hf contents, and the maximum ESD was 46 J cm^{-3} at the 4.5 MV cm^{-1} electric field [12]. The ESD and energy efficiency of the $Hf_{0.3}Zr_{0.7}O_2$ capacitor did not decrease with increasing temperature up to 448 K, which shows its robust thermal stability [12]. This property of the material makes it appropriate for practical application in a wide temperature range. Such promising properties also have high reliability; the large ESD decreased by only ~ 4.5 % even after 10^9-time field cycling at a repetitive pulse height of 3.26 MV cm^{-1} [12]. Furthermore, the HZO films could be generally well deposited using the thermal atomic layer deposition (ALD) technique

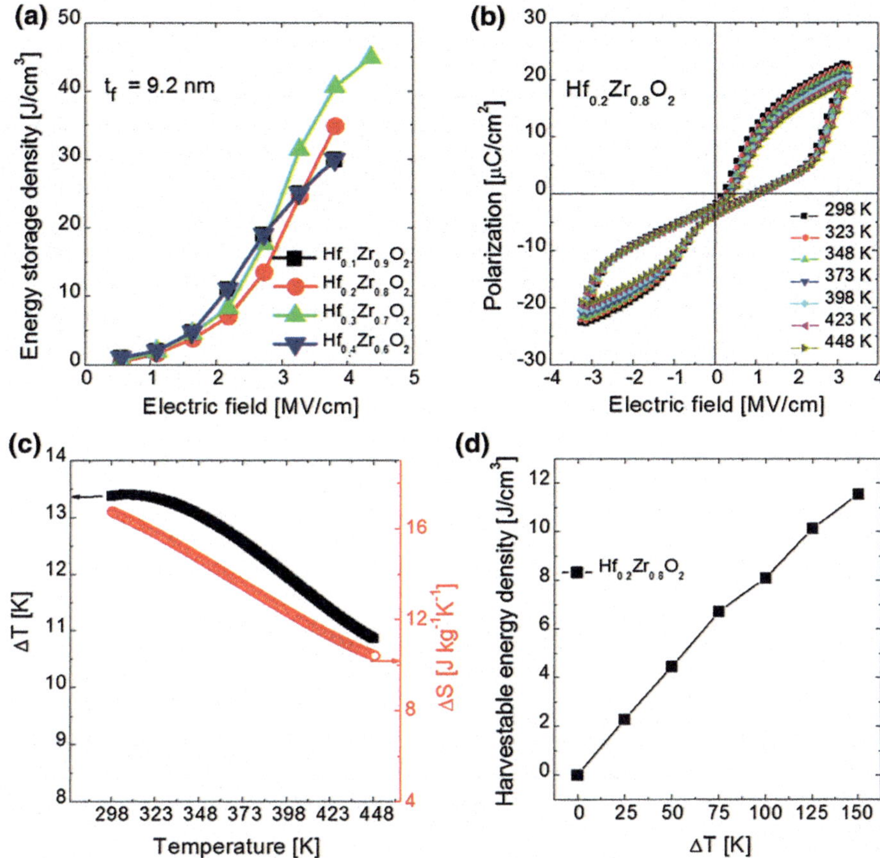

Fig. 14.2 **a** Variations in the energy storage densities of the $Hf_xZr_{1-x}O_2$ films with various compositions (x = 0.1–0.4) as a function of the electric field. Reproduced with permission [12]. Copyright 2014, John Wiley & Sons, Inc. **b** Polarization-electric field hysteresis curves of the ~9.2-nm-thick $Hf_{0.2}Zr_{0.8}O_2$ thin film sandwiched by TiN top and bottom electrodes at various temperatures (298–448 K). Reproduced with permission [20]. Copyright 2015, Elsevier. **c** Changes in ΔT (*black solid dots*) and ΔS (*red open dots*) as a function of temperature of the ~9.2-nm-thick $Hf_{0.2}Zr_{0.8}O_2$ film. **d** Change in the harvestable energy density calculated from the polarization-electric field curves of the ~9.2-nm-thick $Hf_{0.2}Zr_{0.8}O_2$ film. The data in panels (**c**, **d**) are reproduced with permission [13]. Copyright 2015, John Wiley & Sons, Inc.

at substrate temperatures ranging from 500 to 600 K [12]. This demonstrates the compatibility of the material and the process with nanostructured devices.

Until now, the energy storage performances of the AFE HfO_2-based films with dopants other than Zr have not been reported. The AFE Si- and Al-doped HfO_2 films, however, are also expected to be used as electrostatic supercapacitors with large ESD values. Especially, the smaller hysteresis loss of the AFE Si-doped HfO_2 film is expected to increase the efficiency of energy storage.

14.3 Electrocaloric Cooling

The development of a new eco-friendly refrigerating or cooling technology is one of the most urgent issues in the modern society. Especially, the development of a new solid-state cooling technology is required because of the increasing usage of electric devices with integrated computer chips. To resolve this problem, ECE has been intensively studied since the first report on the giant ECE of $PbZr_{0.95}Ti_{0.05}O_3$ films by Mischenko et al. [21]. Before their work, the ECE of bulk electrocaloric materials was too weak that such materials could not be seriously considered feasible solid-state cooling technologies. The cooling performance of an electrocaloric material has been indirectly estimated by calculating the ΔT and ΔS based on the well-known Maxwell's relation of $(\partial S/\partial E)_T = (\partial P/\partial T)_E$. The ΔT and ΔS can be calculated using (14.2) and (14.3), respectively [21, 22].

$$\Delta T = -\frac{1}{\rho C} \int_{E_1}^{E_2} T \left(\frac{\partial P}{\partial T}\right)_E dE, \tag{14.2}$$

$$\Delta S = -\frac{1}{\rho} \int_{E_1}^{E_2} \left(\frac{\partial P}{\partial T}\right)_E dE, \tag{14.3}$$

where ρ and C are the density and specific heat of the electrocaloric material, respectively. ECE has been intensively reviewed by many researchers, and a detailed theoretical background and experimental researches can be found in these review papers [52–55].

Park et al. [20] recently reported the electrocaloric behavior of 9.2-nm-thick HZO films. The ΔT and ΔS of $Hf_{0.2}Zr_{0.8}O_2$ films were the largest among the films with different Hf:Zr ratios. Figure 14.2b shows the P-E characteristics of the $Hf_{0.2}Zr_{0.8}O_2$ films at the temperature range of 298–448 K [20]. The polarization values generally decreased with increasing temperature whereas the electric field needed for the AFE-to-FE and FE-to-AFE transition decreased [20]. For the whole temperature range (298–448 K) of the measurements, the field-induced phase transition could be clearly observed [20]. Figure 14.2c shows the change in the ΔT and ΔS as a function of temperature [20], and the large ΔT (10.9–13.4 K) and ΔS (10.7–16.7 J kg^{-1}K^{-1}) values in a wide temperature range (298–448 K) could be confirmed [20]. For the conventional electrocaloric materials, on the other hand, such large ΔT and ΔS values could be observed only within a narrow temperature range near the Curie temperature [21–28]. The reason that $Hf_{0.2}Zr_{0.8}O_2$ films can have large ΔT and ΔS values in such a wide range of temperature has yet to be elucidated. It is believed, however, that the field-induced phase transition in a wide temperature range is strongly related with it because the phase transition generally occurs with a large change in configurational entropy [13, 20]. There should be a large amount of ΔS due to the field-induced phase transition, and the $(\partial P/\partial T)_E$

Table 14.1 Electrocaloric effect of various materials

	Material	T (K)	ΔT (K)	ΔE (kV cm^{-1})	ΔT/ΔE (K cm kV^{-1})	ΔS (J K^{-1} kg^{-1})
Lead-containing	$PbZr_{0.95}Ti_{0.05}O_2$ [21]	495	12	481	0.025	8
	$Pb_{0.8}Ba_{0.2}ZrO$ [24]	290	45.3	598	0.076	46.9
	$PbSc_{0.5}Ta_{0.5}O_3$ [26]	341	6.2	774	0.008	6.3
	$0.9PbMg_{1/3}Nb_{2/3}O_3-0.1PbTiO_3$ [27]	348	5	895	0.006	5.6
Lead-free	P(0.55VDF–0.45TrFE) [22, 23]	353	12.6	2090	0.006	60
	$SrBi_{2.1}Ta_2O_9$ [29]	500	4.9	600	0.008	N/A
	$Hf_{0.2}Zr_{0.8}O_2$ [20]	298	13.4	3260	0.004	16.7

Reproduced with permission from [20]. Copyright 2015, Elsevier

values were also the largest near the electric field where the phase transition occurs [11]. Table 14.1 summarizes the electrocaloric behavior of various materials [21–28]. It is believed that these materials could be made to have even larger ΔT and ΔS values by optimizing their film orientation, increasing the portion of the AFE phase, and applying different dopants to the HfO_2 films. Especially, the AFE Si- or Al-doped films are also believed to have a giant ECE with large ΔT and ΔS values.

14.4 Pyroelectric Energy Harvesting

The large amount of waste heat generated is a serious problem today [56]. Especially, electric mobile devices are accounting for an ever-growing portion of the waste heat generated, among the various sources of waste heat. Therefore, harvesting electric energy from the waste heat from electric devices should have a huge impact. PEH is one of the promising methods for this energy conversion due to its high potential efficiency [57]. Among the various mechanisms of PEH, Sebald et al. reported that the Olsen cycle was the most promising, with a large harvestable energy density (HED) per single cycle [38]. It should be noted that HED refers to the internal area of the Olsen cycle in Fig. 14.1c [20]. The details of the Olsen cycle were presented in the introduction and in the section on ECE, and it should be noted that PEH is the inverse cycle of ECE.

There have been many reports on the harvesting of energy from waste heat using the Olsen cycle. As the large pyroelectric constant, $(\partial P/\partial T)_E$, is beneficial for PEH, Pb-containing perovskite materials with a large pyroelectric coefficient have been studied to increase the HED using the Olsen cycle [20]. Sebald et al. [14] and Kandilian et al. [18] reported HED values of 0.186 and 0.100 J cm^{-3} cycle^{-1} for a $(PbMg_{1/3}Nb_{2/3}O_3)_{0.9}(PbTiO_3)_{0.1}$ ceramic and a $(PbMg_{1/3}Nb_{2/3}O_3)_{0.68}(PbTiO_3)_{0.32}$ single crystal, respectively. Lee et al. [19] reported that the HED of (Pb,La)

$(Zr_{0.65}Ti_{0.35})O_3$ ceramics can be as high as 1.014 J cm^{-3} cycle^{-1} due to the ergodic relaxor-FE phase transition. Vats et al. [15] reported that $(Bi_{0.5}Na_{0.5})_{0.915}$–$(Bi_{0.5}K_{0.5})_{0.05}Ba_{0.02}Sr_{0.015}TiO_3$ ceramic can have an HED of ~ 1.523 J cm^{-3} cycle^{-1}, which, to the authors' knowledge, is the highest energy density reported before the first report on AFE $Hf_{0.2}Zr_{0.8}O_2$ [20]. Its complicated chemical composition, however, might significantly restrict its practical applications. For the polymeric P(VDF-TrFE) thin films, Navid et al. [17] and Lee et al. [16] reported large HEDs of 0.521 and 0.155 J cm^{-3} cycle^{-1}, respectively. Even though FE polymers are cost-effective and appropriate for mass production, their relatively low melting point of 443 K is a critical barrier to further increasing their HED.

Furthermore, the HED needs to be improved even at a relatively lower temperature fluctuation frequency. Generally, the T fluctuation frequency (or the rate of the irregular temperature fluctuation pattern) of a waste heat source is too low (even lower than 1 Hz) for the general pyroelectric energy harvesters to harvest sufficient power density [58]. Nanoscale pyroelectric systems with thin pyroelectric films or nanostructures, however, might be the simplest and most powerful solutions to the aforementioned problem [58]. Yang et al. [59] experimentally proved that the pyroelectric nanogenerators can harvest thermal energy more effectively.

Consequently, the conditions needed for the eco-friendly and efficient PEH could be summarized as follows: (1) a large HED; (2) a lead-free nature; (3) a thin film or nanostructure form; (4) stability against T fluctuation; and (5) a high electrical breakdown field [20].

The PEH behavior of thin $Hf_{0.2}Zr_{0.8}O_2$ films was also examined by Park et al. [20]. Figure 14.2d shows the change in the HED of the $Hf_{0.2}Zr_{0.8}O_2$ film as a function of the operating temperature range used for the Olsen cycle [20]. The HED of the $Hf_{0.2}Zr_{0.8}O_2$ film was the largest among the films with various compositions, and its HEDs were 2.3, 4.4, 6.7, 8.1, 10.1, and 11.5 J cm^{-3} for the temperature spans of 25, 50, 75, 100, 125, and 150 K, respectively [20]. The maximum HED value of the $Hf_{0.2}Zr_{0.8}O_2$ film was ~ 7.6 times larger than the largest HED value reported by Vats et al. [15, 20]. Table 14.2 summarizes the PEH performances of various materials using the Olsen cycle [14–20]. Until now, only the PEH properties of HZO films have been examined, but it is believed that the AFE HfO_2-based films with other dopants, such as Si and Al, are also promising for PEH.

14.5 IR Sensing

The pyroelectric materials can also be used for the IR sensor, for thermal imaging [30, 31, 60]. The array-based IR sensors can be used for various applications, such as for people counting, health care, security, thermal imaging, and military application [31]. For this application, the development of low-cost materials with a high

Table 14.2 Pyroelectric energy harvesting properties of various materials

	Material	Form	HED (J cm^{-3} cycle^{-1})	ΔE (kV cm^{-1})	ΔT_{HL} (K)	Method
Lead-containing	$0.90PbMg_{1/3}Nb_{2/3}O_3$–$0.10PbTiO_3$ [14]	Ceramic	0.186	35	323	Direct
	$0.68PbMg_{1/3}Nb_{2/3}O_3$–$0.32PbTiO_3$ [18]	Single crystal	0.100	7	363	Direct
	$(Pb,La)(Zr_{0.65}Ti_{0.35})O_3$ [18]	Ceramic	1.014	68	443	Direct
Lead-free	P(VDF-TrFE) [17]	Thin film	0.521	300	358	Direct
	P(VDF-TrFE) [16]	Thin film	0.155	150	358	Direct
	$(Bi_{0.5}Na_{0.5})_{0.915}$–$(Bi_{0.5}K_{0.5})_{0.05}Ba_{0.02}Sr_{0.015}TiO_3$ [15]	Ceramic	1.523	39	413	Indirect
	$Hf_{0.2}Zr_{0.8}O_2$ [20]	Thin film	11.549	3260	423	Indirect

Reproduced with permission from [20]. Copyright 2015, Elsevier

figure of merit (F_v) is highly required. To achieve a low-cost IR sensor, materials compatible with the templates commonly used in the industry, such as Si substrates, are beneficial [30, 31]. It has been suggested that the AFE HfO_2-based film can be a promising candidate for this application, with its Si compatibility. For IR sensor application, the F_v can be formulated as (14.4).

$$F_v = \frac{p}{C \varepsilon_r \varepsilon_0},$$
(14.4)

where p, T, C, ε_r, and ε_0 are the pyroelectric coefficient, operation temperature, heat capacity, dielectric permittivity, and permittivity of vacuum, respectively [60]. The most promising candidates for this application are known to be $LiNbO_3$ and $LiTaO_3$, due to their moderate p and low ε_r, and the F_v values of $LiNbO_3$ and $LiTaO_3$ are 10.8 and 11.6 $\times$ 10^{-2} m^2C^{-1}, respectively, within a single crystal form [60]. Park et al. [20] reported that the F_v of the $Hf_{0.2}Zr_{0.8}O_2$ film could be as large as 32.0×10^{-2} m^2 C^{-1}, which is 2.8 times larger than that of $LiTaO_3$.

14.6 Perspectives

To summarize the energy-related applications of AFE materials, the large ESD, ΔT (or ΔS), HED, and F_v values of the AFE HZO-based films prove that such materials can be promising for the next-generation EES, electrocaloric cooling, PEH, and thermal imaging system [13, 20]. Even better properties are their lead-free, thermally stable, and highly Si-compatible nature with the mature deposition techniques, which make them more fascinating for practical application. Evidently, the EES, ECE, PEH, and IR sensing performances of the HZO films observed in the previous studies are still in the pioneering stage and could be improved further with various strategies, including controlling the film orientation, suppressing the formation of the second phase, and adopting various dopants.

Based on these promising properties and the three-dimensional compatibility of HZO-based thin films, Park et al. suggested a novel multifunctional monolithic device [20]. This monolithic device is based on the idea that the PEH, ECE, and EES can be used in any device that needs electrical energy, where the electrical energy can be produced from waste heat (such as heat from Si chips). Figure 14.3 shows the schematics of the HZO-based monolithic device for PEH, ECE, EES, a high-charge capacitor, and IR sensing integrated on Si substrates. The upper center panel shows the schematics of a PEH device based on microelectromechanical systems, which was originally designed by Oak Ridge National Laboratory [58, 61]. A cantilever is positioned between a hot surface and a cold surface. The resonant motion of the cantilever can be produced due to the difference in the

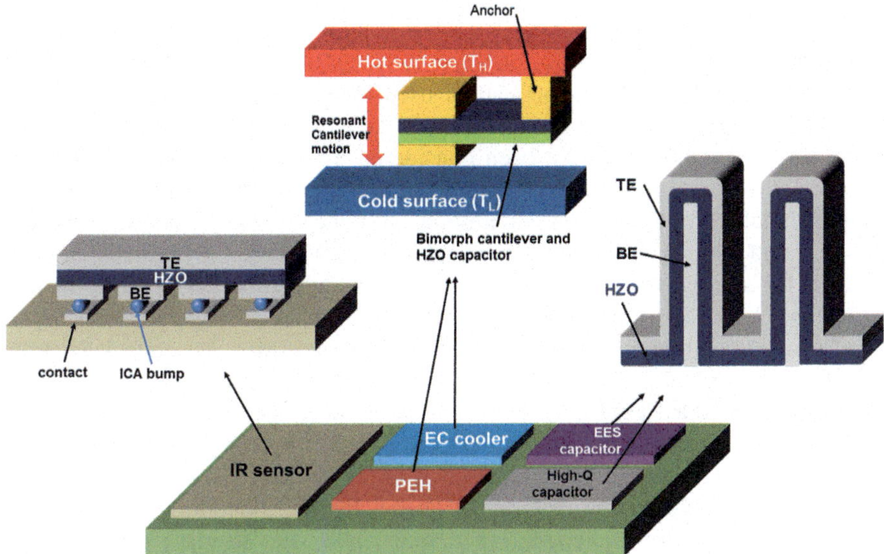

Fig. 14.3 Schematics of the proposed monolithic device with various devices for pyroelectric energy harvesting/electrocaloric cooling (*upper center panel*), high-charge capacitor/electrostatic energy storage (*upper right panel*), and IR sensing (*upper left panel*), respectively (*ICA* interconnect conducting adhesive). Reproduced with permission [20]. Copyright 2015, Elsevier

thermal expansion coefficient of the materials in the bimorph part. Moreover, the appropriate combination of the cycling of the electrical field with the resonant motion can use the PEH or the ECE with careful control. As previously mentioned, the PEH and the ECE are inverse processes of each other, and as such, they can be easily used in an equivalent structure by simply changing the sequence of the applying field, and changing the temperature. The upper right panel in Fig. 14.3 shows the structure of an electrostatic supercapacitor or a high-charge capacitor, which are based on a simple metal-insulator-metal capacitor structure with HZO-based films sandwiched by top and bottom metal electrodes. For the three-dimensional nanostructure using HZO, TiN might be the best electrode material, according to the previous studies [13], and the three-dimensional nanoscale structure was used to increase the ESD and the stored charges by increasing the area of the capacitors. The EES capacitors with an HZO dielectric layer might store the energy harvested by a PEH device using another HZO film with the appropriate external circuit. The upper left panel shows the schematics of an array-based pyroelectric IR sensor [31, 62]. It is believed that the highly Si-compatible HZO-based films are expected to make the IR sensors more appropriate for mass production. These functional devices can be produced on a single Si wafer and can be stacked with Si processors or memory chips as the sources of waste heat, or placed near the chips on the printed circuit board.

14.7 Conclusions

In conclusion, AFEs and relaxors can be used for various applications, including EES, ECE, PEH, and IR sensing, based on their pyroelectricity. The PZT and PVDF-based FE materials have been considered for the aforementioned applications owing to their performances, which are suitable for these applications. The compositional complexity and environmental concern for Pb-based materials, however, and the inappropriate thermal stability of PVDF-based materials, invokes a search for alternatives. Recently, Park et al. examined the performance of thin (~ 10 nm) HZO-based thin films and concluded that they are highly promising for the aforementioned applications [12, 13, 20]. Their performances, calculated from the double hysteretic P-E curves over a wide temperature range, showed the possibility that the doped HfO_2-based AFE films can overwhelm the performances of the conventional materials, which are too thick, difficult to fabricate, not thermally stable, and even not environmentally benign. These properties show that the new AFE materials could be the material of choice for the monolithic integration of computer chips with various passive components with energy storing, energy harvesting, solid-state cooling, and imaging functionalities in an environmentally benign manner.

Acknowledgments This work was supported by the Global Research Laboratory Program (2012K1A1A2040157) of the Ministry of Science, ICT, and Future Planning of the Republic of Korea, and by a National Research Foundation of Korea (NRF) grant funded by the South Korean government (MSIP) (2014R1A2A1A10052979).

References

1. C. Kittel, Phys. Rev. **82**, 729 (1951)
2. G. Burns, F.H. Dacol, Phys. Rev. B **28**, 2578 (1983)
3. G. Xu, Z. Zhong, Y. Bing, Z.-G. Ye, G. Shirane, Nat. Mater. **5**, 134 (2006)
4. L.E. Cross, Ferroelectrics **76**, 241 (1987)
5. R.A. Cowley, Adv. Phys. **60**, 229 (2011)
6. A.A. Bokov, Z.-G. Ye, J. Mater. Sci. **41**, 31 (2006)
7. S. Zhang, F. Li, X. Jiang, J. Kim, J. Luo, X. Geng, Prog. Mater Sci. **68**, 1 (2015)
8. X. Hao, J. Zhai, L.B. Kong, Z. Xu, Prog. Mater Sci. **63**, 1 (2014)
9. K. Yao, S. Chen, M. Rahimabady, M.S. Mirshekarloo, S. Yu, F.E.H. Tay, T. Sritharan, L. Lu, IEEE T. Ultrason. Ferr. **58**, 1968 (2011)
10. J. Li, S. Tan, S. Ding, H. Li, L. Yang, Z. Zhang, J. Mater. Chem. **22**, 23468 (2012)
11. S.E. Reyes-Lillo, K.F. Garrity, K.M. Rabe, arXiv:cond-mat/1403.3878 (2014)
12. M.H. Park, H.J. Kim, Y.J. Kim, T. Moon, K.D. Kim, C.S. Hwang, **4**, 1400610 (2014)
13. M.H. Park, Y.H. Lee, H.J. Kim, Y.J. Kim, T. Moon, K.D. Kim, J. Muller, A. Kersch, U. Schroeder, T. Mikolajick, C.S. Hwang, Adv. Mater. **27**, 1811 (2015)
14. G. Sebald, S. Pruvost, D. Guyomar, Smart Mater. Struct. **17**, 015012 (2008)
15. G. Vats, R. Vaish, C.R. Bowen, J. Appl. Phys. **115**, 013505 (2014)
16. F.Y. Lee, A. Navid, L. Pilon, Appl. Therm. Eng. **37**, 30 (2012)
17. A. Navid, L. Pilon, Smart Mater. Struct. **20**, 025012 (2011)

18. R. Kandilian, A. Navid, L. Pilon, Smart Mater. Struct. **20**, 055020 (2011)
19. F.Y. Lee, S. Goljahi, I.M. McKinley, C.S. Lynch, L. Pilon, Smart Mater. Struct. **21**, 025021 (2012)
20. M.H. Park, H.J. Kim, Y.J. Kim, T. Moon, K.D. Kim, C.S. Hwang, Nano Energy **12**, 131 (2015)
21. A.S. Mischenko, Q. Zhang, J.F. Scott, R.W. Whatmore, N.D. Mathur, Science **311**, 1270 (2006)
22. B. Neese, B. Chu, S.-G. Lu, Y. Wang, E. Furman, Q.M. Zhang, Science **321**, 821 (2008)
23. S.-G. Lu, Q. Zhang, Adv. Mater. **21**, 1983 (2009)
24. B. Peng, H. Fan, Q. Zhang, Adv. Funct. Mater. **23**, 2987 (2013)
25. X. Luo, W. Zhou, S.V. Ushakov, A. Nevrotsky, A.A. Demkov, Phys. Rev. B **80**, 134119 (2009)
26. T.M. Correia, S. Kar-Narayan, J.S. Young, J.F. Scott, N.D. Mathur, R.W. Whatmore, Q. Zhang, J. Phys. D Appl. Phys. **44**, 165407 (2011)
27. A.S. Mishenko, Q. Zhang, R.W. Whatmore, J.F. Scott, N.D. Marthur, Appl. Phys. Lett. **89**, 242912 (2006)
28. D. Saranya, A.R. Chaudhuri, J. Parui, S.B. Krupanidhi, Bull. Mater. Sci. **32**, 259 (2009)
29. P.F. Liu, J.L. Wang, X.J. Meng, J. Yang, B. Dkhil, J.H. Chu, New J. Phys. **12**, 023035 (2010)
30. R.W. Whatmore, J. Electroceram. **13**, 139 (2004)
31. A.J. Holden, IEEE T. Ultrason. Ferr. **58**, 1981 (2011)
32. S.B. Lang, Phys. Today **58**, 31 (2005)
33. M. Mackaym, D.E. Schuele, L. Zhu, L. Flandinm, M.A. Wolak, J.S. Shirk, A. Hiltner, E. Baer, Macromolecules **45**, 1954 (2012)
34. J. Müller, T.S. Böscke, U. Schröder, S. Mueller, D. Bräuhaus, U. Böttger, L. Frey, T. Mikolajick, Nano Lett. **12**, 4318 (2012)
35. S.G. Lu, B. Rozic, Q.M. Zhang, Z. Kutnjak, X.Y. Li, E. Furman, L.J. Gorny, M.R. Lin, B. Malic, M. Kosec, R. Blinc, R. Pirc, Appl. Phys. Lett. **97**, 162904 (2010)
36. S.G. Lu, B. Rozic, Q.M. Zhang, Z. Kutnjak, R. Pirc, M.R. Lin, X.Y. Li, L. Gorny, Appl. Phys. Lett. **97**, 202901 (2010)
37. X.Y. Li, X.S. Qian, S.G. Lu, J.P. Cheng, Z. Fang, Q.M. Zhang, Appl. Phys. Lett. **99**, 052907 (2011)
38. G. Sebald, E. Lefeuvre, D. Guyomar, IEEE T. Ultrason. Ferr. **55**, 538 (2008)
39. S.A. Sherrill, P. Banerjee, G.W. Rubloff, S.B. Lee, Phys. Chem. Chem. Phys. **13**, 20714 (2011)
40. L.C. Haspert, E. Gillette, S.B. Lee, G.W. Rubloff, Energy Environ. Sci. **6**, 2578 (2013)
41. K. Yao, S. Chen, M. Rahimabady, M.S. Mirshekarloo, S. Yu, F.E.H. Tay, T. Sritharan, L. Lu, IEEE T. Ultrason. Ferr. **58**, 1968 (2011)
42. J.H. Klootwijk, K.B. Jinesh, W. Dekkers, J.F. Verhoeven, F.C. van den Heuvel, H.-D. Kim, D. Blin, M.A. Verheijen, R.G.R. Weemaes, M. Kaiser, J.J.M. Ruigrok, F. Roozeboom, IEEE Electron Device Lett. **29**, 740 (2008)
43. P. Banerjee, I. Perez, L. Henn-Lecordier, S.B. Lee, G.W. Rubloff, Nat. Nanotechnol. **4**, 292 (2009)
44. C.C.B. Bufon, J.D.C. González, D.J. Thurmer, D. Grimm, M. Bauer, O.G. Schmidt, Nano Lett. **10**, 2506 (2010)
45. X. Hao, Y. Wang, L. Zhang, L. Zhang, S. An, Appl. Phys. Lett. **102**, 163903 (2013)
46. B. Ma, D.-K. Kwon, M. Narayanan, U. Balachandran, J. Mater. Res. **24**, 2993 (2009)
47. J.F. Scott, *Ferroelectric Memories* (Springer, Heidelberg, 2000), p. 53
48. J. Ge, X. Dong, Y. Chen, F. Cao, G. Wang, Appl. Phys. Lett. **102**, 142905 (2013)
49. J. Ge, D. Remiens, J. Costecalde, Y. Chen, X. Dong, G. Wang, Appl. Phys. Lett. **103**, 162903 (2013)
50. J. Li, S. Tan, S. Ding, H. Li, L. Yang, Z. Zhang, J. Mater. Chem. **22**, 23468 (2012)
51. K. Kim, Y.J. Song, Microelectron. Reliab. **43**, 385 (2003)
52. J.F. Scott, Annu. Rev. Mater. Res. **41**, 229 (2011)
53. X. Moya, S. Kar-Narayan, N.D. Mathur, Nat. Mater. **13**, 439 (2014)

54. X. Li, S.-G. Lu, X.-Z. Chen, H. Gu, X. Qian, Q.M. Zhang, J. Mater. Chem. C **1**, 23 (2013)
55. M. Valant, Prog. Mater Sci. **57**, 980 (2012)
56. US Energy Flow Trends, Lawrence Livermore National Laboratory, https://flowcharts.llnl.gov/#2009. Accessed: September 2014
57. G. Sebald, D. Guyomar, A. Agbossou, Smart Mater. Struct. **18**, 125006 (2009)
58. C.R. Bowen, H.A. Kim, P.M. Weaver, S. Dunn, Energy Environ. Sci. **7**, 25 (2014)
59. Y. Yang, W. Guo, K.C. Pradel, G. Zhu, Y. Zhou, Y. Zhang, Y. Hu, L. Lin, Z.L. Wang, Nano Lett. **12**, 2833 (2012)
60. R.V.K. Mangalam, J.C. Agar, A.R. Damodaran, J. Karthickm, L.W. Martin, A.C.S. Appl, Mater. Interfaces **5**, 13235 (2013)
61. S.R. Hunter, N.V. Lavrik, T. Bannuru, S. Mostafa, S. Rajic, P.G. Datskos, Proc. SPIE **8035**, 80350V (2011)
62. S. Porter, M. Mansi, N. Sumpter, L. Galloway, Sensor Rev. **21**, 283 (2001)

Chapter 15
Adaptive-Learning Synaptic Devices Using Ferroelectric-Gate Field-Effect Transistors for Neuromorphic Applications

Sung-Min Yoon and Hiroshi Ishiwara

Abstract An adaptive-learning ferroelectric neuron circuit is proposed and fabricated on a silicon-on-insulator structure, which is composed of a metal-ferroelectric-semiconductor field-effect transistor (MFSFET) and an oscillation circuit as an artificial synapse and neuron devices, respectively. Typical oxide ferroelectric $SrBi_2Ta_2O_9$ thin film is selected as a ferroelectric gate insulator for the MFSFET. The synapse MFSFET show good memory operations and gradual learning effect. The drain current is gradually modulated with increasing the number of input pulses with a sufficiently short duration. The output pulse frequency of the fabricated neuron circuit is also confirmed to gradually increase as the number of input pulses increased. The weighted sum operation is realized by constructing the synapse array composed of the MFSFETs. The output pulse performance including the pulse amplitude and time-dependent stability are improved by employing Schmitt-trigger oscillator and metal-ferroelectric-metal-oxide-semiconductor gate stack structure, respectively.

15.1 Introduction

Artificial neural networks, which execute a distributed parallel information processing and an adaptive learning function, have attracted much attention for the future highly-developed information-oriented society. In a human brain, a huge quantity of information is processed in parallel and stored as one's past experience. In this system, neurons accept many weighted input signals and generate output pulses when the total value of input signals exceeds a threshold value. The weighting

S.-M. Yoon (✉)
Department of Advanced Materials Engineering for Information and Electronics,
Kyung Hee University, Yongin, Gyeonggi-Do 446-701, Republic of Korea
e-mail: sungmin@khu.ac.kr

H. Ishiwara
Tokyo Institute of Technology, 4259 Nagatsuta, Midori-ku
Yokohama 226-8503, Japan

© Springer Science+Business Media Dordrecht 2016   311
B.-E. Park et al. (eds.), *Ferroelectric-Gate Field Effect Transistor Memories*,
Topics in Applied Physics 131, DOI 10.1007/978-94-024-0841-6_15

operation for input signals is conducted by synapses which are attached to the neurons. Thus, synapses and neurons can be realized using memory devices and processors in artificial neuromorphic systems. However, the hardware implementation of a large-scale network is rather difficult, since the number of synaptic connections becomes huge as the number of neurons increases. One feasible solution to this hardware problem is to use nonvolatile analog memories, by which electrically-modifiable synapse array can be implemented in a small size. Actually, floating-gate MOS devices have been used for this purpose [1–5]. In these devices, since the data are stored as an amount of electrical charge injected through a tunnel oxide into the floating-gate, precise control of the quantity of injected carriers is rather difficult, unless the well-designed control circuit is used.

The nonvolatile analog memory operations can be achieved by realizing a metal-ferroelectric-semiconductor field-effect transistor (MFSFET), in which gate insulator is replaced with a ferroelectric film [6]. In implementation of novel synaptic connections using an array of MFSFETs, we can exploit two specific features much superior to other information processing systems. One is their 'adaptive-learning' capability which means that the electrical properties of a device are changed partially or totally by applying a certain number of usual signals to the device. The other is their flexible 'electrically modifiable function' where the different values of synaptic weight can be programmed and modified by usual electrical signals. However, the difficulties in integrating MFSFETs into Si circuitry have been critical obstacles for fabrication of ferroelectric neuron circuit on a single wafer. From these backgrounds, the main purpose of this study is aimed to fabricate the ferroelectric neuron circuit using the ferroelectric-gate FETs as synapse device by optimizing the fabrication processes and to establish the adaptive-learning function of the ferroelectric neuron circuit.

15.2 Operation Principles of Adaptive-Learning Neuron Circuits

15.2.1 Pulse Frequency Modulation-Type Ferroelectric Synaptic Device Operations

Figure 15.1 shows the basic neuron circuit proposed as an elementary component of the pulse frequency modulation (PFM) type adaptive-learning neural networks [6]. In this circuit, the adaptive-learning function can be realized by controlling the amounts of polarization of the ferroelectric gate in MFSFET. In other words, the channel resistance of MFSFET can be gradually changed as the polarization state of ferroelectric film is partially reversed by pulse signals applied to the gate terminal. Consequently, the synaptic values stored in MFSFETs can be gradually changed by applying an adequate number of input signals. For this reason, the duration of input pulses must be sufficiently shorter than the switching time for the polarization reversal of ferroelectric gate insulator. This feature explains that the proposed

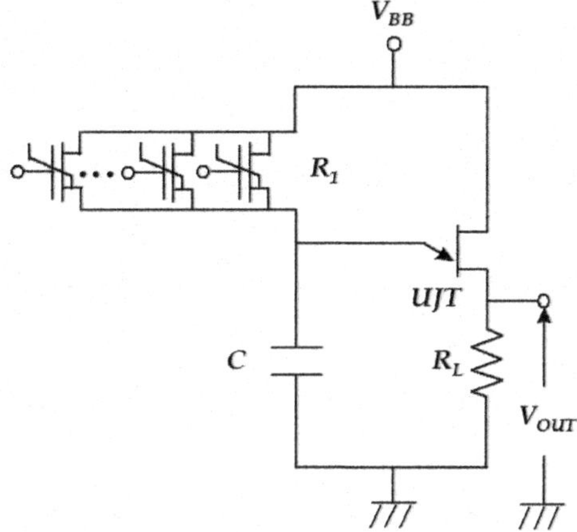

Fig. 15.1 Ferroelectric neuron circuit composed of an MFSFET and a CUJT oscillation circuit

neuron circuit is desirable to be implemented as the PFM system. In this circuit, complementary unijunction transistor (CUJT) is used as a switching component to discharge the capacitor C, which corresponds to the threshold processing in a neuron. Since the interval of output pulse generated from the circuit is proportional to the product of C and S-D resistance (R_{SD}) of MFSFET, the output pulse frequency can be gradually changed during signaling the input pulses. This is similar to the information processing in a human brain, in which current pulses generated in neurons propagate through nerve membranes and axons.

15.2.2 *Multiple-Input Neuron Circuit and Electrically Modifiable Synapse Array*

In neural networks, each neuron has many synapses and they are connected to the neurons in the previous layer. Figure 15.2 shows the schematic diagram of a two-layered neural network, in which the outputs of m neurons are fully connected to the n neurons in the next layer. In this neural network, $m \times n$ synapses are required, which can be realized by parallel connection of the MFSFETs. In this structure, each MFSFET is differently programmed and accepts pulse signals from different neurons. Therefore, the total drain current summed up for all MFSFETs determines the output behaviors of the neuron circuit. The 'weighted-sum' operation of synaptic values in neurons is performed in this way. The prototype layout of the synapse array fabricated on an SOI structure is shown in Fig. 15.3, where Si stripes with a lateral *npn* structure are placed on an insulating layer and then covered with a ferroelectric film, and common metal stripes for gate electrode are placed on the film perpendicular to Si stripes [7]. Since there is no via-holes across the ferroelectric film in this structure, the packing density of synapses is expected to be very high. Furthermore, the synapse array fabricated on SOI structure can be electrically isolated completely from one another, which enables us to give different weight values to the individual synapses with ease.

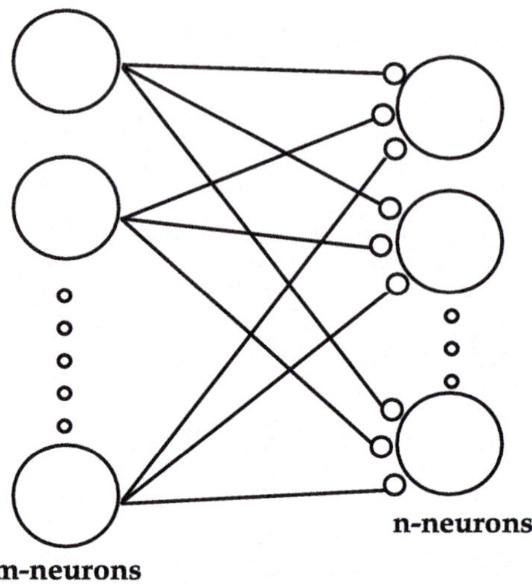

Fig. 15.2 Schematic diagram of a two-layered neural network

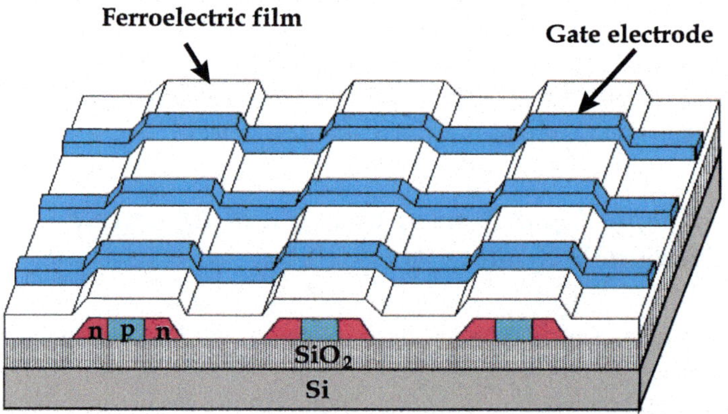

Fig. 15.3 Schematic diagram of a prototype layout of the synapse array fabricated on SOI substrate

15.3 Fundamental Characteristics Ferroelectric Synapse FETs

15.3.1 Fundamental Characteristics of Ferroelectric SrBi$_2$Ta$_2$O$_9$ (SBT) Thin Films

Switching characteristics of SBT film were experimentally investigated in MFM capacitors, based on the Ishibashi and Takagi's theory [8, 9]. In their theory, the reversal current response and reversed polarization are given by following equations.

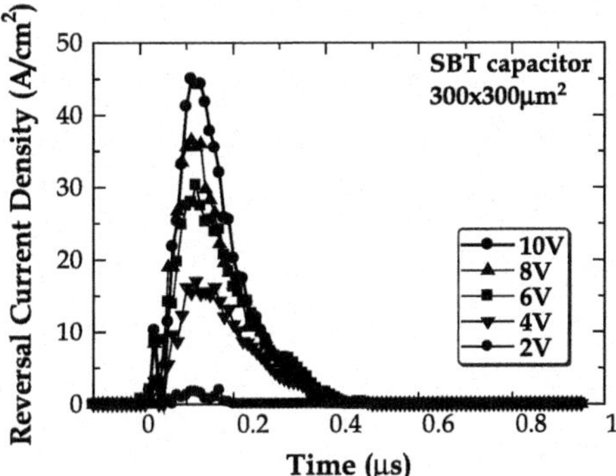

Fig. 15.4 Reversal switching current density calculated by using double-pulse method for the SBT capacitor with the size of 300 × 300 μm²

$$j = 2P_r n/t_s (t/t_s)^{n-1} \exp[-(t/t_s)^n]$$
(15.1)

$$P = P_r(1 - 2\exp[-(t/t_s)^n]),$$
(15.2)

where P_r, t_s, and n denote remnant polarization, switching time, and the dimensionality factor, respectively. Figure 15.4 shows the reversal switching current responses of the SBT capacitor (300 × 300 μm²) estimated by using double-pulse measurement when the input pulse voltage was changed from 2 to 10 V. The values of t_s and n can be deduced by fitting these data using (15.1). It was found that n is about 2.0 and insensitive to the applied voltage. The t_s increases from 158 to 182 ns with decreasing the applied voltage from 10 to 6 V. Assuming that the t_s obeys the exponential law as

$$t_s = t_{s0} \exp(E_a/E),$$
(15.3)

values of t_{s0} and E_a were estimated to be approximately 127 ns and 72.4 kV/cm, respectively. Compared with those for the Pb(Zr,Ti)O₃ (PZT) thin film, the t_s for SBT is shorter and less dependent on the applied electric field [10, 11]. These results provide us a lot of basic properties of the SBT thin films for adaptive-learning MFSFET and ferroelectric neuron circuit.

15.3.2 Basic Device Characteristics of MFSFETs

Next, MFSFETs were fabricated on an SOI structure, in which SBT/Si was chosen as a gate stack structure for its simplicity, even if it is not perfectly promising in its interface property. Figure 15.5 shows the drain current-gate voltage (I_D-V_G) characteristic of the fabricated MFSFET. A counterclockwise hysteresis was obtained and the memory window was approximately 0.45 V for a V_G sweep from 0 to 6 V, which was owing to the ferroelectric nature of the SBT. In order to further examine the memory effect, the drain current-drain voltage (I_D-V_D) characteristics were

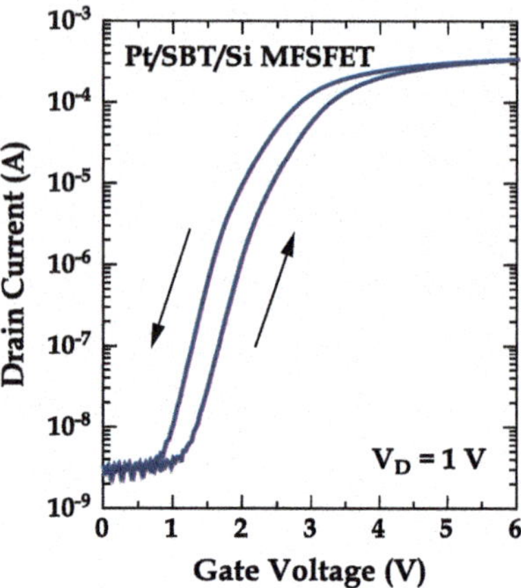

Fig. 15.5 Drain current—gate voltage characteristic of the fabricated MFSFET

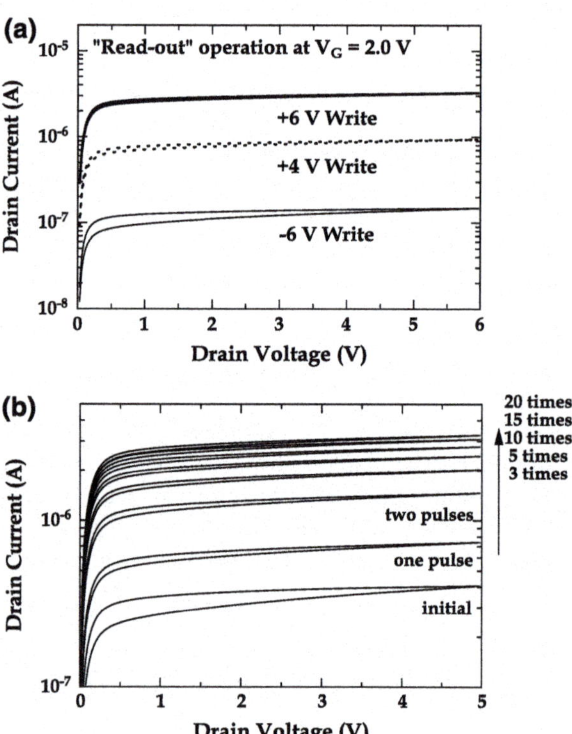

Fig. 15.6 **a** Write and read-out operations in drain current—drain voltage characteristics of the fabricated MFSFET. **b** Gradual variation of drain current in read-out operations of the MFSFET as the increase in applied pulse numbers

measured for the same FET, as shown in Fig. 15.6a. First, 'write' pulse signal of -6 V, $+4$ V, or $+6$ V was applied to the gate terminal. Then, the V_G of 2 V for the 'read-out' operations was applied and I_D was measured. The I_D changed from 'off' to 'on' state by changing the 'write' voltage from -6 to 6 V. From these results, it can be found that the MFSFET was successfully fabricated on an SOI structure with good memory operations. Next, in order to examine the gradual learning-effect, the variation of I_D-V_D characteristics was measured by increasing the number of input pulses applied to the gate terminal of the MFSFET. The width and height of applied pulses were 20 ns and 6 V, respectively. The value of I_D increased as the number of applied pulses increased even if the V_G for the 'read-out' operations was equally adjusted to 2.0 V, as shown in Fig. 15.6b. This indicates that the polarization of ferroelectric SBT gate insulator is gradually reversed by input pulses, and that the channel resistance of MFSFET is changed by the number of input pulses. This analog-like change of I_D's in 'read-out' operations is essential in realizing the adaptive-learning function in the proposed ferroelectric neuron circuit.

15.4 Electrically Modifiable Synapse Array Using MFSFET

In implementation of neuromorphic systems, it is very important to realize a specified synapse device which act as an electrically-modifiable and nonvolatile analog-like memory and stores the synaptic weights. The array structure composed of adaptive-learning MFSFETs was proposed as a novel electrically-modifiable synaptic connection. In this section, this promising candidate for synaptic connection, an MFSFET array, is fabricated using a ferroelectric SBT film and the weighted sum operation is demonstrated [12].

15.4.1 Device Design of Ferroelectric Synapse Array

The prototype synaptic connection was designed to be 3 × 3 MFSFET array structure. The ferroelectric material was chosen as SBT and 5 μm design rule was employed. Use of an SOI structure (Si stripes on an insulating substrate) is essential in giving different synaptic weight values to the synapses connected along a common gate stripe [7]. The designed layout for MFSFET array with 3 × 3 structure is schematically shown in Fig. 15.7. The Si islands are connected to the source terminals of FETs through the body contacts, in order to prevent the floating body effect of SOI substrate.

15.4.2 Fabrication Process

First, the device region was separated in islands with a rectangular shape using a plasma etching system. Then, in order to form highly-doped n source and drain (S-D) regions with low resistance, phosphorous ions were repeatedly implanted. Since the S-D regions are used as conductors connecting FET's in the array

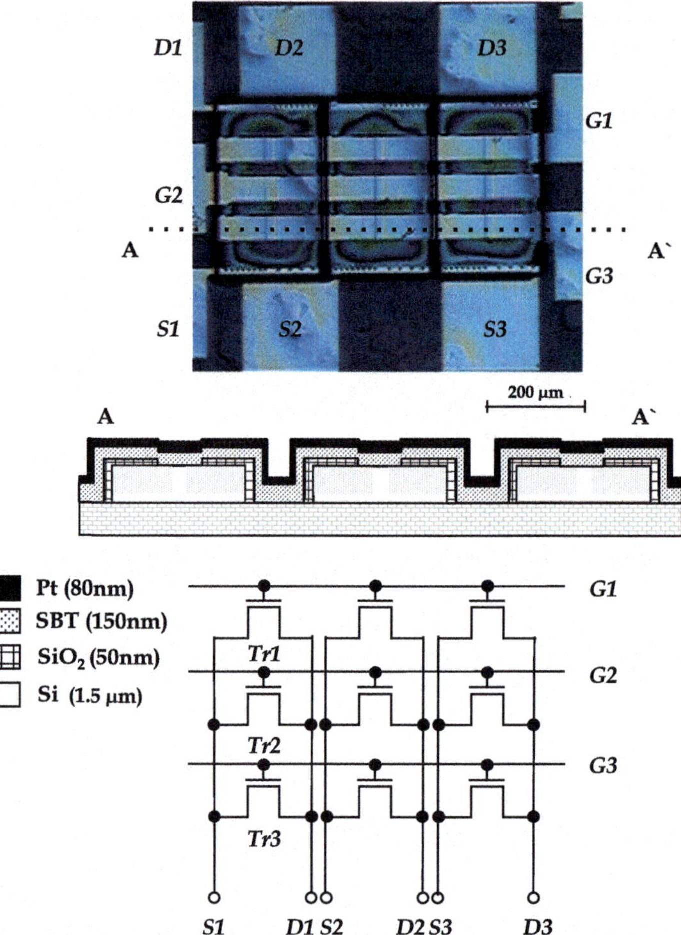

Fig. 15.7 **a** Microscopic photo image, **b** schematic cross-sectional view, and **c** equivalent circuit diagram of the MFSFET synapse array fabricated on SOI structure

structure, the resistance of these regions is necessary to be made as low as possible. The sheet resistance of this region was designed to be about 25 Ω/sq. Following dry oxidation of the Si islands for passivation, gate windows for the deposition of ferroelectric gate film were formed by wet chemical etching. The ferroelectric SBT film was deposited using liquid source misted chemical deposition (LSMCD) method for expecting better step coverage the surface steps. The final thickness of SBT gate film was about 150 nm. They were annealed for crystallization at 800 °C for 30 min in an O$_2$ atmosphere. The Pt gate electrodes were patterned by lift-off process. Contact holes for the source and drain were formed by dry etching in a reactive ion etching (RIE) system using the gas mixture of Ar/Cl$_2$. Finally, Al electrodes and metal interconnections were formed by lift-off process. The channel length and width of fabricated MFSFET's are 5 and 50 µm, respectively. A photograph of the fabricated 3×3 MFSFET array structure is shown in Fig. 15.7, in which three FET's are connected in parallel on each Si island and three FET's located on three different islands have a common gate metal stripe, as shown in the middle (cross section) and bottom (equivalent circuit) parts.

15.4.3 Weighted Sum Operation of Synapse Array

Figure 15.8a shows the I_D-V_G characteristics of three FETs located on the same Si island. Three FETs show similar characteristics with counterclockwise traces. The variation of threshold voltage of MFSFETs on the same Si island was about 0.2 V. Although the small variation was also observed in the measurement of I_D's in 'read-out' operations of I_D-V_D characteristics, the MFSFETs with array structure was confirmed to be successfully fabricated on an SOI substrate without large

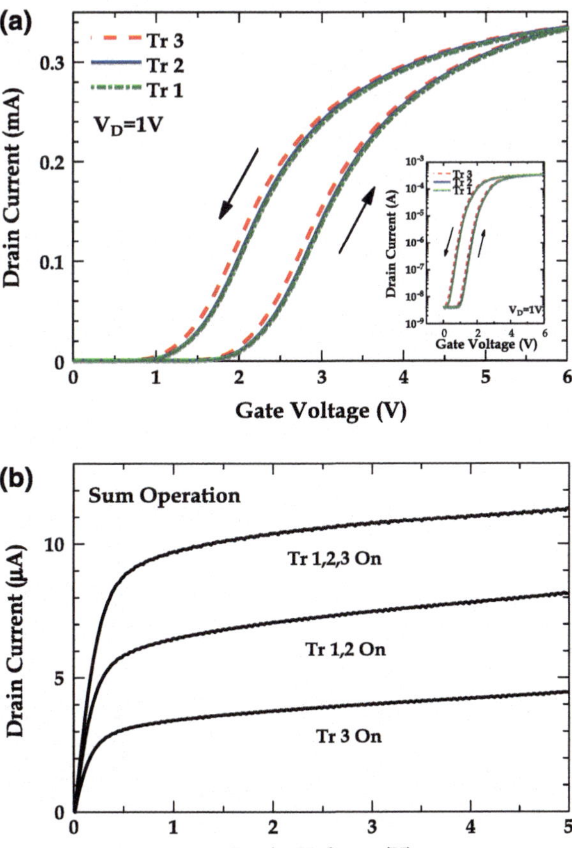

Fig. 15.8 **a** Drain current—gate voltage characteristics of three MFSFETs fabricated with 3×3 array structure. **b** Sum operation of stored data in drain currents for the MFSFET array structure

fluctuations in their electrical characteristics. Using this MFSFET memory array, we can simply conduct the 'sum' operation of stored data. First, the pulse signals of +6 V were applied to $G1$, $G2$, and $G3$ to write the data and the I_D's between $D1$ and $S1$ were measured under the different conditions, as shown in Fig. 15.8b; (1) $V_{G3} = 1.4$ V, $V_{G1} = V_{G2} = 0$ V, (2) $V_{G1} = V_{G2} = 1.4$ V, $V_{G3} = 0$ V, and (3) $V_{G1} = V_{G2} = V_{G3} = 1.4$ V. As can be seen in the figure, the I_D is almost doubled when two FETs are turned-on, and it is roughly three times larger than that of one FET when three FETs are turned-on. From this result, it can be found that the stored data in MFSFET array can be summed up with non-volatility. Next, the 'weighted-sum' operation was demonstrated in the fabricated MFSFET array. The different values of synaptic weight were stored by applying input pulses as programming signals. The I_D's in 'read-out' operations varied with an approximate ratio of 1:2:4 when the height of 'write' pulses was adjusted to 2, 3, or 6 V, as shown in Fig. 15.9a, which corresponds to the configuration of different initial synaptic values. In order to add up the stored data, I_D's between $D1$ and $S1$ were measured for 3-bit input signals, as shown in Fig. 15.9b. In this figure, the I_D's in 'read-out' operations clearly vary with eight different values. In other words, the 3-bit analog-to-digital conversion was successfully demonstrated as an example of the weighted-sum operation. From these results, it can be concluded that the electrically modifiable functionality of MFSFET array structure is very promising for implementing the high-density synaptic connections.

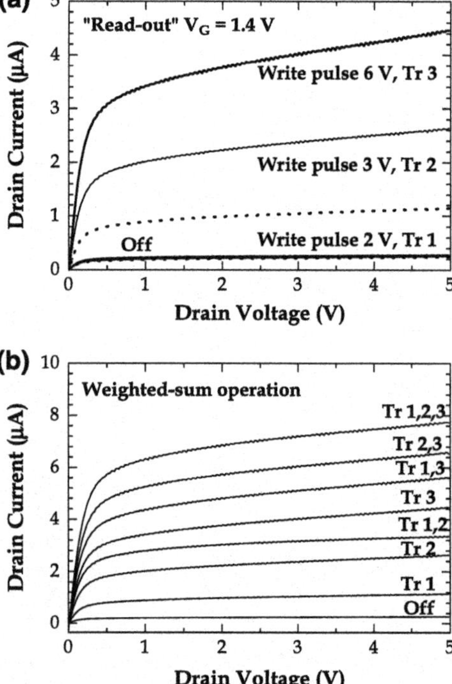

Fig. 15.9 **a** Configuration of initial synaptic values into the MFSFETs in the array structure by applying different amplitude voltage pulses to Tr1, Tr2, and Tr3, respectively. **b** Demonstration of weighted-sum operation in MFSFET array structure

15.5 Adaptive-Learning Neuron Circuit Composed of an MFSFET and a CUJT Oscillation Circuit

In this section, an adaptive-learning ferroelectric neuron circuit is fabricated by integrating an MFSFET and a CUJT oscillation circuit on an SOI structure with a 3 µm-thick p-type Si layer [13–15].

15.5.1 Device Design and Circuit Layout

All devices of the neuron circuit were designed by a 5-µm design rule. The capacitors were designed to be 3, 10, and 30 pF and fabricated with the structure of $Pt/SiO_2/n^+$-Si. R_L was designed to be 60–80 Ω. The fabrication procedures are as follows. First, the device regions were separated into islands of rectangular shapes using plasma etching system. The reaction gas and their ratio were CF_4:O_2 and 45:5, respectively. The ion implantation processes for forming the active regions of device and contact regions were performed. After the Si islands were oxidized by dry oxidation for passivation, gate windows for deposition of SBT films were formed by wet chemical etching. SBT films were deposited by LSMCD method. The deposition process by LSMCD method was repeatedly performed until the desired film thickness was obtained and they were annealed for crystallization at 750 °C for 30 min in an O_2 atmosphere using a rapid thermal annealing (RTA) system. The final thickness of SBT gate was about 150 nm. Then, a Pt gate film was deposited by *e*-beam evaporation method for forming the gate electrode and it was patterned by lift-off process, which can also be acted as a protection layer for the SBT gate insulator during subsequent fabrication processes. SBT film was patterned by the selective etchant, NH_4F:HCl solution. Contact holes were easily formed by wet chemical etching. Finally, Al interconnection and electrode pads were formed by lift-off process. 10 sheets of photo-mask were used in fabrication of this neuron circuit. A photograph of the integrated ferroelectric neuron circuit is shown in Fig. 15.10.

15.5.2 Adaptive-Learning Function of Ferroelectric Neuron Circuit

R_L, C, and CUJT showed normal device operations. The memory operations of the fabricated MFSFET were investigated. In this measurement, in order to verify that the threshold voltage shift shown in Fig. 15.11 is really caused by the ferroelectric nature of SBT film, the dependence of the memory window width on the sweep rate of V_G was measured. Figure 15.11a shows the I_D-V_G characteristics for the fastest sweep rate case (6×10^4 V/s) was compared with that for the normal case (0.5 V/s).

Fig. 15.10 A microscopic photograph of the integrated ferroelectric neuron circuit on an SOI structure, in which MFSFET and CUJT oscillation circuit are worked as synapse and neurons, respectively

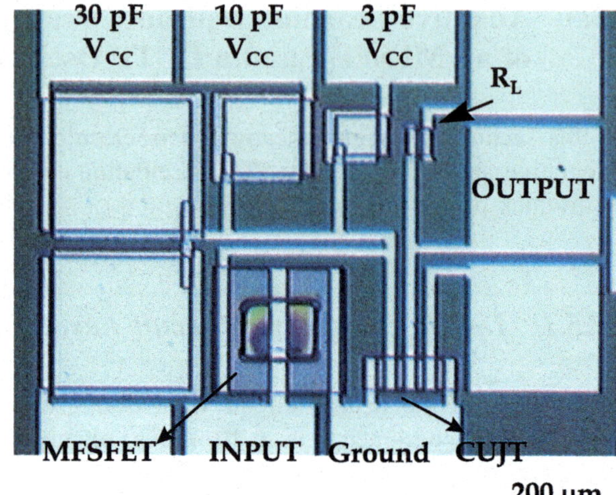

In the fastest case, 5 kHz triangular wave voltage from 0 to 6 V was applied using a virtually grounded circuit shown in Fig. 15.11b. The measured value of memory window is practically independent of the V_G sweep rate, as shown in Fig. 15.11c, which clearly shows that the obtained hysteresis is not due to mobile ions but due to ferroelectricity of SBT film. The adaptive-learning effect resulted from the gradual polarization reversal was similarly confirmed, as shown in Fig. 15.12a, in which the I_D gradually increased as the number of applied pulses increased, when the 'read-out' voltage was fixed at 1.4 V. On the basis of obtained device characteristics, the normal operation of the ferroelectric neuron circuit was examined. First, the oscillation frequency was measured as a function of DC input voltage applied to gate of MFSFET, as shown in Fig. 15.12b. The output pulse frequency changed with a hysteretic characteristics, which reflects the ferroelectric memory operation of MFSFET. In order to realize the adaptive-learning function in the PFM-type circuit, it is necessary to change the number of input pulses, each of which has the same width and height (20 ns, 6 V). Typical output waveforms are shown in Fig. 15.13, in which the output waveform after application of a single pulse is compared with that after sixty pulses. During this measurement, a constant DC voltage of 1.65 V was applied to the gate terminal, but the circuit did not oscillate before the first pulse was given to the gate. Figure 15.13c shows the variation of output pulse frequency in the circuit as a function of input pulses numbers, which clearly demonstrates the adaptive-learning function of the ferroelectric neuron circuit. It is concluded from these results that the neuron circuit changes its output characteristics (response) by the past experience imposed by the input pulses (stimulus). Although the adaptive-learning function of the ferroelectric neuron circuit was successfully obtained, there remained some problems. In other words, small output pulse height of CUJT oscillation circuit and short memory retention time of MFSFET should be improved for the practical applications in the circuit.

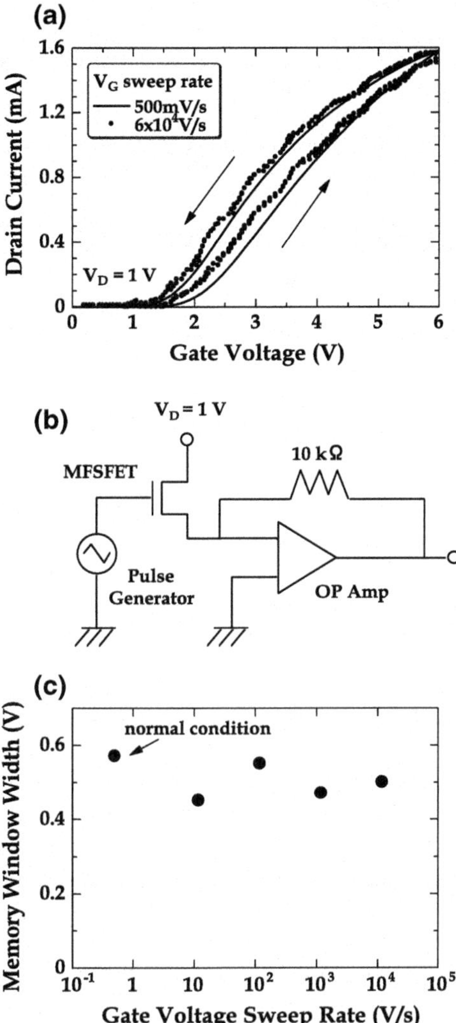

Fig. 15.11 **a** Comparisons of drain current—gate voltage characteristics for slow sweep rate of gate voltage (0.5 V/s) with that for fast sweep rate of gate voltage (6 × 10⁴ V/s). **b** Circuit diagram of a virtually grounded circuit using an OP-amp. **c** Variations in the memory window width as a function of the sweep rate of gate voltage

15.6 Improvement of Output Characteristics in Ferroelectric Neuron Circuit Using CMOS Schmitt-Trigger Oscillator

15.6.1 Device and Circuit Designs

In the proposed neuron circuit, the small output pulse height of CUJT oscillation circuit is still a problem. The height of output pulses must be high enough to reverse the ferroelectric polarization of MFSFET, since the output pulse of a neuron is used as an input signal to the next layer neuron. However, the pulse height obtained in

Fig. 15.12 **a** Gradual
learning effect in read-out
operations for the synapse
MFSFET integrated into the
ferroelectric neuron circuit.
b Modulations in output pulse
frequency as a function of DC
input signals applied to the
MFSFET in the ferroelectric
neuron circuit

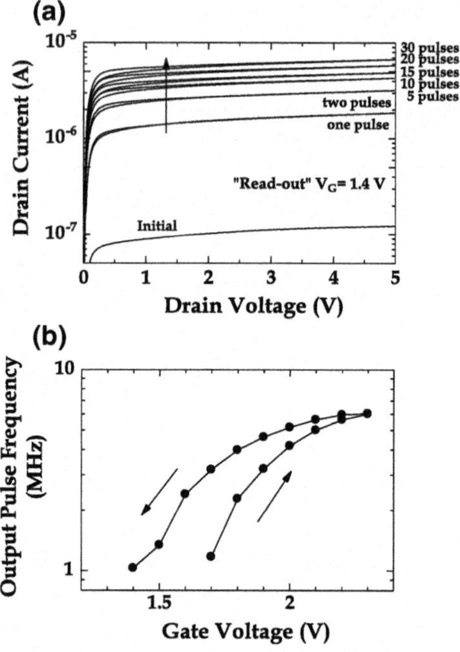

the CUJT oscillation circuit was as small as 0.1 V. This value is too small to be used as input signals. To solve this problem, a new configuration of circuit, CMOS Schmitt-trigger oscillator, was proposed as a switching component of ferroelectric neuron circuit [16]. The diagram of the circuit loaded with a CMOS Schmitt-trigger is shown in Fig. 15.14. The PFM-type oscillation operation is basically identical to that of neuron circuit using the CUJT. The CMOS Schmitt-trigger, enclosed by the dotted line, has the hysteretic behavior in input-output transfer characteristic. Hence, charging and discharging of a capacitor C can be performed through p-ch FET connected in parallel. The threshold voltages for increasing and decreasing input signals can be changed by varying the ratio of two feedback resistors, $R1/R2$. The ferroelectric neuron circuit was newly designed and fabricated by integrating the CMOS Schmitt-trigger oscillator with MFSFET. In order to minimize the process damage to the ferroelectric film, the conventional CMOS processes were conducted except for the interconnection process prior to deposition of the SBT film, though MFSFET was fabricated in the same way as employed in the fabrication of neuron circuit using CUJT. For the full fabrication process, 12 sheets of photo-mask were used. A photograph of the fabricated circuit is shown in Fig. 15.15a.

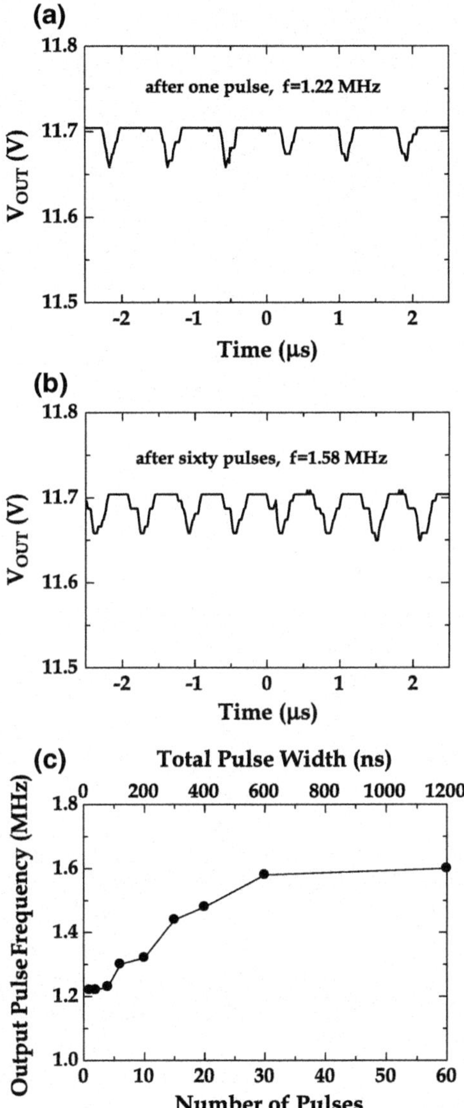

Fig. 15.13 Output pulse waveforms of the integrated ferroelectric neuron circuit when **a** one and **b** sixty pulses with 6-V amplitude and 20-ns duration were applied as input signals. **c** Variation in output pulse frequency as the increase in the number of input signals

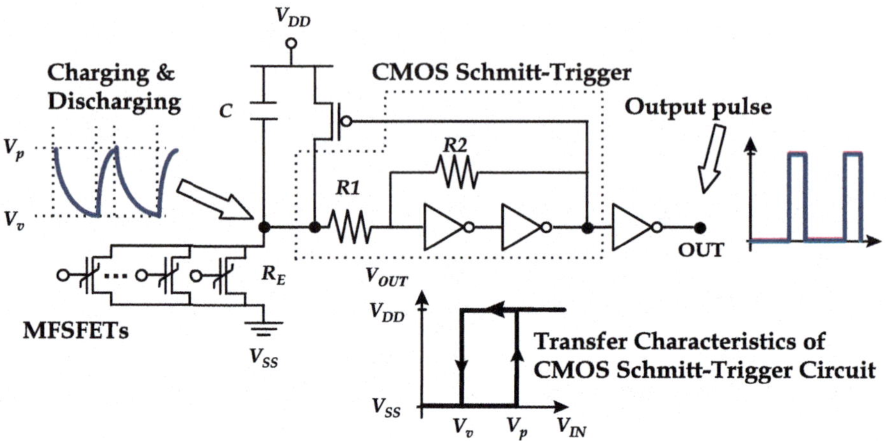

Fig. 15.14 Circuit diagram of the ferroelectric neuron circuit using a CMOS Schmitt-trigger oscillator

15.6.2 Adaptive-Learning Functions with Improved Output Characteristics

All the fabricated devices in the circuit were confirmed to normally operate. Especially, the fabricated MFSFET showed a relatively good memory operations in I_D-V_G measurement, and its I_D's in 'read-out' operations gradually increased when the number of pulses was applied to the gate as input signals, which were not so different from the case of MFSFET in the neuron circuit using CUJT. To examine the improved behavior of newly fabricated ferroelectric neuron circuit using CMOS Schmitt-trigger oscillator, the measurements similar to those carried out in Sect. 15.5 were typically performed. The power supply voltage (V_{DD}) was 5 V. Figure 15.15b shows typical output pulse waveforms after a single pulse and sixty pulses were applied to the gate. It can be confirmed that the circuit started to oscillate by application of a single pulse and the oscillation frequency increased as the number of applied pulses increases. It is noticeable that the height of output pulse was almost the same as V_{DD}. This feature is completely different from the case of the neuron circuit using CUJT. It is concluded that the output pulse signals generated from CMOS Schmitt-trigger oscillator can be used as input signals for neuron in the next-layer without connecting an additional amplifier.

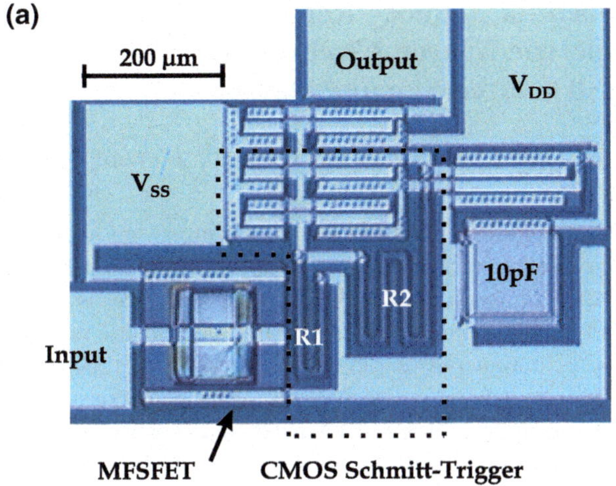

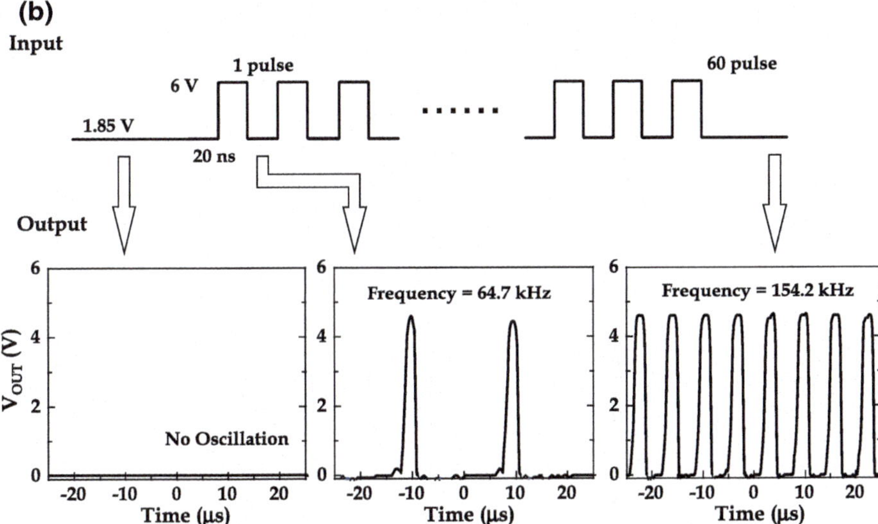

Fig. 15.15 **a** Microscopic photo image of the ferroelectric neuron circuit composed of an MFSFET and a CMOS Schmitt-trigger oscillator. **b** Adaptive-learning output waveforms with improved pulse height

15.7 Improvement of Memory Retention in Ferroelectric Neuron Circuit Using MFMIS-Structured Synapse Device

15.7.1 Device Designs for MFMIS Synapse Device

Although the reliable nonvolatile memory operation with a long retention time of a synapse device is another important issue in this application, the memory retention characteristic of MFSFET with a Pt/SBT/Si gate structure is unsatisfactory. Figure 15.16 shows typical output waveforms of the ferroelectric neuron circuit, in which the frequency of output pulses rapidly decreases with time, and the oscillation operation of the circuit stops in about 40 min. This behavior is well explained from the retention characteristics of MFSFET used as a synapse device. Therefore, in order to improve the memory retention characteristics of the ferroelectric neuron circuit, it is necessary to modify the device structure and fabrication process of MFSFET. In this study, MFMOS-FET (O: oxide) with a Pt/SBT/Pt/Ti/SiO$_2$/Si structure was proposed in order to improve the memory retention time of MFSFET, in which the excellent interface property and the small gate leakage current can be expected by introducing the SiO$_2$ buffer layer with a good quality in MOS structure [17]. Furthermore, this structure has a good structural merit that MFM and MOS

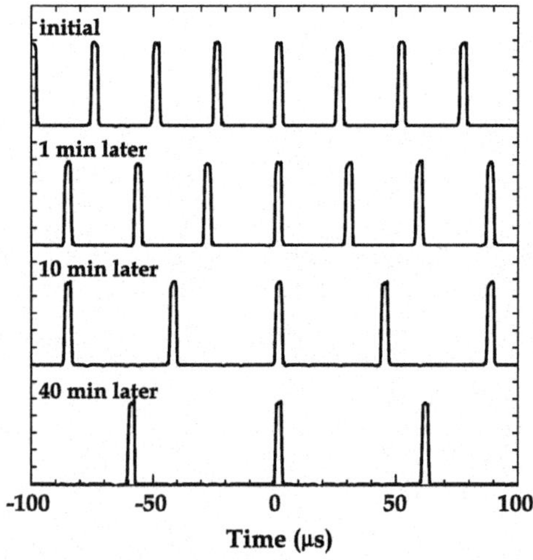

Fig. 15.16 Variations **a** in output pulse waveforms in the ferroelectric neuron circuit using MFSFET and **b** in output pulse frequency with retention time

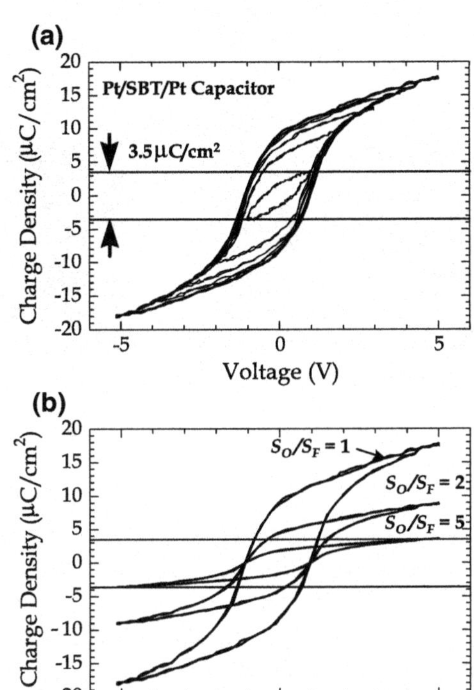

Fig. 15.17 **a** Typical Q-V hysteresis of SBT capacitor. **b** Variations in Q-V hysteresis by changing the area ratio of MFM (S_F) and MOS (S_O) capacitors

capacitors can be independently designed, which is very promising to improve the retention characteristic. In other words, it is noticeable that the available charge for controlling the channel conductance of MOSFET is not determined by the remnant polarization (P_r) of ferroelectric film, but determined by the maximum induced charge of MOS capacitor. For example, the maximum induced charge density in MOS capacitor is only 3.5 μC/cm^2, assuming that the breakdown field of SiO$_2$ is 10 MV/cm. Since this value is much smaller than the P_r of SBT capacitor, as shown in Fig. 15.17a. Therefore, we can use only a minor hysteresis loop of the SBT capacitor. However, it is evident that the use of this minor loop has such demerits that the coercive field is small and the polarization direction is easily reversed due to the depolarization field generated in ferroelectric film during the memory retention period. Consequently, excellent characteristics as a nonvolatile memory cannot be expected in an FET with a simple MFMOS structure. An MFMOS structure with a small area (S_F) MFM capacitor on a larger area (S_O) MOS capacitor

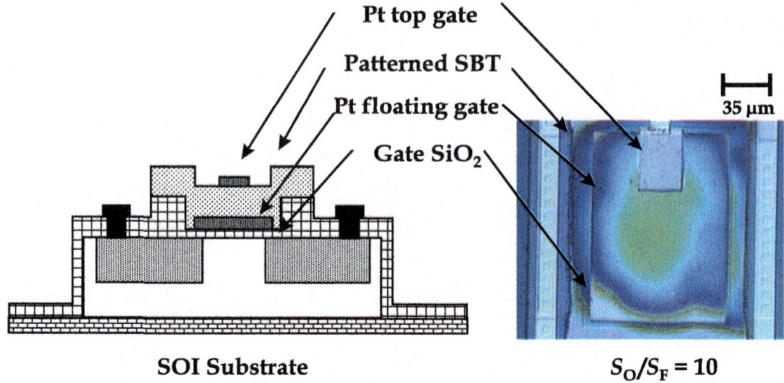

Fig. 15.18 Schematic cross-sectional view and microscopic photo image of the fabricated MFMOS-FET with the S_O/S_F ratio of 10

is a good solution to this problem, in which the P_r of MFM capacitor can be equivalently reduced, as schematically shown in Fig. 15.17b. From these discussions, the MFMOS-FET was designed and fabricated on an SOI structure. The thickness of SiO$_2$ layer (9 nm) and the area ratio of S_O/S_F were so chosen from the estimations of optimum operating points analysis when the operating voltage was given to be 5 V. Figure 15.18 shows a schematic cross-section and a photograph of the fabricated MFMOS-FET with $S_O/S_F = 10$, in which the S_O/S_F ratio was changed from 3 to 15 by changing the size of top Pt gate electrode.

15.7.2 Adaptive-Learning Functions with Improved Memory Retention Characteristic

The memory operations of fabricated MFMOS-FETs are given in Fig. 15.19a. In the I_D-V_G characteristics, the memory window increases from 0.35 to 1.40 V as the S_O/S_F ratio is changed from 3 to 15, and it is almost saturated at S_O/S_F ratio larger than 10. Next, in order to measure the retention characteristics, a programming input pulse of +6 or −4 V was applied for 100 ms to write 'on' and 'off' state, respectively, which was preceded by a 'reset' pulse of −4 and 6 V to initialize the ferroelectric polarization. Then, I_D was continuously measured by keeping V_G at 0.8 V and V_D at 0.1 V. The memory retention characteristics of the fabricated MFMOS-FETs are shown in Fig. 15.19b. In the case that $S_O/S_F = 3$, the initial on/off ratio of I_D is as small as 7.0, because of the narrow memory window. In the case that $S_O/S_F = 7$, the initial on/off ratio is still small (52) and the ratio decreases

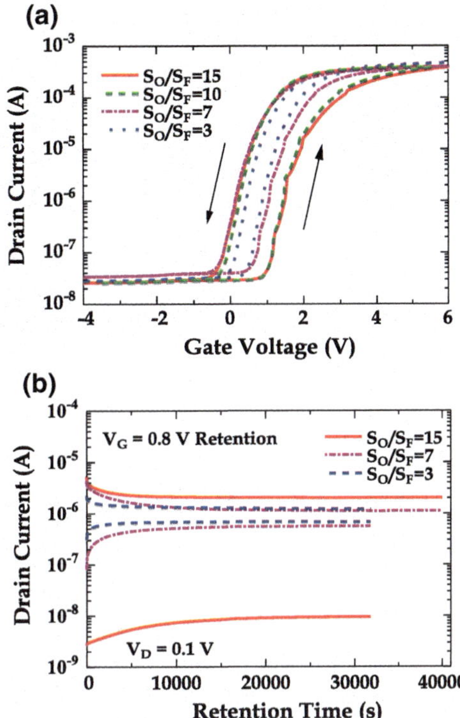

Fig. 15.19 **a** Drain current—gate voltage characteristics and **b** memory retention characteristics of the MFMOS-FETS fabricated with various S_O/S_F ratios

to less than 10 within 1000 s. On the other hand, in the case that $S_O/S_F = 15$, the on/off ratio of I_D is larger than 1800 at first, and it is still larger than 200 after about 10 h has passed. These results are considered to be mainly due to the improvement of interface quality by use of SiO_2 as an insulating buffer layer and due to the optimization of S_O/S_F ratio in the MFMOS structure. Using the MFMOS-FET with a relatively good characteristics as a synapse device, the ferroelectric neuron circuit was fabricated, as shown in Fig. 15.20a, in which the CMOS Schmitt-trigger was used as an oscillation component. Since the fabricated circuit showed the normal oscillation operation, variation of output pulse frequency was monitored by keeping V_G at 1.85 V, after sixty input pulses (20 ns, 6 V) were applied to the gate of MFMOS-FET. A typical result is given in Fig. 15.20b, in which the result in Fig. 15.15b is also plotted for comparison. The output pulse frequency was almost constant up to 3000 s. This feature is completely different from the case of neuron circuit using MFSFET. It is concluded from this result that the MFMOS-FET with improved memory retention characteristic is very promising for the synapse device in ferroelectric neuron circuit.

Fig. 15.20 a Microscopic photo image of the fabricated ferroelectric neuron circuit using the MFMOS-FET with the S_O/S_F ratio of 15. **b** Improvement of retention characteristics for the ferroelectric neuron circuit using the MFMOS-FET

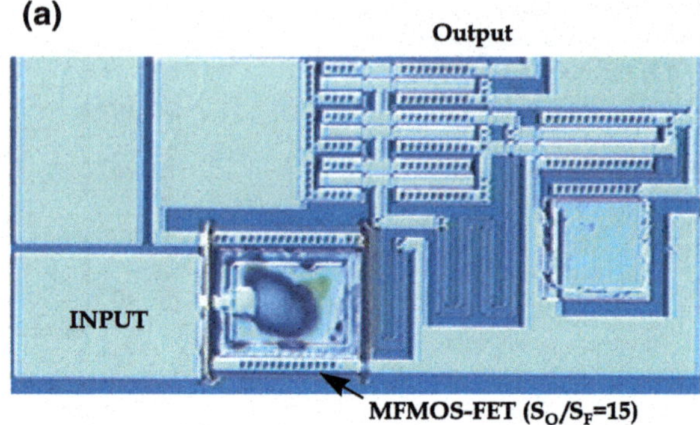

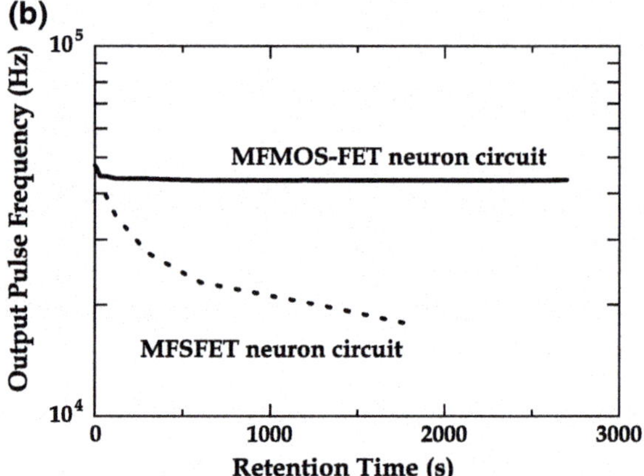

15.8 Conclusions and Outlooks

Ferroelectric devices and related functional circuits have been energetically researched for realization of new memory concepts. Similarly, the artificial neuromorphic systems have been drawing a considerable attention due to their ability to carry out the distributed parallel information processing and the adaptive-learning function. The object of this work was the realization of 'ferroelectric neuron circuit' with adaptive-learning capability. In this work, in order to integrate the synapse device of ferroelectric-gate FETs with the neuron oscillation circuit, a number of approaches were performed in fabrication processes and evaluation methods, so that the appropriate fabrication processes and related technologies were proposed and established for the ferroelectric neuron circuit. The flexible information processing schemes such as weighted-sum operation of MFSFET synapse array and adaptive-learning function of the ferroelectric neuron circuit were experimentally verified in the fabricated circuits. Furthermore, the intrinsic problems of the firstly proposed circuit, such as the small output pulse height and short retention time, were also successfully solved by replacing a CUJT with a CMOS Schmitt-trigger and by employing an MFMOS-FET as a synapse device, respectively. All the results briefed above are the first demonstrations in the integrated neuron circuit using a ferroelectric thin film. It can be concluded from these results that this novel ferroelectric neuron circuit with an adaptive-learning function is very promising candidates for the next-generation large-scale neural networks.

References

1. T. Shibata, T. Ohmi, IEEE Trans. Electron Devices **39**, 1444 (1992)
2. T. Shibata, H. Kosaka, H. Ishii, T. Ohmi, IEEE J. Solid State Circuits **30**, 913 (1995)
3. O. Fujita, Y. Amemiya, IEEE Trans. Electron Device **40**, 2029 (1993)
4. K. Nakajima, S. Sato, T. Kitaura, J. Murota, Y. Sawada, IEICE Trans. Electron. **E78-C**, 101 (1995)
5. C. Diorio, P. Hasler, B.A. Minch, C.A. Mead, IEEE Trans. Electron. Devices **43**, 1972 (1996)
6. H. Ishiwara, Jpn. J. Appl. Phys. **32**, 442 (1993)
7. H. Ishiwara, T. Shimamura, E. Tokumitsu, Jpn. J. Appl. Phys. **36**, 1655 (1997)
8. M. Avrami, J. Chem. Phys. **7**, 1103 (1939)
9. Y. Ishibashi, Y. Takagi, J. Phys. Soc. **31**, 506 (1971)
10. E. Tokumitsu, N. Tanisake, H. Ishiwara, Jpn. J. Appl. Phys. **33**, 5201 (1994)
11. E. Tokumitsu, R. Nakamura, K. Itani, H. Ishiwara, Jpn. J. Appl. Phys. **34**, 1061 (1995)
12. S.M. Yoon, E. Tokumitsu, H. Ishiwara, IEEE Electron Device Lett. **20**, 229 (1999)
13. S.M. Yoon, Y. Kurita, E. Tokumitsu, H. Ishiwara, Jpn. J. Appl. Phys. **37**, 1110 (1998)
14. S.M. Yoon, E. Tokumitsu, H. Ishiwara, Jpn. J. Appl. Phys. **38**, 2289 (1999)
15. S.M. Yoon, E. Tokumitsu, H. Ishiwara, IEEE Electron Device Lett. **20**, 526 (1999)
16. S.M. Yoon, E. Tokumitsu, H. Ishiwara, IEEE Trans. Electron Device **47**, 1630 (2000)
17. S.M. Yoon, E. Tokumitsu, H. Ishiwara, Jpn. J. Appl. Phys. **39**, 2119 (2000)

Chapter 16
Applications of Oxide Channel Ferroelectric-Gate Thin Film Transistors

Eisuke Tokumitsu and Tatsuya Shimoda

Abstract In this chapter, recent topics on oxide-channel ferroelectric-gate transistors are presented. First, nonvolatile memory circuit application of ferroelectric-gate transistors is discussed by comparing it with Flash memory. It is pointed out "read disturb" may become serious in the memory circuits using ferroelectric-gate transistors in contrast to Flash memory. To solve the read disturb problem, two transistor memory cell structures are presented for both NAND and NOR configurations. In NAND configuration, one memory cell consists of parallel connection of a memory transistor and a pass transistor, whereas one memory cell consists of series connection of a memory transistor and a cut-off transistor in NOR configuration. Next, two-transistor cell NAND memory arrays using oxide-channel ferroelectric-gate transistors for both memory and pass transistors have been fabricated. It is confirmed that the stored "off" data in unselected cells remain almost intact during the readout procedures of a selected cell. Next, solution process is demonstrated to fabricate oxide channel ferroelectric gate transistors. All-oxide, all-solution-processed ferroelectric-gate transistors are demonstrated. In addition, newly developed nano-rheology printing (n-RP) technology, which utilize direct nanoimprint of oxide gel films, is used to fabricate oxide-channel ferroelectric thin film transistors without using conventional lithography process.

E. Tokumitsu (✉) · T. Shimoda
Green Devices Research Center, Japan Advanced Institute of Science
and Technology, 1-1 Asahidai, Nomi, Ishikawa 923-1292, Japan
e-mail: e-toku@jaist.ac.jp

T. Shimoda
School of Materials Science, Japan Advanced Institute of Science
and Technology, 1-1 Asahidai, Nomi, Ishikawa 923-1292, Japan

© Springer Science+Business Media Dordrecht 2016
B.-E. Park et al. (eds.), *Ferroelectric-Gate Field Effect Transistor Memories*,
Topics in Applied Physics 131, DOI 10.1007/978-94-024-0841-6_16

16.1   Introduction

A straightforward application of ferroelectric-gate field effect transistors is, of course, nonvolatile memory applications [1, 2]. As described in other chapters in this book, there are two types in ferroelectric random access memory (FeRAM); capacitor-type and transistor-type. For transistor-type FeRAM, mainly two configurations can be considered as in Flash memory, which are NOR and NAND configurations. In this chapter, we first describe features of ferroelectric-gate transistors in comparison of floating-gate Si-MOSFETs used in Flash memory, in terms of nonvolatile memory circuit applications. Then, we present fabrication of NAND structure using oxide channel ferroelectric-gate thin film transistors. Next, we introduce solution process as a new fabrication process for oxide-based ferroelectric-gate TFTs.

16.2   Memory Circuit Application Using Ferroelectric Gate Transistors

Ferroelectric materials exhibit hysteresis loops in charge (polarization)-electric filed characteristics (P-E) and its polarization can be switched at low voltage. This is usually considered as an advantage of FeRAM, because low voltage switching will lead to low power consumption. However, this can be a disadvantage when we construct memory circuits using ferroelectric-gate field-effect transistors. Figure 16.1 schematically shows drain current–gate voltage (I_D–V_G) characteristics, or transfer curve of ferroelectric-gate transistor. To write "1", a positive voltage, $+V_{WR}$, is applied to the gate, and after the application of $+V_{WR}$, the threshold voltage becomes low, which makes the device on-state. To write "0", a negative voltage $-V_{WR}$, is applied to the gate, and then the threshold voltage becomes high, which makes the device off-state. The polarity of the programming is opposite for Flash memory. Floating-gate MOSFETs used in the Flash memory have similar transfer curve as shown in Fig. 16.1. However, when a large positive voltage is applied to the gate to inject electrons to the floating gate by tunneling, its threshold voltage becomes high, because of the electrons in the floating gate. A notable difference between ferroelectric-gate transistor and floating-gate MOSFET in Flash memory is its programming (write) voltage. Much higher voltage is needed for Flash memory to write data than for FeRAM using ferroelectric-gate transistor.

The stored datum can be read out by applying an appropriate read voltage, V_{READ}, to detect "on" or "off" state in both memories. Figure 16.2 shows typical NAND and NOR configurations used in the memory circuits. In the NAND configuration, ferroelectric-gate transistors were connected in series and once one transistor is turned off, the string is open circuit condition. This feature is used to detect the stored information in the selected cell (transistor). In other words, to read

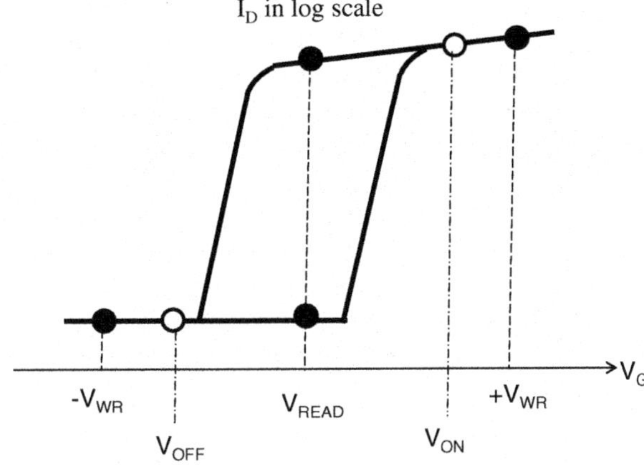

Fig. 16.1 Schematic illustration of transfer curve of ferroelectric-gate transistor

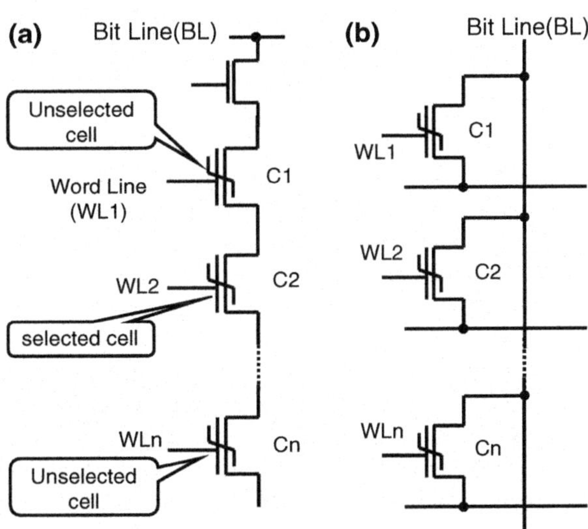

Fig. 16.2 **a** NAND and **b** NOR configurations in transistor-type FeRAM

out the stored information of the selected transistor (for example, C2 in Fig. 16.2a), all other unselected transistors (C1, C3, C4, …Cn in Fig. 16.2a) of the string must be turned on during the readout process. Hence, an appropriate gate voltage (as shown in Fig. 16.1 as V_{ON}) must be applied to the unselected transistors. In floating-gate MOSFETs used in the Flash memory, a programming voltage to write a datum to the selected transistor is high (12–20 V), because tunneling effect is used for electron injection to the floating gate. Since the gate voltage to turn on the unselected transistors, V_{ON}, is sufficiently low compared to the programming voltage, the stored data of the unselected cells remain intact during the readout process even if these unselected transistors stored "off" data. This is also the case for recent SONOS (silicon-oxide-nitride-oxide-silicon) type memory transistors.

On the other hand, the programming voltage, $+V_{WR}$, of the ferroelectric-gate transistors is low (less than 10 V), which results in low-voltage write operation. However, since the voltage to turn on the unselected transistors, V_{ON}, is close to $+V_{WR}$, when we use ferroelectric-gate transistors in NAND configurations, the stored data of the unselected transistors may be affected by the gate voltage applied during the readout process. This is called "readout disturb" effect.

In the NOR configuration, the phenomena is contrastive. As shown in Fig. 16.2b, transistors are connected in parallel to a bit line in the NOR configuration, and once one transistor is turned on, a bit line is pulled down to the plate line (or ground level). If an unselected transistor is on-state during the read out process, we cannot read the stored information from the selected transistor correctly. Hence, all unselected transistors must be turned off to read out the datum of the selected transistor. The situation is easily realized in Flash memory, when enhancement-type floating-gate MOSFETs are used. If the lower threshold voltage is adjusted positive, the transistor is off at $V_G = 0$ V, regardless of the stored information. In addition, since an absolute value of equivalent negative gate voltage to extract electrons from the floating gate is very large, there is no effect on stored data of unselected transistors during the readout process. On the other hand, when NOR configuration consists of ferroelectric-gate transistors, "read disturb" will be one of the problems as found in NAND configuration. Most of transfer curves reported for ferroelectric-gate transistors show that the devices are on state even when $V_G = 0$ V after positive programming voltage is applied due to remanent polarization of the ferroelectric gate insulator. Hence, if such devices are used for NOR configuration, in order to read out the stored datum of the selected transistor, unselected transistors must be turned off by applying an appropriate negative voltage to the gates during the readout process. As shown in Fig. 16.1, this voltage, V_{OFF}, is close to negative write voltage, $-V_{WR}$. Hence, the readout process may partly or totally destroy the stored data of the unselected transistors if they are programmed to on-state.

Readout disturb problem can be effectively solved by using a two-transistor cell in both NAND and NOR configurations. For NAND configuration, as shown in Fig. 16.3a, addition of one pass transistor which parallel connected to ferroelectric-gate transistor will make the memory cell to be turned on during the readout procedure, without turning on the ferroelectric-gate transistor. The details of

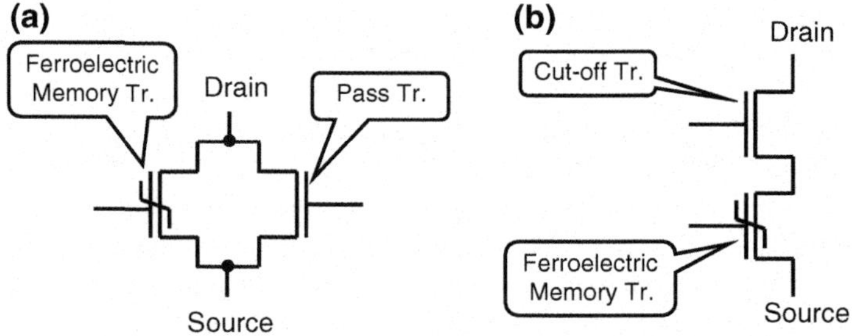

Fig. 16.3 Two-transistor memory cell structure for **a** NAND and **b** NOR configuration

read operation will be described in the next section. A disadvantage of two-transistor memory cell is it needs large area, compared to standard one-transistor memory cell. To decrease the footprint, the two-transistor memory cell can be realized by stacking two transistors as Kaneko et al. reported [3]. In this stacked structure, the upper and bottom surfaces of the oxide semiconductor act as channels of two different transistors. For NOR configuration, addition of one series connected cut-off transistor, as shown in Fig. 16.3b, will make it possible to turn off the non-selected memory cells during the readout process, without turning off the ferroelectric-gate transistors. This also can be realized by one transistor configuration using a split-gate structure [4].

16.3 Fabrication of NAND Memory Cell Arrays Using Oxide-Channel Ferroelectric Gate Transistors with 2-Tr Memory Cell Configuration

We have fabricated two-transistor configuration NAND memory cell arrays using oxide channel ferroelectric-gate transistors (FGTs). Figure 16.4 shows (a) circuit and (b) photograph of the fabricated memory cell array [5]. In this configuration, a memory cell consists of two ferroelectric-gate transistors, one ferroelectric-gate transistor (memory transistor; m-Tr) is used for data storage and the other is used as a pass transistor (p-Tr). When ferroelectric-gate transistors are used as pass transistors, it is not necessary to turn on the devices during the readout process if these transistors are programmed to maintain on- state. Instead, a ferroelectric-gate pass transistor of selected cell must be turned off during the readout process. The operation procedure is described as follows,

1. Write procedure

Suppose we write a datum to C12. At the stand-by state, all ferroelectric pass transistors are programed to on-state, hence there is a current path through these pass transistors. To write a datum to a memory cell of C12, word line 2 (WL2) which connected to the gate of ferroelectric memory transistor is selected. To write "1" or "0", positive or negative voltages is applied to WL2. Other WLs are kept at 0 V. At this time, bit line 1 (BL1) is 0 V, whereas the same voltage as WL2 is applied to other BLs. Pass lines, which are connected to the gate of pass transistors, are kept 0 V. Since all pass transistors are programed at on state, source and drain of pass transistors of BL1 is 0 V and those of pass transistors of other BLs are at the same voltage as WL2. Hence, the write voltage is applied only to selected cell, C12, effectively.

2. Read procedure

To read out the data stored in C12, a negative voltage is applied to PL2 to turn off the pass transistor with applying an appropriate small voltage to BL1. At this time,

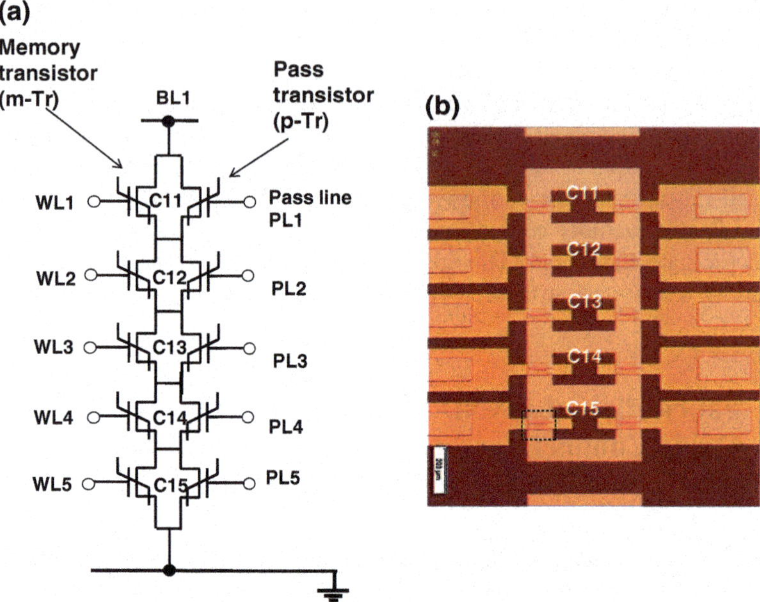

Fig. 16.4 **a** Circuit and **b** photograph of the fabricated NAND memory cell array using oxide channel ferroelectric-gate transistors [5]

all WLs and other PLs are kept at 0 V. When the pass transistor of C12 is turned off, we can read out the stored data of memory transistor of C12, because other pass transistors connected to BL1 are all at on state. After this procedure, positive voltage is applied to PL2 to turn on the pass transistors of PL2 again, which protects the stored data of the memory transistor.

Figure 16.4b shows a photograph of fabricated NAND memory cell array, which consists of ten (5 memory transistors (m-Tr) and 5 pass transistors (p-Tr)) oxide-channel ferroelectric-gate transistors, which results in 5 memory cells, C11–C15. Fabrication process is as follows, first a Pt bottom gate electrode was deposited by sputtering and patterned. Then, ferroelectric stacked gate insulator (20 nm-thick $(Bi,La)_4Ti_3O_{12}$(BLT)/160-nm-thick $Pb(Zr,Ti)O_3$ (PZT)) was prepared by the sol-gel technique. The thin top BLT layer prevents Pb-diffusion and create better interface with oxide channel. Next, Pt source/drain electrodes are formed prior to the deposition of ITO channel (20 nm) layer. The channel length and channel width of each device is 4 and 60 μm, respectively.

Figure 16.5 shows measurement circuits and corresponding transfer curves of one memory cell, which consists of one memory transistor and one pass transistor. Transfer curves were measured for (a) both memory and pass transistors connected in parallel, (b) one transistor (memory transistor) with the other transistor (pass transistor) on, and (c) one transistor (memory transistor) with the other transistor (pass transistor) off. The gate of pass transistor was connected to ground during the

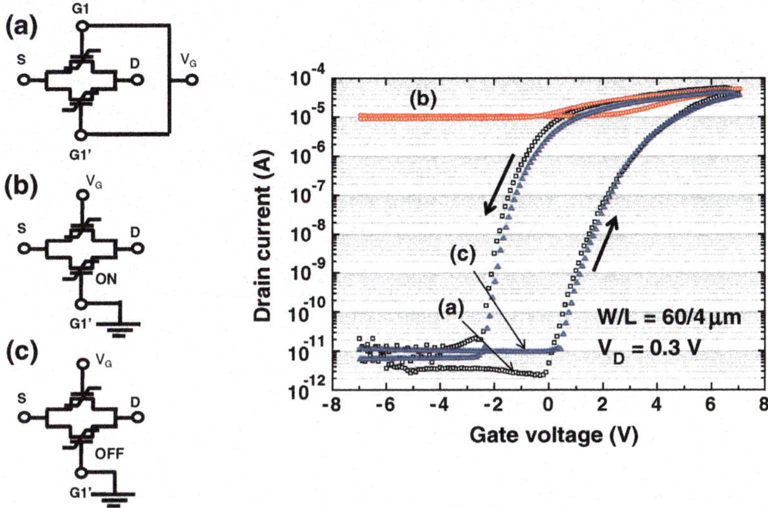

Fig. 16.5 Measurement circuits and corresponding transfer curves of one memory cell

measurements for cases (b) and (c). When both memory and pass transistors are operated simultaneously, transfer curve shows a clear hysteresis with a memory window of 3 V, an on/off ratio of 10^6, which indicates normal operation of ferroelectric-gate transistor. On the other hand, when the pass-transistor is turned on, a large current level of 10^{-5} A was observed even for negative gate bias region, which indicates the memory cell is shortened regardless of the state of the memory transistor. When the pass transistor is turned off, (case (c)), the transfer curve is similar to that of case (a), which indicates that current-voltage characteristics of the memory transistor can be readout by turning off the pass transistor.

Since we confirmed nonvolatile memory operation of one memory cell, we next evaluated operation of memory cell array using C11 and C12 cells. First, all pass transistors were programmed to be on-state so that they can flow current even for V_G = 0 V. Secondly, we programmed memory transistors of C11 off and C12 on. Then the datum of selected C12 was readout by turning off the pass transistor of C12. At this time, the pass transistor of C11 was kept on state, since no bias voltage was applied to pass line 1 (PL1). A large on current of 10^{-5} A was observed, which shows the on-datum of C12 was successfully readout. Thirdly, after the several readout operations of C12, a pass transistor of C11 was turned off and drain current of the bit line was detected, to check the read disturb effect of C11, which are not selected during the readout procedures. The detected current was as small as 10^{-10} A, which indicates memory cell C11 remained off during the C12 readout procedures. Note that if we do not use pass transistors, ferroelectric-gate memory transistor of C11 must be turned on during the C12 readout procedures. According to transfer curve shown in Fig. 16.5, we need to apply 4–5 V to turn on the device, which definitely affects the off state of the transistor. Although the detected current was slightly larger than that of off current of Fig. 16.5, it is more than 5 orders of magnitude less than the on level current. This indicates that the off datum stored in unselected C11 cell remained almost intact during the readout procedures of C12.

16.4 Solution Process for Oxide-Channel Ferroelectric-Gate Transistors

16.4.1 All Oxide Ferroelectric-Gate TFTs by Total Solution Process

An interesting approach for oxide-based device fabrication is the use of solution process, which will lead to lithography-less patterning by the printing technologies. In addition, meal oxides exhibit variety of electrical properties from semiconductors and conductors to insulator, as well as ferroelectric material. Hence, ferroelectric-gate transistors can be fabricated using only oxide materials. We demonstrated nonvolatile ferroelectric-gate TFTs using all oxide materials including electrodes. In addition, all layers of the device were fabricated by solution process [6]. Figure 16.6 shows schematic illustration of the fabricated device. At first, $LaNiO_3$ (LNO) film was fabricated by solution process on $SrTiO_3$ (STO) substrate as a bottom gate electrode. LNO has the same perovskite crystalline structure as STO and PZT. We confirmed that a low resistivity around 10^{-4} Ω cm was obtained when the film was annealed above 600 °C. Then, a ferroelectric $Pb(Zr,Ti)O_3$ (PZT) gate insulator was formed by solution process at an annealing temperature of 625 °C. Figure 16.7 shows transmission electron microscope (TEM) cross section of PZT/LNO interface and diffraction patterns (TED) of PZT and LNO layers. It is found that abrupt interface without transition layer was observed in PZT/LNO structure. In addition, even local epitaxial relation between LNO and PZT, which have both perovskite structure, was also found, even though the PZT film were fabricated by solution process with relatively high temperature annealing. Remanent polarization of the PZT film prepared on LNO electrode by solution process is approximately 30 $\mu C/cm^2$. Next, 200-nm-thick ITO source/drain electrodes were formed also by solution process with high annealing temperature which results in high carrier concentration and low resistivity. Finally, 20-nm-thick

Fig. 16.6 Schematic illustration of all-oxide FGT. All layers were fabricated by solution process

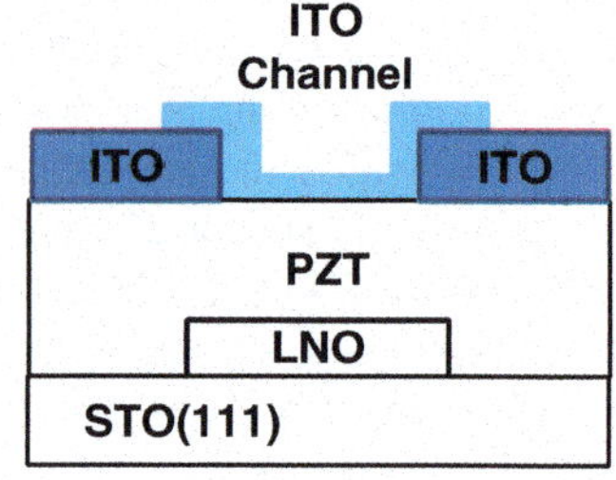

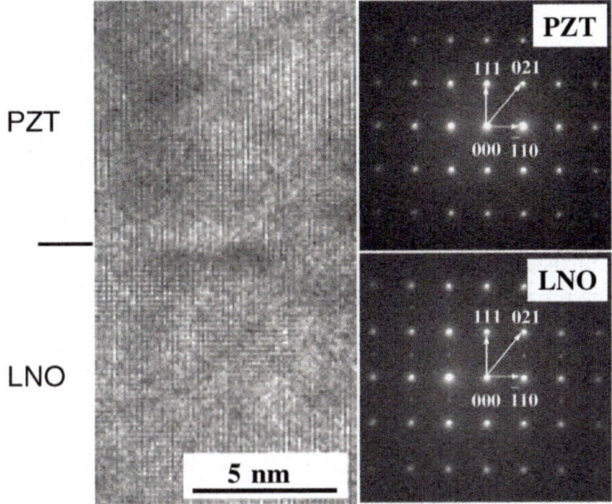

Fig. 16.7 TEM cross section and TED patterns of PZT/LNO structure fabricated by solution process [6]

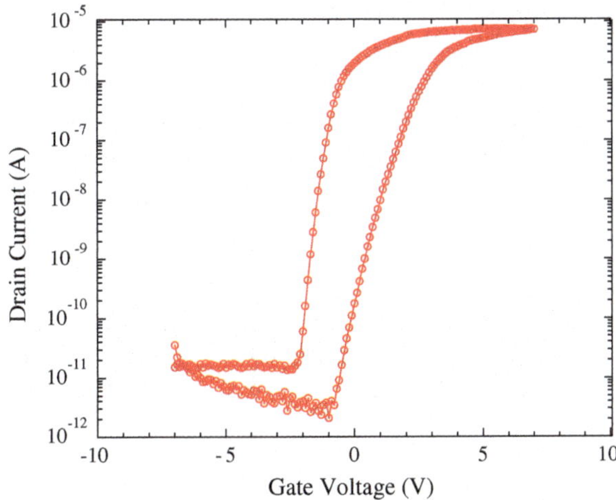

Fig. 16.8 Transfer curve of all-oxide, all-solution-processed ferroelectric-gate TFT [6]

ITO channel layer was formed by solution process. Since ITO was used for both channel and source/drain regions, annealing temperature dependence on electrical properties of ITO films was investigated before the device fabrication and proper annealing condition was selected for channel and source/drain regions. Carrier concentration of the channel layer is 5×10^{19} cm^{-3}. Large induced charge density by ferroelectric gate insulator makes it possible to use such a conductive oxide with high carrier concentration as a channel layer [7, 8].

Transfer curve of fabricated FGT is shown in Fig. 16.8. n-channel transistor operation with a large drain current on/off ratio of 10^7 was obtained. In addition, transfer curve exhibited hysteresis due to the ferroelectric-gate insulator with a memory window of 2 V, which demonstrates non-volatile memory function.

16.4.2   Fabrication of Oxide-Channel Ferroelectric-Gate TFTs by Nano-rheology Printing (n-RP)

Solution processes also enable us to use printing technologies to fabricate various patterns without using conventional lithography process. Ink-jet printing has been utilized to fabricate organic TFTs. However, the minimum line width by the ink-jet printing technology is generally more than 10 μm. Furthermore, to fabricate scaled transistors, precise shape control of the film is required, which is hard to be realized by the ink-jet printing technique. We have proposed a novel printing technology for fabrication of nm-size oxide electron devices with precise shape control, based on direct nano-imprint of oxide gel films. Since the rheological properties of the gel film is important to obtain fine patterns, we call this technology "nano-rheology printing (n-RP)" [9]. In the n-RP process, first, source solution of oxide is spin-coated and dried on the substrate. Then, a mold with precise patterns is pressed to the oxide gel film directly at a temperature around 200 °C. Note that no resist material is used. After the removal of the mold, the residual layer is etched off. At this stage, carbon-related species still remain in the film. Hence, after the pattern formation, the film is annealed at elevated temperatures. To obtain the fine patterns, careful design of source solution along with the optimization of imprint conditions is necessary. By this technique, oxide fine patterns less than 100 nm were demonstrated.

Figure 16.9 shows fabrication process of oxide ferroelectric-gate transistors [9]. First, the conductive oxide was patterned by n-RP to form bottom gate electrodes.

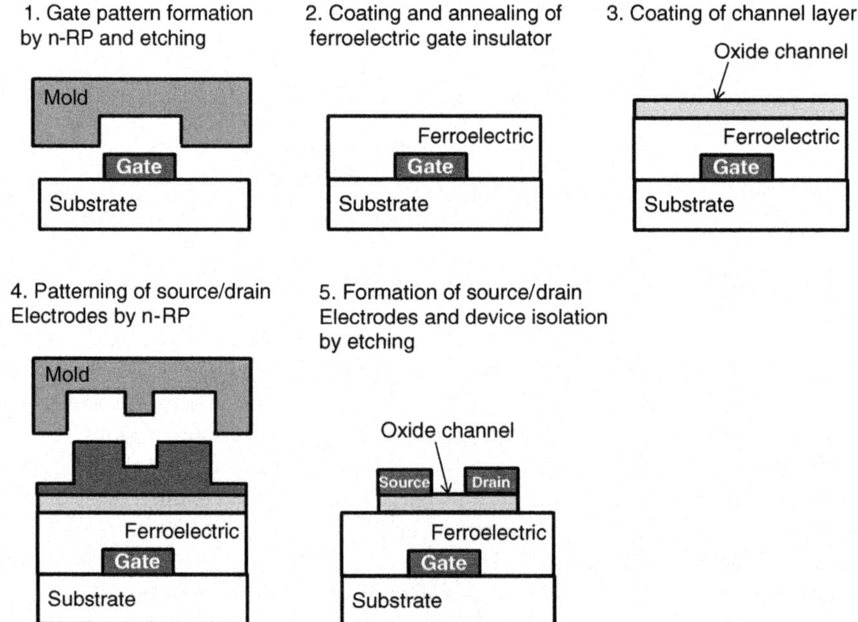

Fig. 16.9 Process flow for fabrication of ferroelectric-gate TFT by nano-rheology printing (n-RP) [9]

After annealing the bottom gate electrodes, a ferroelectric film was spin-coated and annealed as a gate insulator. Then, thin oxide channel layer was formed by solution process, too. After the annealing of the channel layer, ITO solution are spin-coated and patterned. As shown in Fig. 16.9, there is residual film of ITO at this stage. Hence, finally the residual film was removed by dry etching and bottom-gate, top-contact thin film transistor structure was fabricated.

To fabricate ferroelectric-gate TFT, LNO and PZT were used as a bottom gate electrode and ferroelectric-gate insulator. For channel layer, In_2O_3 was used because In_2O_3 has lower carrier concentration and higher mobility than ITO. Figure 16.10 shows (a) schematic cross section and (b) *top-view* photograph with AFM image, of the ferroelectric-gate TFT fabricated by n-RP process. The channel length and width are 500 nm and 28 μm, respectively. (The bottom gate length is 10 μm). Note that the device patterns were fabricated by n-RP only and no conventional lithography process was used and that the channel length is as small as 500 nm. Transfer curve is shown in Fig. 16.10c. Normal n-channel transistor

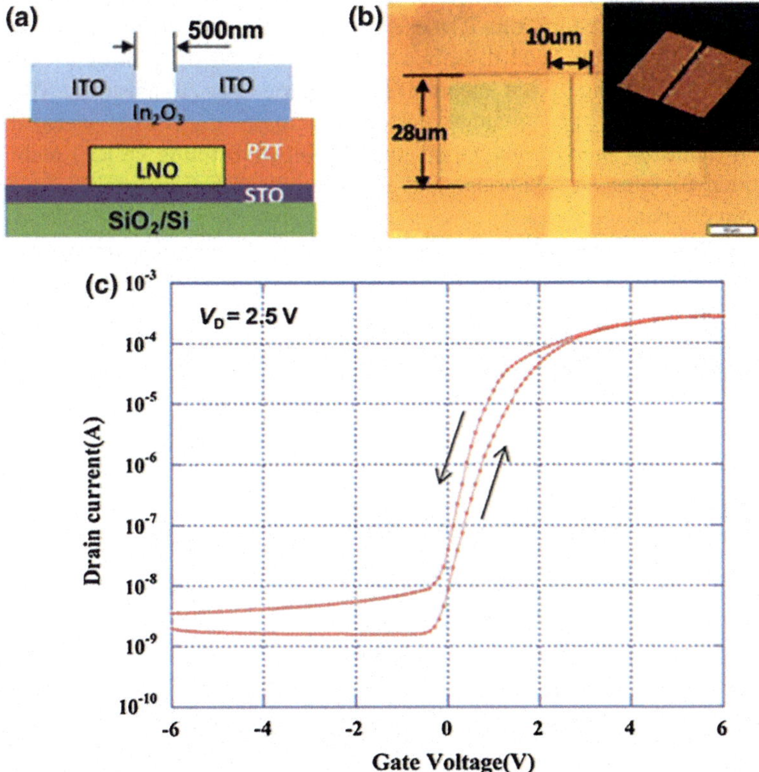

Fig. 16.10 a Schematic cross section and **b** top-view photograph with AFM image, **c** transfer curve, of the ferroelectric-gate TFT fabricated by n-RP without using conventional lithography [9]

characteristics was obtained with an on/off ratio of 10^5. In addition, although the observed memory window is small, hysteresis due to the ferroelectric-gate insulator was confirmed.

16.5 Summary and Conclusion

It is pointed out that "read disturb" is pronounced in nonvolatile memory circuits both in NAND and NOR configurations using ferroelectric-gate transistors. This is because a voltage to turn on the device which is required for unselected memory cells during the readout operation of a selected cell in the NAND configuration, is close to the positive program (write) voltage. To avoid read-disturb problem, two-transistor cell NAND memory arrays using oxide-channel ferroelectric-gate transistors have been fabricated. In the fabricated NAND memory cell, both memory and pass transistors have ferroelectric gate insulator. Hence, it is not necessary to apply a positive voltage to pass transistors of unselected cells to turn on, if these are programmed to be on in advance. Memory storage without read disturb was confirmed. Next, solution process was introduced to fabricate oxide channel ferroelectric gate transistors. All oxide, all-solution-processed ITO-channel ferroelectric-gate transistors were demonstrated. In addition, newly developed nano-rheology printing (n-RP) technology was applied to fabricate oxide-channel ferroelectric thin film transistors without using conventional lithography process. These results show that the solution process is not a substitute of conventional vacuum thin film deposition techniques, but a new technique which produces new device fabrication process and new functionality.

Acknowledgments The author would like to acknowledge members of JST ERATO, "Shimoda Nano-Liquid Process" project.

References

1. H. Ishiwara, M. Okuyama, Y. Arimoto (eds.), *Ferroelectric Random Access Memories: Fundamentals and Applications* (Springer, Berlin, 2004)
2. M. Okuyama, Y. Ishibashi (eds.), *Ferroelectric Thin Films: Basic Properties and Device Physics for Memory Applications* (Springer, Berlin, 2005)
3. Y. Kaneko, H. Tanaka, M. Ueda, Y. Kato, E. Fujii, I.E.E.E. Trans, Electron Devices **58**, 1311 (2011)
4. H. Saiki, E. Tokumitsu, IEICE Trans. Electron. **E87-C**, 1700 (2004)
5. B.N.Q. Trinh, T. Miyasako, T. Kaneda, P.V. Thanh, P.T. Tue, E. Tokumitsu, T. Shimoda, Ext. Abst. 2011 International Conference on Solid State Devices and Materials (SSDM), Nagoya (2011), p. 967
6. T. Miyasako, B.N.Q. Trinh, M. Onoue, T. Kaneda, P.T. Tue, E. Tokumitsu, T. Shimoda, Jpn. J. Appl. Phys. **50**, 04DD09-1-6 (2011)

7. E. Tokumitsu, M. Senoo, T. Miyasako, J. Microelectron. Eng. **80**, 305–308 (2005)
8. T. Miyasako, M. Senoo, E. Tokumitsu, Appl. Phys. Lett. **86**(16), 162902-1–162903 (2005)
9. T. Kaneda, D. Hirose, T. Miyasako, P.T. Tue, Y. Murakami, S. Kohara, J. Li, T. Mitani, E. Tokumitsu, T. Shimoda, J. Mater. Chem. C **2**, 40–49 (2014)

Printed by Printforce, the Netherlands